McGraw-Hill Series in Water Resources and Environmental Engineering

CONSULTING EDITOR
George Tchobanoglous, University of California, Davis

Bailey and Ollis: *Biochemical Engineering Fundamentals*
Bouwer: *Groundwater Hydrology*
Canter: *Environmental Impact Assessment*
Chanlett: *Environmental Protection*
Chow, Maidment, and Mays: *Applied Hydrology*
Crites and Tchobanoglous: *Small and Decentralized Wastewater Management*
Davis and Cornwell: *Introduction to Environmental Engineering*
deNevers: *Air Pollution Control Engineering*
Eckenfelder: *Industrial Water Pollution Control*
LaGrega, Buckingham, and Evans: *Hazardous Waste Management*
Linsley, Franzini, Freyberg, and Tchobanoglous: *Water Resources and Engineering*
McGhee: *Water Supply and Sewerage*
Mays and Tung: *Hydrosystems Engineering and Management*
Metcalf & Eddy, Inc.: *Wastewater Engineering: Collection and Pumping of Wastewater*
Metcalf & Eddy, Inc.: *Wastewater Engineering: Treatment, Disposal, Reuse*
Peavy, Rowe, and Tchobanoglous: *Environmental Engineering*
Sawyer and McCarty: *Chemistry for Environmental Engineering*
Tchobanoglous, Theisen, and Vigil: *Integrated Solid Waste Management: Engineering Principles and Management Issues*
Wentz: *Hazardous Waste Management*
Wentz: *Safety, Health, and Environmental Protection*

Bioremediation Principles

Bioremediation Principles

Juana B. Eweis

Department of Civil and Environmental Engineering
University of California, Davis
Davis, California

Sarina J. Ergas

Department of Civil and Environmental Engineering
University of Massachusetts, Amherst
Amherst, Massachusetts

Daniel P. Y. Chang

Department of Civil and Environmental Engineering
University of California, Davis
Davis, California

Edward D. Schroeder

Department of Civil and Environmental Engineering
University of California, Davis
Davis, California

Boston Burr Ridge, IL Dubuque, IA Madison, WI New York San Francisco
St. Louis Bangkok Bogotá Caracas Lisbon London Madrid
Mexico City Milan New Delhi Seoul Singapore Sydney Taipei Toronto

WCB/McGraw-Hill

A Division of The ***McGraw-Hill*** *Companies*

BIOREMEDIATION PRINCIPLES

This book is printed on acid-free paper.

1 2 3 4 5 6 7 8 9 0 DOC/DOC 9 0 9 8 7

ISBN 0-07-057732-3

Library of Congress Cataloging-in-Publication Data

Bioremediation principles / Juana B. Eweis . . . [et al.].
p. cm.—(McGraw-Hill series in water resources and environmental engineering)
Includes index.
ISBN 0-07-057732-3
1. Soil remediation. 2. Groundwater—Purification. 3. Bioremediation. I. Eweis, Juana B. II. Series.
TD878.B5556 1998
628.5—dc21 97-44577

Publisher: *Tom Casson*
Executive editor: *Eric Munson*
Marketing manager: *John T. Wannemacher*
Project manager: *Kimberly Hooker*
Senior production supervisor: *Madelyn S. Underwood*
Designer: *Felicia McGurren*
Compositor: *Shepherd, Inc.*
Typeface: *10.5/12 Times Roman*
Printer: *R. R. Donnelley & Sons Company*

http://www.mhhe.com

DEDICATION

This book is dedicated to our colleague and close friend Professor Kate M. Scow of the Department of Land, Air, and Water Resources at the University of California, Davis. Professor Scow has been our mentor in the study of soil microbiology, participated with us on numerous research projects, and generously opened her laboratory and graduate seminars to a long line of environmental engineering students. She continues to be an inspirational guide for engineers, as well as scientists, into the wonders of communal life beneath the surface of our planet. She has our continued gratitude, admiration, and most of all, friendship.

Juana B. Eweis
Sarina J. Ergas
Daniel P. Y. Chang
Edward D. Schroeder

ABOUT THE AUTHORS

JUANA B. EWEIS Juana Eweis was born and raised in the old city of Jerusalem. She received a B.S. degree in Civil Engineering from Birzeit University in 1987, and then worked at the municipality of Ramallah as a staff engineer. Her work included design of concrete structures, as well as supervising the construction of new roads and a sewer system in the city. In 1989 Ms. Eweis received a Fulbright scholarship to study International Development Technology at Humboldt State University in Arcata, California. Ms. Eweis received a M.S. degree in Civil and Environmental Engineering from the University of California, Davis in 1994 and will complete requirements for a Ph.D. in 1998. Ms. Eweis has worked on projects involving bioremediation of polynuclear aromatic hydrocarbons, fuels, solvents, and other compounds both in soil matrices and in airstreams. She taught classes in water chemistry and bioremediation. Her most current research involves liquid and vapor phase biodegradation of the gasoline additive methyltert-butyl ether. Ms. Eweis is a member of the Fulbright Association, the Air and Waste Management Association, and the American Chemical Society.

SARINA J. ERGAS Sarina Ergas received a B.S. in Environmental Resources Engineering from Humboldt State University and a M.S. and Ph.D. in Civil Engineering from the University of California, Davis. In 1994 she joined the Department of Civil and Environmental Engineering of the University of Massachusetts, Amherst. Dr. Ergas teaches graduate and undergraduate courses in biological processes, air quality, groundwater hydrology and contamination, and surface water quality modeling. Her research interests are in the areas of biological air pollution control, membrane bioreactors, biostability of drinking water, denitrification, and bioventing. She is a member of the Air and Waste Management Association, the American Society of Civil Engineers, and the Association of Environmental Engineering Professors.

DANIEL P. Y. CHANG Professor Chang was raised in Los Angeles, California. Following a brief Hollywood career as a child, he focused on science and engineering and received his B.S., M.S., and Ph.D. degrees in Mechanical Engineering from the California Institute of Technology. After serving in the U.S. Air Force, in 1973 Professor Chang joined the Department of Civil and Environmental Engineering at the University of California, Davis. He currently chairs the Department. His research interests include the physico-chemical behavior of pollutants; monitoring and generation of aerosols and their effects on health; emission and control of toxic volatile organic compounds; combustion of chlorinated wastes, and biodegradation of vapor-phase organics. Professor Chang has authored or co-authored over 100 papers, reports, and book chapters dealing with air pollution. He is a member of the Air and Waste Management Association, the American Chemical Society, the Combustion Institute, the American Association for the Advancement of Science, and Sigma Xi.

EDWARD D. SCHROEDER Professor Schroeder received his B.S. and M.S. degrees in Civil Engineering from Oregon State University. Following a period at the Robert A. Taft Sanitary Engineering Center of the U.S. Public Health Service, in 1963 he entered the doctoral program in Chemical Engineering at Rice University. At Rice, Professor Schroeder focused on biological wastewater treatment under the supervision of the late Arthur W. Busch. He began teaching at the University of California, Davis in 1966. Professor Schroeder has authored or co-authored over 150 papers, reports, and book chapters in the areas of biological processes and water quality management. At present, his principal research interests are in the areas of biodegradation of petrochemical compounds, biological treatment of vapor-phase contaminants, microbial nitrification and denitrification, and control of struvite precipitation in anaerobic process effluents. He is a member of the International Association for Water Pollution Research, the Water Environment Federation, and the Association of Environmental Engineering Professors. Previous books by Professor Schroeder include *Water and Wastewater Treatment* (McGraw-Hill, 1978) and with George Tchobanoglous, *Water Quality: Characteristics, Modeling, Modification* (Addison-Wesley, 1985).

PREFACE

Biological transformation of contaminants in soil and groundwater is a subject that is drawing increasing interest throughout the environmental science and engineering community. Transformations may occur naturally, although these are usually quite slow and limited by the availability of nutrients, oxygen, or appropriate microorganisms. Altering the reaction environment to increase the rate of biological transformation results in bioremediation. Altering the reaction environment in a manner that is cost effective and results in treatment to a prescribed level requires application of engineering methods and principles. For this reason environmental engineers usually are given the responsibility for the development, design, operation, and management of bioremediation processes and systems. Unfortunately, environmental engineering programs rarely include course work dealing with subsurface systems. Since 1990, a large number of books have been published on soil and groundwater remediation, including many specifically focused on bioremediation. However, few attempts have been made to provide a text for engineering students in which quantitative methods are emphasized.

This text is intended for use in an undergraduate or introductory graduate engineering course on remediation of contaminated soil and groundwater. Students using the book are assumed to have taken fluid mechanics and an introductory course on environmental engineering. Because few undergraduate engineering students have the opportunity to take course work in soil science, groundwater hydrology, and microbiology, introductions to these areas of study are presented. As indicated by the title, emphasis is placed on the principles of bioremediation. Although considerable design related information is given, particularly in the chapters on bioremediation processes, the intent is to provide students with a fundamental understanding of the underlying mechanisms associated with bioremediation. Such understanding allows students to address the complex problems resulting from varying contamination site and soil characteristics, contaminant properties, and microbial interactions. Worked examples are given in each chapter and study problems, which can be used for homework assignments, are provided at the end of each chapter. Numerous literature references are provided to allow students to investigate issues in more depth.

The first three chapters deal with the general nature of soil and groundwater contamination. An introduction to the subject, including sources of contamination, contaminant characteristics of importance, and current remediation approaches and methods, is presented in the first chapter. An introduction to soil characteristics of particular importance in contaminant transport and remediation is given in chapter two. In chapter three mechanisms of contaminant transport in soil and groundwater are discussed, as well as discussions on sorption, desorption, volatilization, dispersion and other reactions. Transport calculations are limited to simple, one-dimensional models.

Chapters four through six focus on microbial processes. In chapter four, Microbial Ecology, a general introduction to microbial communities is presented.

Emphasis is placed on the bacteria and their growth requirements because these microorganisms are the principal agents of contaminant transformation in soil. A detailed consideration of bacterial metabolism is presented in chapter five where the thermodynamic basis for microbial transformations is developed and bacterial metabolic processes of particular importance in bioremediation are discussed. Biodegradation of selected groups of contaminants (e.g., aromatic compounds) is discussed in chapter six. This chapter has two principal foci, hydrocarbons and halogenated organics.

The final four chapters of the book focus on methods of bioremediation currently in use. In situ treatment of soil and groundwater is discussed in chapter seven. Topics discussed include three methods of remediation that are not necessarily biological; pump and treat, air sparing, and soil vapor extraction. Pump and treat and soil vapor extraction are methods of contaminant removal that are often segments of bioremediation systems. In the case of pump and treat systems, the methods of above-ground water treatment used are identical to those used for conventional wastewater treatment and are not discussed in this book. Biological treatment of soil vapor extract is discussed in chapter ten. Air sparging is used to strip contaminants from groundwater and to provide oxygen required for microbial respiration. The extent to which air sparging enhances bioremediation is not well documented, but the technique is commonly applied in conjunction with other methods. Bioremediation of soils by land treatment (landfarming) and soil composting is discussed in chapter eight. Both methods usually ex situ, that is excavation of the contaminated soil and treatment is carried out at a prepared site or in a reaction vessel. Another ex situ method of bioremediation, slurry-phase treatment, is discussed in chapter nine. In slurry-phase treatment, water is mixed with contaminated soil to form a fluid that can be mixed, aerated, and pumped. Contaminant desorption is enhanced, which potentially increases microbial degradation rates. The addition of required nutrients is also simpler in slurry-phase systems. Biodegradation of vapor phase organics using biofilters and biotrickling filters is discussed in chapter ten. Vapor-phase organics result from partitioning between soil and air or water and air. The most common sources of such organics are soils contaminated with volatile compounds that are treated by soil vapor extraction and/or air sparging. Thus, application of vapor-phase treatment usually is associated with in situ soil and groundwater remediation. Vapor-phase microbial degradation processes are commonly used for treatment of off-gases from wastewater treatment and manufacturing facilities. However, application of vapor-phase biological treatment for treatment of gas from soil vapor extraction processes has been very successful in terms of contaminant transformation and cost effectiveness.

This text was developed for use in a senior elective course offered since 1994 by the Department of Civil and Environmental Engineering at the University of California, Davis. Students in the class suggested many improvements in the presentation and found many errors in the manuscript. We are extremely grateful for their forbearance and hope their contributions to the clarity and precision of the text make learning easier for future students of bioremediation.

Juana B. Eweis
Sarina J. Ergas
Daniel P. Y. Chang
Edward D. Schroeder

CONTENTS

CHAPTER 1

Introduction

The use of biological methods and processes for the remediation of contaminated soils and aquifers is the focus of this book. Biological processes are most readily applied to the transformation of organic contaminants, and consequently emphasis will be placed on the characteristics of organic compounds and factors making organics amenable to biological treatment. Bioremediation is a rapidly changing and expanding area of environmental engineering. Engineers working in soil and groundwater remediation in general, and bioremediation in particular, must deal with exacting regulations, serious public health issues, and a great deal of uncertainty related to knowledge of actual conditions in subsurface systems. Professional judgment and integrity are the most important tools in solving most remediation problems. This book is intended to provide a technical foundation for understanding and applying bioremediation concepts rather than to serve as a comprehensive handbook for system design.

The use of bioremediation in the treatment of hazardous waste is a relatively new concept, yet it is a rapidly growing trend in environmental management. A significant factor in the development of bioremediation has been the enactment of environmental laws and regulations that favor waste treatment rather than waste disposal. Treatment of contaminated sites has been increasingly emphasized in most industrialized nations since 1970 (Caplan, 1993), and biological treatment methods have generally been found to be less expensive than chemical or physical methods. The market for bioremediation is quite large. For example, out of approximately 750,000 existing underground storage tank (UST) facilities in the United States, over 300,000 have leaked and approximately 30,000 new releases occur each year. Over half of these facilities are used to store petroleum hydrocarbons, contaminants which have been proved to biodegrade relatively easily.

The number of bioremediation sites is not entirely clear at this time. For example, in California, leaking underground storage tanks are believed to be the largest source of contaminated soils and groundwater. Over 17,000 leaking underground storage tanks had been identified up to Jan. 1, 1991 (State Water Resources Control Board, 1991). In the large majority of the leaking underground storage tank sites (UST sites) petroleum products were involved. Solvents and pesticides are the materials in question in a relatively small number of cases. However, because of the toxicity of many solvents and pesticides, and the resulting low allowable concentrations in soil and water, these cases present difficult problems. Remediation programs are most often under the supervision of local agencies. Reports of the type of remediation program are either not well documented or not up to date in many cases. Bioremediation programs are often found to be incorrectly described in terms of both the methods used and their effectiveness in decontamination of a site.

Organic contaminants found in soils can be classified as naturally occurring or of anthropogenic origin. Naturally occurring organics resulting from the decay of plant and animal tissues, termed humic materials, are normal constituents of soils and groundwaters. Humic materials are structurally complex and are resistant to further biological transformation. Soils generally contain less than 3 percent humic material on a mass basis and groundwaters usually contain less than 5 mg/L of total humic materials. Organics of anthropogenic origin range widely in characteristics. Of particular significance are the biodegradability, polarity, solubility, volatility, and toxicity of anthropogenic organic contaminants.

Biodegradability. Readily metabolizable, nontoxic (to the microorganisms) organic compounds are normally oxidized very rapidly by microorganisms in soil. The principal problems in remediation of soils contaminated with degradable compounds are (1) providing inorganic nutrients required for microbial growth, (2) matching the rate of biodegradation with the rate of application, and (3) preventing plugging of soil pores with excessive microbial growth. Soils and groundwaters contaminated with compounds resistant to biological transformation are difficult to remediate.

Polarity and solubility. Nonpolar compounds tend to be hydrophobic and are most likely to concentrate (partition) into the organic material found in soils. The result is that nonpolar compounds are generally less mobile in soil and groundwater and the spread of nonpolar compounds in groundwater aquifers is usually slower than the spread of polar compounds. In practical terms, the time required for nonpolar compounds to travel from the site of contamination to nearby wells will be longer than for polar compounds. Solubility is similar in significance to polarity but is more directly related to the potential availability of compounds in the liquid phase. Bioremediation is carried out by microorganisms, principally bacteria and fungi, and contaminants must be in solution for the transformation processes to occur.

Volatility. Volatile compounds tend to partition into the vapor phase and can be removed from soil and groundwater by stripping. The contaminated vapor can then be treated above ground surface. Volatile compounds are often quite mobile in unsaturated soils, and emissions from soils may constitute a hazard at the contamination site or during excavation.

Toxicity. The key factor driving the need for remediation of contaminated soil and groundwater is human toxicity. Disposal or discharge of toxic chemicals to

TABLE 1.1
Maximum concentration limits (MCLs) of selected organic compounds in drinking water

Compound	Empirical formula	Molecular weight	Solubility,* mg/L	MCL, mg/L
Benzene	C_6H_6	78.1	1,800	0.005
Toluene	C_7H_8	92.2	500	2
Xylenes	C_8H_{10}	106.2	198†	10
Ethylbenzene	C_8H_{10}	106.2	150	0.7
Pentachlorophenol	C_6OHCl_5	266.3	20	0.2
Carbon tetrachloride	CCl_4	153.8	800	0.005
Trichloroethylene	C_2HCl_3	131.4	1.1	0.005
Methylene chloride	CH_2Cl_2	84.9	20,000	0.005
Ethylenedibromide (EDB)	$C_2H_4Br_2$	187.9	4,000	0.00005
Polychlorinated biphenyls (PCBs)**	$C_{12}H_nCl_{10-n}$	varies	varies	0.0005
Tetrachlorodibenzo(*p*)dioxin	$C_{12}H_2O_2Cl_4$	322		0.00000005
Vinyl chloride	C_2H_3Cl	62.5	2,792	0.002

*in water at 20°C.
†*p*-Xylene, solubility of *m*-xylene is 175 mg/L and *o*-xylene is 130 mg/L.
**Polychlorinated biphenyls are a group of contaminants that share the same biphenyl molecular structure; *n* ranges from 0 to 9.

soils presents a difficult problem because (1) toxic materials are often resistant to biodegradation, (2) once materials are in the soil environment less control exists with respect to their transport and fate, and (3) the risk to water supplies is very high because maximum concentration limits (MCLs) for toxic materials in water supplies are often extremely low.

Deposition of nontoxic wastes in soil remains a viable disposal option. Septic tank–leach field systems are used by over 20 million people in the United States, agricultural residues are often used as soil amendments, and land disposal is in common use for both agriculture and industry. Groundwater contamination is a concern whenever land disposal is used, primarily because of nitrate (NO_3^-) and other inorganic ions.

Remediation of contamination is often required because of the very low concentrations of specific toxic chemicals that are allowable in water supplies. The maximum concentration limits (MCLs) of selected toxic compounds in water supplies are given in Table 1.1.

EXAMPLE 1.1. QUANTITY OF FREE PRODUCT REQUIRED TO EXCEED DRINKING WATER MCLS FOR SELECTED COMPOUNDS. Determine the volume of trichloroethylene (TCE) required to exceed the drinking water MCL of 5 μg/L in 1 L of water and the volume of water that can be contaminated by 1 L of TCE. Density of pure TCE at 20°C is 1.46 kg/L.

Solution

1. Determine the volume of TCE having a mass of 5 μg.

$$V_5 = \frac{\text{mass TCE}}{\rho_{\text{TCE}}} = \frac{5\ \mu\text{g}}{(1.46\ \text{kg/L})(10^9\ \mu\text{g/kg})}$$

$$= 3.4 \times 10^{-9}\ \text{L of TCE will contaminate 1 L of drinking water}$$

2. Determine the volume of water contaminated by 1 L of TCE

$$V_w = \frac{1}{3.4 \times 10^{-9}} = 2.9 \times 10^8 \text{ L} = 290{,}000 \text{ m}^3$$

$$= 77{,}000{,}000 \text{ gal of water}$$

SOURCES OF CONTAMINATION

Large volumes of contaminants most commonly enter the soil from leaking underground storage tanks (USTs), landfills, and waste-disposal ponds. Placing fuel liquid storage tanks underground became common to save space and for the safety of occupants of nearby buildings. Over time metal tanks corrode and small leaks develop. Leaks also occur at pipe joints and connections as the result of settling and ground movement. Usually the leakage rates are small and difficult to measure (see Example 1.2). Similarly, storage ponds often develop small leaks in the liners over time. Prior to 1970, few controls existed on design and operation of waste-disposal ponds. Often ponds were unlined and much of the deposited liquid material quickly percolated into the soil. Current regulations set design standards for liners, distance to groundwater, and monitoring for leakage, but thousands of older, less well designed and monitored sites exist. Small amounts of toxic contaminants can pollute very large quantities of water, as demonstrated in Example 1.1. Leakage from tanks and ponds generally forms plumes, as indicated in Figure 1.1. In the case of leaks from USTs, the plumes may be *free product,* while plumes from waste-disposal ponds are usually composed mostly of water and contain a mixture of contaminants. Contamination at landfills results from leachate, liquids which collect in the bottom of the landfills and migrate through liners, as shown in Figure 1.2. The plumes may be sat-

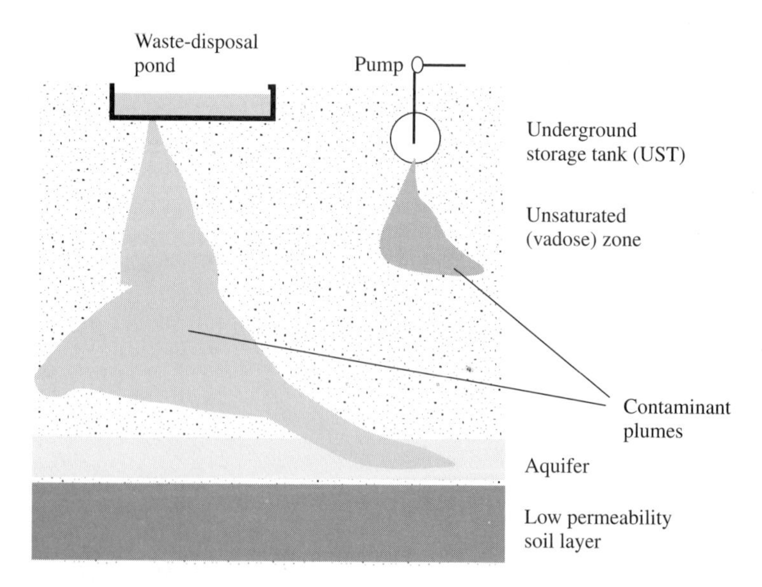

FIGURE 1.1
Schematic of plumes generated from small leaks in waste-disposal ponds and underground storage tanks.

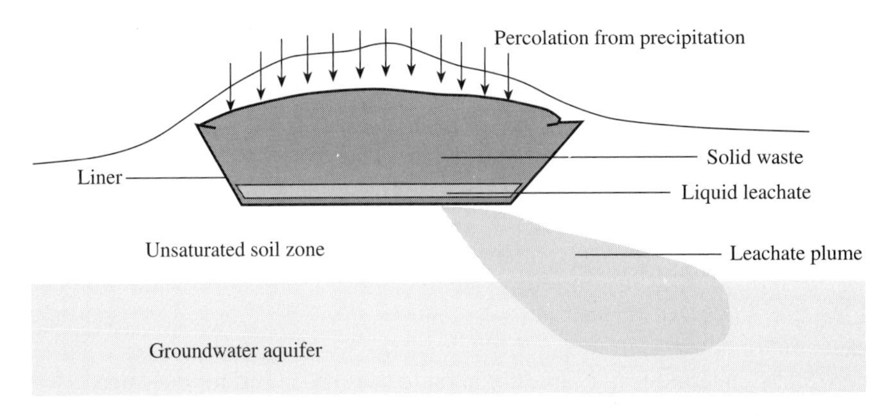

FIGURE 1.2
Contamination of soil and groundwater by leachate from a landfill.

urated or unsaturated, but the flow is generally downward and spreads through available channels, following the paths of minimum resistance. Interaction of contaminants with the soil occurs, principally through adsorption and desorption of nonpolar materials. Chemical reactions, including ion exchange, may also occur. Plumes tend to travel slowly in the unsaturated (vadose) zone but eventually will reach groundwater if leakage continues. Mixing with water in the aquifer is a function of relative density, with light plumes tending to float, dense plumes tending to sink, and neutral plumes tending to mix well. Cases of gasoline and jet fuel leakage have often resulted in pools of free product accumulating over an unconfined aquifer.

EXAMPLE 1.2. RATE OF LEAKAGE FROM AN UNDERGROUND STORAGE TANK. An underground storage tank used to store gasoline (specific gravity ≈ 0.8) has developed a 0.5-mm-diameter hole due to corrosion near the invert. The tank has a diameter of 2 m and a length of 4 m, and an average of 1,500 L is pumped from the tank each day. Estimate (*a*) the rate of leakage from the tank as a function of tank depth, (*b*) the volume, and (*c*) the fraction of each tank full of gasoline that is lost to leakage.

Solution

1. Determine the rate of leakage

$$q_L = A_h v_h = \frac{\pi}{4} d^2 C_D (2gz)^{1/2}$$

where q_L = volumetric leakage rate, m^3/s
A_h = area of hole, m^2
v_h = velocity through hole, m/s
d = diameter of hole, m
C_D = coefficient of discharge ≈ 0.8
g = acceleration due to gravity = 9.8 m/s^2
z = depth of gasoline in tank, m

$$q_L = \frac{\pi}{4}(5 \times 10^{-4})^2 (0.8)[2(9.8)]^{0.5} z^{0.5}$$

$$= 6.96 \times 10^{-7} z^{0.5} \text{ m}^3/\text{s}$$

2. Write a mass balance on the tank assuming that the unsaturated zone in the soil is at atmospheric pressure.

$$\rho_{\text{gas}} \frac{dV}{dt} = -\rho_{\text{gas}} \left(Q_p + q_L \right)$$

$$= -\rho_{\text{gas}} \left(1.5 \frac{\text{m}^3}{\text{day}} + 6.96 \times 10^{-7} z^{0.5} \frac{\text{m}^3}{\text{s}} \right)$$

where ρ_{gas} = mass density of gas, kg/m^3
Q_p = rate of pumping

$$\frac{dV}{dt} = -1.74 \times 10^{-5} \text{ m}^3/\text{s} - 6.96 \times 10^{-7} z^{0.5} \text{m}^3/\text{s}$$

The volume versus time relationship can be most easily solved numerically using the hydraulic elements given below and assuming a constant pumping rate.

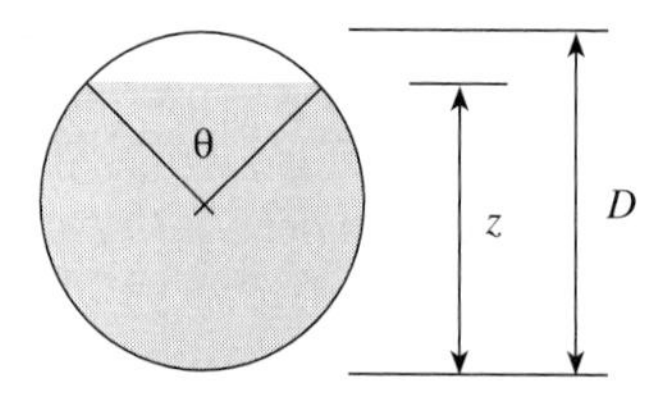

$$\text{Empty volume } = V = L \frac{D^2}{8} \left(\frac{\pi \theta}{180} - \sin \theta \right)$$

$$z = \frac{D}{2} \left(1 + \cos \frac{\theta}{2} \right)$$

Results of the solution propagated by determining the time increments required for specified changes in depth are given below.

Time, days	***z*, m**	**Vol leaked, m^3**	**Vol pumped, m^3**	**Vol remain, m^3**
0.00	2.0	0.00	0.00	12.56
0.41	1.8	0.03	0.62	11.91
1.13	1.6	0.09	1.70	10.78
2.01	1.4	0.15	3.02	9.40
2.98	1.2	0.22	4.47	7.87
3.99	1.0	0.29	6.00	6.28
5.01	0.8	0.34	7.53	4.69
5.99	0.6	0.39	9.00	3.17
6.88	0.4	0.43	10.35	1.79
7.62	0.2	0.45	11.46	0.65
8.05	0.0	0.46	12.10	0.00

3. Determine the volume fraction lost through leakage.

$$V_L = 0.46 \text{ m}^3$$

$$\text{Fraction leaked } = \frac{0.46 \text{ m}^3}{12.56 \text{ m}^3} = 0.037$$

The fraction lost is actually quite high. Assuming gasoline costs \$0.30/L (\$1.14/gal) the dollar loss per tank is \$138. Over a year the cost would be \$6,257. However, the cost of removing the gasoline from the soil would be much greater. Note that the loss rate would be difficult to detect without very careful monitoring.

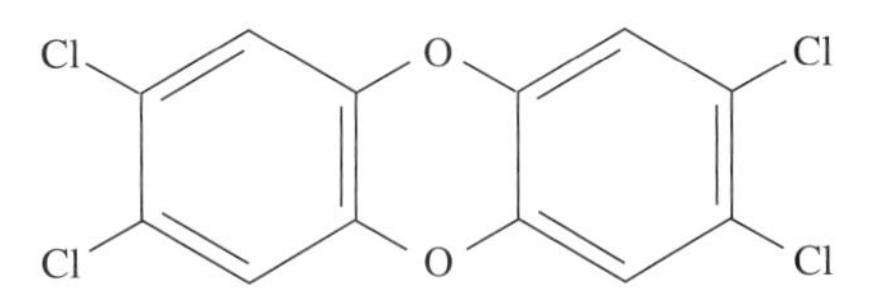

FIGURE 1.3
Tetrachlorodibenzo(*p*)dioxin

Serious soil and groundwater contamination problems also result from spills and improper disposal of toxic materials. For example, accidents during transport of chemicals may result in spillage of large quantities of pure products on small areas of ground. Left untreated, the chemicals can percolate into the soil and eventually may contaminate local groundwaters. Disposal of chemicals in homes and commercial enterprises is often done in a manner that results in soil and groundwater contamination. Waste oil and cleaning solvents from garages, agricultural chemical residuals on farms and at airfields used by crop dusters, paints, and cleaning supplies have often been dumped onto soil or buried on private land. Manufacturing plants in rural locations have often disposed of waste materials "out back," with major soil and groundwater contamination resulting. Such a case occurred in the small community of Lathrop, California, where tank rinsate from manufacture of the soil fumigant dibromochloropropane (DBCP) was disposed of in a field behind the facility. Groundwater concentrations as high as 1,700 mg/L were found in investigations following determination that DBCP was both a male sterilant and a carcinogen. Misuse of toxic materials can also result in soil contamination, as was the case in Times Beach, Missouri, where a mixture of chemical wastes and used crankcase oil was sprayed on roads and other unpaved areas as a dust-control measure. Following a number of animal deaths in 1971, high concentrations of tetrachlorodibenzo(*p*)dioxin (TCDD) (Figure 1.3) were found in the soil. Removal of the upper 6 in of soil proved unsatisfactory as a remediation measure, and eventually the community had to be evacuated.

Since the begining of the 20th century environmental engineers and scientists have steadily broadened their focus from the provision of clean, safe, and aesthetically pleasing water for consumption and treatment of wastewaters for the protection of human health, to include the improvement of the aquatic and air environments, and most recently to remediation of polluted soils and groundwater aquifers. Environmental protection and remediation of environmental damage became an issue of public concern following the publication of Rachel Carson's *Silent Spring* in 1962. Prior to publication of *Silent Spring* few members of the technical community had interest in ecological effects of waste discharges beyond the relationships to waterborne disease transmission, aesthetic damage resulting from algal blooms, fish kills, and floating debris, and the acute impacts on fisheries. Since 1962, engineers, scientists, and the general public have become increasingly well educated about the subtle effects to all forms of life resulting from chronic exposure to toxic materials in the environment, the importance of maintaining a diverse genetic pool, and potential global impacts of environmental damage in relatively small regions. A number of notable contributions have been made to public understanding of the importance of good environmental stewardship over the past 35 years. Scholarly books, such as E. O. Wilson's *The Diversity of Life*

(1992), have been written in a manner and style available to nonspecialists. The discovery that depletion of the stratospheric ozone was a result of emissions of chlorofluorocarbons has resulted in widespread, and initially contentious, discussion in the popular press, passing of a considerable body of control legislation and a number of international treaties, and the spawning of a large research program. During the same period we have found that many sites exist where chemicals accidentally or intentionally leaked, deposited, or were disposed of in soils have contaminated groundwater aquifers or rendered large tracts of land unusable and dangerous to humans and other forms of life.

Large-scale problems of soil and groundwater contamination are the direct result of the development of our modern industrial society. Waste production is associated with nearly every aspect of modern life. Manufacturing requires the extraction and processing of natural resources—minerals, forest products, and petroleum. Resource extraction and development generally results in the production of large amounts of waste materials. Abandoned extraction sites often become sources of large-scale pollution. Until relatively recently, wastes associated with mineral extraction and lumber harvesting were usually stored at the site or disposed of in lakes or rivers. Cost to the processor was the principal factor in selecting among disposal options. Controls and regulations were established in response to events or conditions that became intolerable.

Since 1900, the production of organic chemicals has increased immensely. More important are the changes in the types of compounds produced and the uses of these chemicals. At the beginning of the 20th century chemicals used in agriculture were derived principally from animal wastes, plant residues, and minerals. Today a wide assortment of petroleum-derived compounds are used as pesticides and herbicides. Many compounds, in some cases after extensive worldwide use, have been found to be toxic to wildlife and fish (e.g., DDT), highly toxic to mammals (e.g., phosphate-based pesticides such as Parathion), or carcinogenic (e.g., halogenated compounds such as polychlorinated biphenyls and dibromochloropropane). The complexity of the ecosystem makes determination of the effects of chemical use very difficult. Disturbance of population balances through the use of a chemical to control a particular pest may result in undesirable increases in other organisms.

Petroleum products provide a particularly good example in that the principal use of petroleum is for fuel. However, the range of petroleum-based chemicals used in modern society is staggering: plastics, pharmaceuticals, pesticides, herbicides, and detergents. The most ubiquitous soil and groundwater contamination problem in the world is undoubtedly gasoline leakage from underground storage tanks. The U.S. EPA estimates that over 200,000 leaking underground storage tank (UST) sites exist in the United States. Virtually as common is the spillage of motor oil on soil due to improper handling. Major contamination occurs as the result of accidents during transport of petroleum. Releases from wrecked tankers have cause significant ecological damage throughout the world. Huge quantities of pesticides and herbicides are used commercially and by individuals in the United States. Disposal of pesticide and herbicide containers is a serious problem, particularly in the case of the small containers used for homes and garden applications. Most of these containers end up in municipal landfills or incinerators and comprise an uncontrolled release into the environment. Disposal of industrial chemicals in landfills

TABLE 1.2
Classification of soil and groundwater contamination

Type of compounds	Typical locations	Mobility	Toxic effects
Agricultural chemicals	Manufacturing plants, chemical distributors, farms, crop duster airports	Generally low	Diseases of the nervous system, cancers
Gasoline and diesel	Service stations, military bases, refineries	Low to moderate	Carcinogens are included among a wide variety of petroleum compounds
Paints	Municipal landfills	Moderate to high	Heavy metal poisoning, nervous system damage, cancer
Solvents	Electronics plants, garages, military bases	Moderate to high	Carcinogens are included, nerve damage, toxicity
Polyaromatic Hydrocarbons (PAHs)	Coal gas manufacturing	Low to moderate	A number of PAHs are known or suspected carcinogens
PCBs	Electrical transformers	Low	Cancers
Dioxins	Chemical manufacture, vehicle exhaust, waste combustion	Low	Tumor promotion, chloracne

became a nationally recognized problem when the Love Canal case became a public issue in the 1970s. The Love Canal, an industrial chemical waste landfill site, was obtained by a school district in Niagara, New York, through eminent domain. The district sold the site to a real estate developer and a subdivision was constructed on the Love Canal. Reports of landfill leachate surfacing and seepage into basements were followed by suggestions that miscarriages and birth abnormalities were unusually high in the subdivision. A class action lawsuit resulted in abandonment of the subdivision and the beginning of a major new phase of environmental awareness worldwide.

Remediation of contaminated soil and groundwater has become a major industry throughout the world. The principal types of problems are classified in Table 1.2.

Polluted soils and groundwaters can be reclaimed through application of a variety of physical, chemical, and biological methods. In bioremediation inorganic and/or organic materials are removed from soils and groundwater through the action of microorganisms. Target materials for bioremediation include a wide range of organic compounds, heavy metals, such as mercury, and potentially toxic ions, such as cyanide (CN^-) and nitrate (NO_3^-). Many polluting materials that are deposited in the soil or groundwater are transformed to a nonpolluting state under normal or ambient conditions. For example, if a glass of orange juice is poured onto the soil surface, most of the organic components will be decomposed in a relatively short period of time by naturally occurring soil bacteria. Time required for decomposition will be a function of the soil characteristics, the temperature, and the presence of nutrients required for microbial growth, but the organic compounds in the juice will be microbially degraded to their lowest oxidation states. In situations where bioremediation is applied, the polluting materials are unlikely to be degraded naturally, or the time involved will be unacceptably long. In most cases of concern the target materials are anthropogenic and xenobiotic. Often the organic compounds are toxic and/or hazardous and their presence prevents use of the polluted soil or groundwater.

In recent years biological treatment systems have been employed to remove contaminants from all three media in which they occur: soil, water, and vapor. The term *bioremediation* is used to describe a large variety of engineered systems that use microorganisms to degrade organic chemicals. In the literature, bioremediation typically refers to systems used in the treatment of soils and water. Biofiltration, a variant of bioremediation, refers to the treatment of contaminated vapors. The degradation of organic contaminants by microorganisms is a naturally occurring process that is limited by chemical, physical, and environmental constraints. Examples of such constraints are the molecular structure of the contaminant and its resistance to biodegradation, the lack of contact between the contaminants and the degrading culture (bioavailability), the presence of the microbial culture that is capable of degrading that contaminant, and the appropriate environmental conditions for the microorganisms. In bioremediation systems, the constricting conditions are modified and the degrading action of microorganisms is enhanced such as by correcting some of the environmental factors that limit bioactivity.

The ultimate goal of bioremediation is to mineralize the contaminants, that is, to transform a harmful chemical into harmless compounds such as carbon dioxide or some other gas or inorganic substance, water, and cell material for the degrading organisms. Most microorganisms use oxygen to oxidize and biodegrade organic matter (aerobic biodegradation); others use nitrate, sulfate, methane, or other electron acceptors (anaerobic biodegradation). In the past, most bioremediation processes have relied on aerobic biodegradation. That is generally because many contaminants are easy to degrade under aerobic conditions, because anaerobic degradation usually occurs at a slower rate, and because it is harder to maintain anaerobic conditions in an engineered bioremediation process. Anaerobic processes are favored for certain groups of contaminants which are easier to break down under anaerobic conditions, such as highly chlorinated compounds. In this book, the focus will be on aerobic biodegradation since it is more commonly applied. Anaerobic biodegradation is discussed to a limited extent.

CURRENT BIOREMEDIATION PRACTICE

Concepts of bioremediation have evolved from management and treatment of municipal and industrial wastewater and solid wastes. Land disposal of wastewaters on *sewage farms,* which began in the late 19th century, involved the use of soil bacteria in decontamination processes. More sophisticated methods of contaminant treatment, such as trickling filters, activated sludge, and anaerobic fermentation, were developed in the first half of the 20th century. Since 1960, biological treatment processes have continued to include new methods of land treatment and processes for biodegradation of particular types of compounds. Developments in wastewater and solid-waste treatment have been transferred to the treatment of contaminated soils and groundwaters, that is, bioremediation. In recent years, most of the work published on bioremediation has been on the treatment of soils contaminated with petroleum products. That is partly because most petroleum hydrocarbons are relatively easy to degrade and as such are amenable to bioremediation, and partly because of the large number of sites contaminated with petroleum hydrocarbons from leaking underground storage tanks. Bioremediation has been successfully used to treat petroleum-contaminated soils for the last 30 years (Ryan et al., 1991).

Recently, bioremediation has become increasingly important in the field of hazardous-waste management. Some of the chemicals that were once thought to be resistant to degradation, including chlorinated species such as trichloroethylene and certain polychlorinated biphenyls (PCBs), have been shown to be biodegradable, at least under laboratory conditions. Other compounds that are currently being targeted by bioremediation include (1) solvents such as acetone and alcohols; (2) aromatics such as benzene, toluene, ethylbenzene, and xylenes which are collectively known as BTEX, as well as polycyclic aromatic hydrocarbons (PAHs), and chlorobenzene; (3) nitro- and chlorophenols; and (4) pesticides (Skladany, 1992). Among the most commonly encountered contaminants in soil and groundwater are aromatic hydrocarbons such as BTEX, resulting from spills or leakage, and chlorinated aliphatics such as tetrachloroethylene or perchloroethylene (PCE), trichloroethylene (TCE), and 1,1,1-trichloroethane, used in industry for degreasing (McCarty, 1991).

Bioremediation is a young field of technology, and this fact must be kept in mind when reading the literature. A large fraction of the applications of bioremediation to date have been experimental and the work reported has been focused on testing the applicability of a bioremediation method for specific site conditions and specific contaminants. Most of the published literature cites cases in which bioremediation has been successful. Cases in which bioremediation has met with failure have not been documented to the same extent.

In the literature, the success of bioremediation is often measured by the percent reduction in contaminant concentration in the soil or groundwater. Such criteria are weak since bioremediation may achieve high percent removals yet fail to meet cleanup goals. At the same time, the contaminants may be transported out of the soil or water, or transformed abiotically, through other processes, such as volatilization, migration, or photooxidation. In such a case, the goals of bioremediation, which are to detoxify and immobilize the contaminants, would not necessarily be met. A successful bioremediation process should include controls to account for contaminant transport, such as a cover to collect volatile material or monitoring wells to detect contaminant migration. At the same time, "proof" that biodegradation has occurred needs to be obtained. That can be in the form of increased microbial activity, increased release of carbon dioxide, increased uptake of oxygen, or presence of metabolic products.

The term bioremediation is applied to any system or process in which biological methods are used to transform or immobilize contaminants in soil or groundwater. One result of this very broad definition is that little knowledge is gained from saying that bioremediation has been or will be applied in a particular case. The physical conditions involved in bioremediation processes can be classified as in situ, solid phase, or bioreactors. Treatment processes used in the bioremediation of contaminated soils, gases, and water differ considerably. A brief description of approaches and processes in current use is presented here. More detailed descriptions are given in Chapters 7 through 10.

Groundwater Bioremediation

Methods used for groundwater bioremediation can be categorized as pump and treat or in situ. In pump and treat systems groundwater is pumped to the surface, treated, and either used directly or returned to the aquifer. Surface treatment often involves aeration, the addition of nutrients, and in some cases seeding with microbial cultures capable of degrading contaminants known to be in the groundwater. The processes used are similar to what is involved in waste-water treatment, such as activated sludge and rotating contactors. In some cases the surface treatment is physical as it may involve adsorption onto activated carbon or air stripping. Pump and treat operations have generally not been successful in remediating groundwater back to drinking water standards.

The term in situ comes from Latin and means "in its original place." That is, treatment occurs in the subsurface. The activity of the degrading microorganisms in an aquifer is stimulated by the introduction of oxygen and nutrients. Oxygen, which is often found to be the most rate-limiting factor in subsurface biodegradation, can be introduced in a number of ways such as air sparging or through the use of hydrogen peroxide. Nutrients are usually introduced in a water solution through infiltration galleries (for shallow depths) or injection wells. Subsurface conditions are not easily controlled, and problems associated with in situ treatment include difficulty in delivery of oxygen or an alternate electron acceptor at satisfactory rates or to the necessary locations, difficulty in delivering microorganisms to contaminated zones, isolation of contaminants in noncontiguous interstitial openings, and inability to prevent migration of contaminants beyond the treatment zone.

Soil Bioremediation

Bioremediation of contaminated soils can be carried out in situ, or the material can be excavated and treated on-site or at a separate treatment facility. In situ treatment methods include soil venting, in cases where volatile contaminants are dominant, and bioventing where semivolatile and nonvolatile contaminants are involved. Ex situ processes include land treatment (usually a form of landfarming), composting, and slurry-phase bioreactors. Both in situ and ex situ treatment often involve a combination of biological and nonbiological processes and operations. For example, soil washing is often used to concentrate contaminated materials and reduce the volume that must be treated. In soil washing surfactants are used to remove sorbed materials from larger components such as rock and gravel. Washing usually has little effect on materials sorbed to soil organic material and fines (silt and clay), but separation of larger materials makes biological treatment considerably simpler.

In situ treatment

In situ soil bioremediation requires transporting oxygen, and possibly nutrients, through the contaminated volume. In some cases the indigenous microbial population is unsatisfactory in terms of species present and an enriched microbial culture needs to be added as well. Even highly porous soils present relatively severe limitations on the transport of liquids and particles. For this reason, addition of nutrients and microorganisms is difficult. In cases where the contaminants are volatile (e.g., a gasoline spill) soil venting and ex situ gas treatment can be applied. In soil venting air is drawn through the polluted soil zone and the off-gases are treated. Some biodegradation of the contaminants may occur in the soil, but usually nutrient limitations exist and the amount of biodegradation is minimal. If the off-gas treatment is biological the soil venting operation becomes a type of bioremediation system. In the past, soil venting operations have been operated as physical treatment processes, where the main objective was to recover as much of the volatile contaminants as possible. Bioventing is a form of in situ treatment applicable to less volatile contaminants in soil. Air (or oxygen) is drawn into the contaminated zone and nutrients are added by infiltration or injection. If successful, bioventing is a cost-effective alternative to excavation and treatment on the surface.

Landfarming

Landfarming involves aeration and mixing of contaminated soil by tilling, addition of nutrients (and in some cases microorganisms), and control of moisture content by periodic addition of water. In most cases contaminated soils are excavated and treated at a site where migration can be controlled by construction of leaching barriers (compacted clay or plastic liners). In some cases the contaminated soils are near enough to the surface to make the excavation step unnecessary and the treatment is effectively in situ. Degradation processes in landfarming are principally biological. Photochemical oxidation may be significant in some cases. Emission of contaminants to the atmosphere through volatilization is often a limiting constraint on the application of landfarming.

Composting

In composting, the contaminated material is mixed with organic bulking agents, such as manure, and formed into piles or windrows. The bulking agents help increase porosity to facilitate airflow, and energy released during organic degradation results in temperature elevation of the pile. Water is added periodically and the piles or windrows are mechanically mixed at regular (typically weekly) intervals.

Static piles are a form of composting in that bulking agents, nutrients, and water are added. However, static piles are not mixed and temperatures are usually near ambient. Aeration can be passive (resulting from thermal density gradients between the ambient air and air in the pores) or forced by applying vacuum and sucking air through the pile as shown in Figure 1.4. Bulking agents used are usually made up of manure, which supports a larger microbial population than soil and provides inorganic nutrients, and relatively inert materials such as sawdust, wood chips, or compost. Water is added periodically, as needed to sustain the microbial population.

Bioreactors

Bioreactors are slurry-phase operations in which the contaminated soil is placed in a containment vessel and enough water is added to allow continuous mixing. Oxygen can be added as required and off-gas controls are often used to prevent loss of volatile organic compounds through stripping. Off-gas controls include gas recycling, use of the off-gases in combustion processes, and, potentially, microbial gas cleaning (biofiltration).

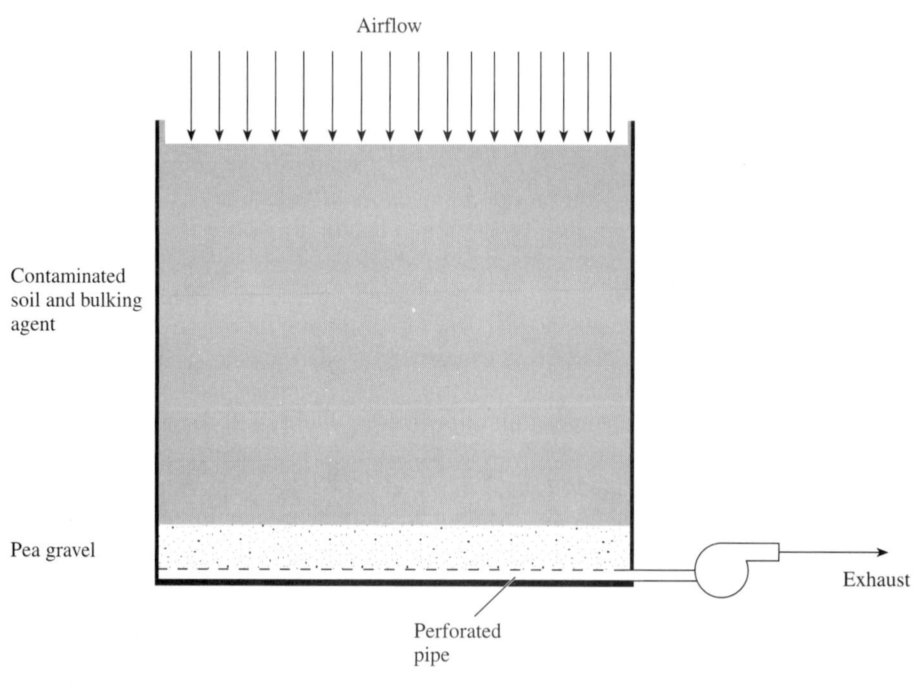

FIGURE 1.4
Schematic diagram of a static pile bioremediation process. Contaminated soil is mixed with an organic-containing bulking agent such as manure and placed over a layer of gravel or other coarse support material. Air is drawn down through the pile by applying a vacuum to the system. Periodic water addition to maintain moist conditions is usually necessary.

Microbial Gas Cleaning

Microbial gas cleaning is accomplished by passing the contaminated air through porous beds called biofilters or biotrickling filters. Packing used in the porous beds is usually composed of either a mixture of compost and an inert bulking agent or porous ceramic pellets constructed of diatomaceous earth or sintered glass. Microbial communities grow attached to the packing surface. The necessary liquid film is provided by maintaining a saturated air stream, by blowing an aerosol through the bed, or by addition of a water stream to the bed. Nutrients necessary for microbial growth are added periodically or with an aerosol or recycled liquid stream depending on the process configuration used. Schematic diagrams of microbial gas cleaning systems are shown in Figure 1.5.

Microbial gas treatment is used for removing volatile organic compounds (VOCs) from bioremediation process off-gases, from nonbiological remediation process off-gases, and from gases generated by both waste treatment and industrial operations (Ottengraf and Van Den Oever, 1983). Biological treatment of gas streams is quite economical relative to alternatives such as combustion or adsorption onto granular activated carbon. However, processes in current use have proved to be somewhat difficult to operate, particularly when mixtures of contaminants are being treated, contaminant concentrations vary widely over short periods of time, and acids, such as HCl and H_2SO_4, are products.

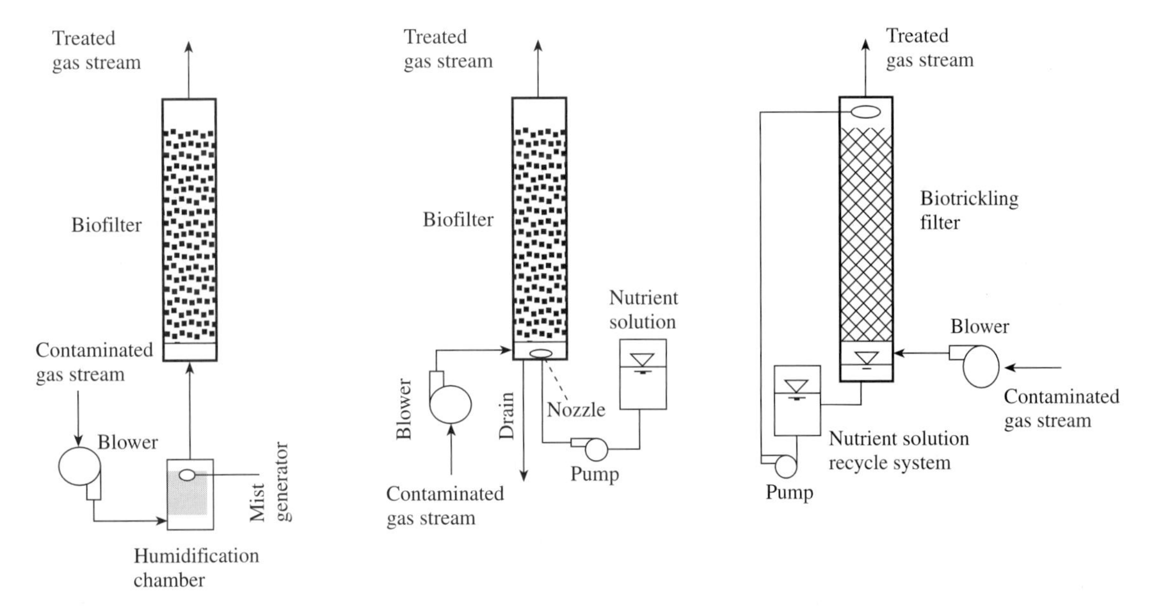

FIGURE 1.5
Schematic diagrams of microbial gas cleaning systems with different configurations.

Emissions of VOCs during bioremediation can occur during loading of bioreactors or operation of bioremediation processes, or may result from the production of volatile materials in the site. In soil treatment systems such as landfarming and static pile processing, volatiles may be emitted during the construction of the windrows or mixing of contaminated soil with bulking agents. In slurry-phase treatment VOCs may be emitted as a result of mixing and aeration. Production of VOCs, followed by emission, may occur as a result of breakdown of the parent compounds, particularly if local anaerobic conditions exist, or because of volatilization when the soil is agitated. Production of VOCs from nonvolatile compounds during biodegradation is a theoretical problem but the issue appears to have little practical significance (Lang et al., 1989), at least for aerobic processes. Microbial transformation of VOCs into more volatile compounds (e.g., trichloroethene into vinyl chloride) may occur under anaerobic conditions, as the result of reductive bond cleavage, and has been observed. Vinyl chloride is a common component of off-gases in sanitary landfills and anaerobic sludge digesters, and the source is believed to be chlorinated solvents (Lang et al., 1987, 1989). Most bioremediation processes are designed to operate aerobically, and therefore VOC emissions will be restricted to the compounds that have been identified at the site.

As noted above, compounds that are easily degraded by soil microorganisms will normally be metabolized without difficulty. For example, simple alcohols, such as ethanol, methanol, and propanol, are among the group of easily biooxidized VOCs. Sorption, biodegradation, and volatilization are essentially competitive processes and would all occur. As suggested in Figure 1.6, VOCs sorbed onto

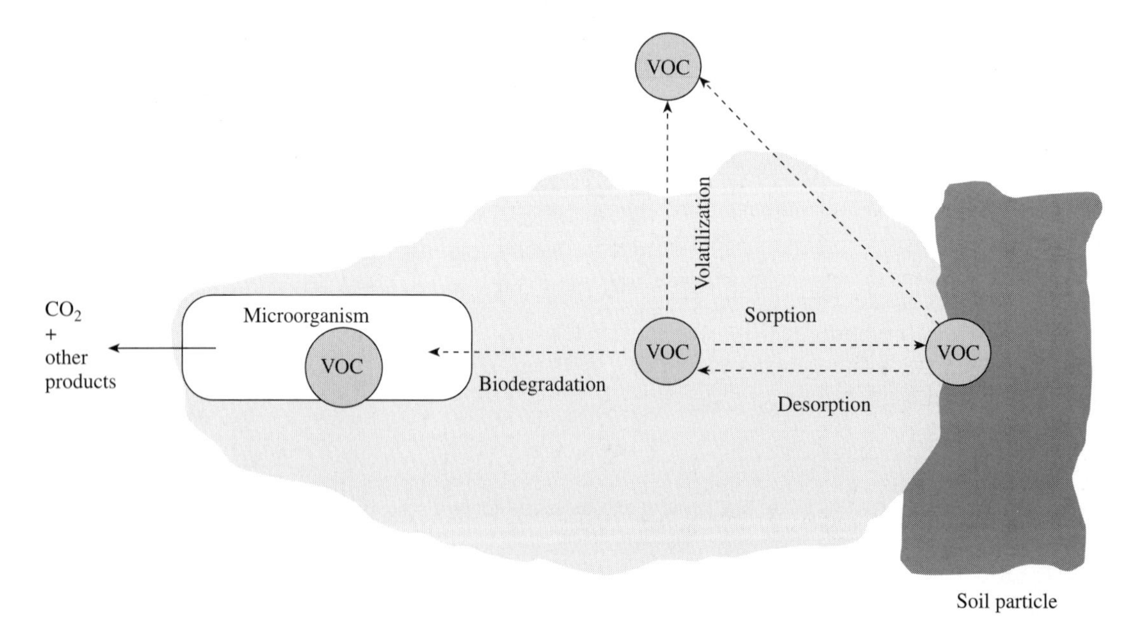

FIGURE 1.6
Mechanisms of removal of VOCs dissolved in bound water within the soil structure. Sorption on soil is not a terminal fate because equilibrium will be established with both the air and water.

soil particles may be volatilized or biodegraded. Emissions would occur from soil contaminated with easily metabolized compounds, but the fraction of available material biodegraded would be large.

ADVANTAGES AND DISADVANTAGES OF BIOREMEDIATION

Bioremediation offers several advantages over physical and chemical treatment processes used to treat contaminated water and soil. Cleanup costs using bioremediation are typically $100 to $250 per cubic meter while more conventional technologies such as incineration or secure landfilling may cost in the range of $250 to $1,000 per cubic meter (Gabriel, 1992). Another advantage achieved by bioremediation is that it is aimed at biodegrading and detoxifying hazardous contaminants, whereas other technologies such as venting, adsorption onto activated carbon, solidification/stabilization, soil washing, and disposal into landfills simply transfer contaminants to a different medium or location. Bioremediation is also a relatively simple technology compared to most others. In situ bioremediation can be carried out with minimal site disruption, volatile compound emission, and health risk to neighboring residents or site occupants.

A number of disadvantages are associated with application of bioremediation. The most important disadvantages are the difficulty in predicting performance and the difficulty of scaling up from laboratory or pilot-plant tests. Success of a bioremediation project is dependent on the ability of process operators to create and maintain environmental conditions necessary for microbial growth. Microorganisms are sensitive to temperatures, pH, contaminant toxicity, and contaminant concentration, moisture content, nutrient concentrations, and oxygen concentration. A decrease in microbial activity will slow down degradation and extend the treatment period. If microbial activity stops (e.g., due to the buildup of toxic metabolites), restarting the process may be extremely difficult. Cleanup goals may not be achievable using bioremediation because some contaminants are nonbiodegradable or only partially biodegradable or because the levels of contaminant removal cannot be attained microbially. As the contaminant levels diminish, biological degradation slows down and the microorganisms may switch to other energy sources or stop growing altogether. In such a case, bioremediation may not be enough to treat a site and another treatment technology may have to be employed. Finally, bioremediation is often relatively time-consuming. The time needed to remediate a site generally depends on the rate at which the contaminants are degraded.

FACTORS INFLUENCING BIOREMEDIATION

Factors influencing the effectiveness of bioremediation can be placed in three principal groups; environmental, physical, and chemical. In designing and operating a bioremediation process consideration must be given to all three areas. Monitoring conditions in operating systems is often a major expense in bioremediation.

Environmental Factors

Environmental factors are those necessary to provide optimum growth conditions for the microorganisms carrying out the bioremediation reactions. Microorganisms are sensitive to changes in temperature, pH, nutrient availability, oxygen, and moisture content. Individual species of organisms have particular optimal ranges in which growth is most rapid. Like nonbiological reaction systems, microbial reaction rates usually increase with increasing temperature. However, microorganisms have optimal temperature regions and temperature values above which microbial activity ceases. Soil bacteria typically function between 5 and 40°C, but the optimal temperature within that range is a function of particular species. Molds and yeasts often perform better at low pH than at neutral or basic pH values. Similarly, bacteria of the genus *Thiobacillus* oxidize sulfide to sulfate (e.g., $H_2S + 2O_2 \rightarrow 2H^+ + SO_4^{-2}$) and often grow very well at pH values as low as 1 but do not grow at pH values above 6. Most microorganisms grow best in the pH range 6 to 9 with optimal growth conditions between pH 7 and 8.

Nutrients are chemicals required for microbial growth that do not provide energy or carbon to the organisms. The most commonly required nutrients are nitrogen and phosphorus, both of which are usually deficient at remediation sites. Oxygen is the most widely used terminal electron acceptor in biological processes and also is required in certain types of enzymatically catalyzed oxidation-reduction reactions. Microorganisms (or mammals, for that matter) oxidize organic or inorganic compounds to obtain energy for growth. The oxidation process releases electrons that are passed through a chain of reactions within the cell and at the end must be discharged to the environment. The terminal electron acceptor is the final recipient of the electrons and, in the case of aerobic metabolism, O_2 is the acceptor and H_2O is the product. Moisture is important in bioremediation because microorganisms obtain all of their nutrients for growth from solution. Unfortunately, the solubility of O_2 in water is very low (9 mg/L at 20°C) and high moisture contents (e.g., where the soil pores are saturated) result in poor oxygen availability.

Physical Factors

The most important physical factors in bioremediation are availability of contaminants to microorganisms, presence of water, and a supply of a suitable electron acceptor, such as oxygen. Availability is a complicated concept involving affinity of contaminants for the solid and gas phases, pore structure of the solid phases, and presence of the necessary microbial communities. All contaminants have some affinity for the soil and gas phases. Many common contaminants have low water solubilities and sorb strongly onto soil particles. For example, petroleum hydrocarbons are generally nonpolar and tend to partition strongly onto the solid phase, which results in very low liquid-phase contaminant concentrations. Contaminants may accumulate in micropores too small for bacteria to colonize (typically less than 1 μm). Because microorganisms obtain nutrients from the liquid phase, the rate of bioremediation may be limited by the desorption rate. Similarly, light hydrocarbons tend to partition strongly into the gas phase and gas-liquid transfer may become the rate-limiting step. In general, in situ type processes do not enhance availability because the contaminants may be trapped inside pores. Ex situ processes offer more control. Through excavation and the mixing that is involved in processes such as land treatment and slurry-phase, aggregates are broken up and the microorganisms have a better chance at coming in contact with the contaminants. Mixing in slurry-phase treatment results in agitation and surface scrubbing that can result in the release of some of the adsorbed contaminants. For in situ treatment, surfactants are sometimes used to desorb and solubilize the chemicals. The effectiveness of surfactants has not yet been established. Some reports claim success at the expense of high dosages of surfactants that may add to the contamination problem (Aronstein et al., 1991).

Water is required because, as noted above, microorganisms obtain organic carbon, inorganic nutrients, and electron acceptors required for microbial growth from the liquid phase. Thus water must be in contact with contaminants and be present in quantities that allow microbial communities to develop. However, water may inhibit airflow and reduce the supply of oxygen necessary for microbial respiration. Optimal moisture contents exist for bioremediation of unsaturated soils with values typically in the range of 150 to 250 g of water per kg of dry soil. In bioremediation of contaminated aquifers, supplying oxygen is often the most difficult problem because of the low solubility of oxygen in water (approximately 7.5 to 9 mg/L in groundwaters).

Oxygen is the most widely used terminal electron acceptor for microbial respiration. In unsaturated soils, supplying oxygen is generally not a major problem. However, the low solubility of oxygen in water and the heterogeneity of groundwater aquifers result in making transport of oxygen throughout contaminated zones or to specific sites extremely difficult. Alternative electron acceptors exist that are more easily supplied to aquifers (e.g., NO_3^-, NO_2^-, SO_4^{2-}, Fe^{3+}, CO_2). The range of microorganisms capable of using each alternative is limited, the extent of the reactions catalyzed is more constrained, and the final products are often less completely oxidized relative to oxygen.

Chemical Factors

The most important chemical factor in bioremediation is the molecular structure of the contaminant, how it influences its chemical and physical properties, and its biodegradability. Biodegradability is related to factors such as solubility, degree of branching, degree of saturation, and the nature and extent of substitution. Solubility is important because microorganisms obtain nutrients from aqueous solution. High solubility results in potentially high availability of a compound. Although saturated carbon-carbon bonds are not difficult for microorganisms to break, the degree of saturation is related to volatility and solubility. Saturated rings such as highly branched alkanes are difficult for microorganisms to degrade (Evans et al., 1988). The effect of branching is seen in the relative degradability of isomers (Gibson, 1984, 1988). For example, *n*-octane (straight chain octane) is more easily degraded than 3-propyl pentane, although both have the same empirical formula, C_8H_{18}. Addition of chlorine, nitrogen, sulfur, and phosphorus to organic molecules tends to make them more stable biologically (Reineke and Knackmuss, 1988; Strand et al., 1991). Amino compounds are an exception, particularly the α-amino acids which are fundamental units of proteins and are easily metabolized by many microorganisms.

Bioremediation is a multidisciplinary field. It combines disciplines such as microbiology, soil science, chemistry, geology or hydrogeology, and process engineering. In subsequent chapters concepts from these disciplines are introduced to provide an understanding of soil and groundwater contamination and approaches to bioremediation. In Chapters 2 and 3 the focus is on the soil environment and the fate and transport of contaminants in the soil. Microbiology is discussed in Chapters 4 and 5. Chapter 6 deals with the chemistry of pollutants, and the factors that make certain groups of contaminants difficult to degrade. Bioremediation process selection, design, and operation are discussed in Chapters 7 through 10.

PROBLEMS AND DISCUSSION QUESTIONS

1.1. Pentachlorophenol (PCP), a compound used in pressure treating of wood to resist rot, has been found to cause cancer in animals and is toxic to humans. Solubility of PCP is 14 mg/L and the maximum concentration limit in drinking water is 0.2 mg/L. Determine the volume of water that will be contaminated by a discharge of 10 kg of PCP.

1.2. Gasoline from a boat dock is accidentally discharged to a small lake (100 ha surface area, average depth = 7 m) at the rate of 1 L/day. Benzene comprises approximately 1 percent of the gasoline on a mass basis (g benzene/g gasoline). Determine the time required before the benzene MCL of 0.005 mg/L is exceeded. Assume (*a*) the density of gasoline is 800 g/L, (*b*) the lake is well mixed, (*c*) benzene is conserved, and (*d*) the flows into and out of the lake are negligible.

1.3. Consider the lake of Problem 1.2: Assume that the average flow into and out of the lake is 1,000 m^3/day. Determine (*a*) the average residence time of water in the lake (θ_H), (*b*) the steady-state benzene concentration resulting from the 1 L/day leak of gasoline described, (*c*) if and when the MCL will be exceeded.

1.4. Consider the lake of Problems 1.2 and 1.3. Assume that the volatilization of benzene can be described as

$$J_B = -K_L C_B$$

where J_B = flux of benzene from the lake surface to the atmosphere, g/$m^2 \cdot$ day
K_L = mass transfer coefficient = 10^{-6} m/s

Determine the steady-state concentration of benzene in the lake and if the MCL will be exceeded.

1.5. A common assumption made when mass transfer rate coefficients are not available is that *local equilibrium* exists. The meaning of local equilibrium is that at the phase interface equilibrium exists and Henry's law or another appropriate equilibrium model can be applied. Consider the case of a cylindrical tube containing a solution of toluene. The tube has a volume of 100 mL and an area of 10 cm^2. Air having 100 percent relative humidity flows across the liquid surface at a rate of 0.1 L/min. Assume that concentration gradients in the liquid are negligible and that gas-liquid equilibrium is achieved very rapidly.

Determine the time required to volatilize 90 percent of the dissolved toluene assuming that Henry's law can be used to describe the gas-liquid equilibrium.

$HC_L = C_g$

$H_{toluene}$ = Dimensionless Henry's law coefficient for toluene

= 0.28 at 25°C

C_L, C_g = liquid- and gas-phase toluene concentrations, mg/L

0.1 L/min

1.6. Mass transfer coefficients for toluene can be expected to be of the order of 2×10^{-4} m/s. For the system of Problem 1.5, estimate the time required to remove 90 percent of the toluene assuming that $C_g << C_L$.

1.7. The adsorption data in the table below were developed using a Langmuir isotherm. Assume a 1-L liquid volume and 1 g of sorbent were used. Fit a Freundlich isotherm and a linear isotherm to the data and consider if the alternative models would be acceptable. You may wish to fit the alternative models over *regions* of the data rather than the entire set.

C_o, mg/L	100	90	80	70	60	50	40	30	20	10	5	1
C_e, mg/L	6.83	6.46	6.33	5.67	5.23	4.76	4.48	3.63	2.92	2.5	1.35	0.5

REFERENCES

Aronstein, B. N., Y. M. Calvillo, and M. Alexander (1991): "Effects of Surfactants at Low Concentrations on the Desorption and Biodegradation of Sorbed Aromatic Compounds in Soil," *Environmental Science and Technology,* vol. 25, no. 10, pp. 1728–1731.

Caplan, J. A. (1993): "The Worldwide Bioremediation Industry: Prospects for Profit," *Trends in Biotechnology,* vol. 11, pp. 320–323.

Carson, R. (1962): *Silent Spring,* Houghton Mifflin, Boston, MA.

Evans, W. C., and G. Fuchs (1988): "Anaerobic Degradation of Aromatic Compounds," *Annual Review of Microbiology,* Ornston, L. N., A. Balows, and P. Baumann (eds.), Annual Reviews, Palo Alto, CA.

Gabriel, P. F. (1992): Innovative Technologies for Contaminated Site Remediation: Focus on Bioremediation, in *Bioremediation: the State of Practice in Hazardous Waste Remediation Operations,* a live satellite seminar, sponsored by AWMA and HWAC, Jan. 9.

Gibson, D. T. (ed.) (1984): *Microbial Degradation of Organic Compounds,* Marcel Dekker, New York.

Gibson, D. T. (1988): "Microbial Metabolism of Aromatic Hydrocarbons and the Carbon Cycle," in *Microbial Metabolism and the Carbon Cycle,* Hagedorn, S. R., R. S. Handson, and D. A. Kunz (eds.), Harwood Academic Publishers.

LaGrega, M. D., P. L. Buckingham, and J. C. Evans (1994): *Hazardous Waste Management,* McGraw-Hill, New York.

Lang, R., D. Herrera, D. Chang, G. Tchobanoglous, and R. Spicher (1987): Trace Organic Constituents in Landfill Gas, report prepared by the Department of Civil Engineering, University of California, Davis, for the California Waste Management Board, Sacramento, CA.

McCarty, P. L. (1991): "Engineering Concepts for in situ Bioremediation," *Journal of Hazardous Materials,* vol. 28, pp. 1–11.

Ottengraf, S. P. P., A. H. C. Van Den Oever (1983): "Kinetics of Organic Compound Removal from Waste Gases with a Biological Filter," *Biotechnology and Bioengineering,* vol. 25, pp. 3089–3103.

Reineke, W., and H. J. Knackmuss (1988): "Microbial Degradation of Haloaromatics," *Annual Review of Microbiology,* Ornston, L. N., A. Balows, and P. Baumann (eds.), Annual Reviews, Palo Alto, CA.

Ryan, J. R., R. C. Loehr, and E. Rucker (1991): "Bioremediation of Organic Contaminated Soils," *Journal of Hazardous Materials,* vol. 28 , pp. 159–169.

Skladany, G. J. (1992): Overview of Bioremediation, in *Bioremediation: the State of Practice in Hazardous Waste Remediation Operations,* a live satellite seminar, sponsored by AWMA and HWAC, Jan. 9.

State Water Resources Control Board (1991): "Report on Releases of Hazardous Substances from Underground Storage Tanks, 91-DCWP," State of California, Sacramento, CA.

Strand, S. E., J. V. Wodrich, and H. D. Stensel (1991): "Biodegradation of Chlorinated Solvents in a Sparged, Methanotrophic Biofilm Reactor," *Research Journal Water Pollution Control Federation,* vol. 63, no. 6, p. 859.

Wilson, E. O. (1992): *The Diversity of Life,* W. W. Norton, New York.

CHAPTER 2

The Soil Environment

The term *soil* refers to the loose material of the earth's surface. Soil provides mechanical support and nutrients for plant and microbial growth. Fertile soils, those that support abundant production of food and fiber, are characterized by both the presence of nutrients and a physical structure amenable to living organisms. A broad range of microorganisms (bacteria, actinomycetes, fungi, algae, and protozoa) are nearly always present in soil, although population densities vary widely. The surface of soil granules is the site of many of the biochemical reactions that take place in the cycling of organic matter, nitrogen, and other minerals; in the weathering of rocks; and in the nutrition of plants (Alexander, 1991).

Physical and chemical properties of soils have a profound influence on aeration, nutrient availability, and water retention and thus on biological activity. The most important of these properties are particle size, porosity, moisture content, aeration status, chemical composition, clay fraction, cation-exchange capacity, and organic fraction. Particle size affects the surface chemistry of soils and the size of the pores. The amount of pore space depends on the texture, structure, and organic content of the soil. In clay soils, where particle size is of the order of micrometers, smaller pore sizes dominate while in sandy soils pores are larger but the total quantity of pores is less. Water moves more quickly through large pores, but little is retained and therefore coarse soils drain rapidly. Plant and microbial growth requires water and thus an optimal pore structure exists in which water is retained but a significant fraction of the pores remain filled with air.

SOIL MAKEUP

The soil matrix is made up of five major components: minerals, air, water, organic material, and living organisms. Mineral materials are the principal structural components of soils and make up over 50 percent of the total volume. Air and water together make up the pore volume, which usually occupies 25 to 50 percent of the total volume. The relative proportion of air to water fluctuates considerably with the moisture content of the soil. Organic material takes up between 3 and 6 percent of the volume on average, while living organisms occupy less than 1 percent.

Two principal processes are responsible for soil formation over time: weathering and breakdown of rocks and parent minerals, and colonization and activity of plants and microorganisms. In the early stages of formation, microbial activity is scarce, principally owing to the deficiency of carbon and nitrogen. The initiation of living processes in soil is dependent on the presence of organisms capable of carbon dioxide and nitrogen fixation. Such organisms almost always use light as their energy source, and for this reason colonization of new soils takes place near the surface. *Cyanobacteria,* also known as blue-green algae, are one of a few groups of microorganisms capable of both photosynthesis and nitrogen fixation and are nearly always among the first groups of microorganisms observed on newly formed soils. Products of microbial metabolism include complex organics and mineral nutrients required for the growth of more fastidious microorganisms, plants, and animals. Thus the initial colonization of soil by *Cyanobacteria* and other microorganisms leads to the establishment of higher vegetation. When plants die, chemoheterotrophic microorganisms break down the tissue. Much of the tissue is biochemically transformed into a stable organic matrix known as humus, and over time accumulation of humus builds up the organic fraction of the soil.

The Mineral Fraction

The dominant mineral in soil is silicon dioxide (SiO_2). Aluminum and iron are also plentiful, while calcium, magnesium, potassium, titanium, manganese, sodium, nitrogen, phosphorus, and sulfur are present in lesser amounts (Alexander, 1991). Chemical composition varies greatly between soils and at different depths within the same soil. Microorganisms obtain a portion of their required nutrients from the mineral portion of soil, and consideration must be given to its chemical composition. Nutrients required by microorganisms include nitrogen, phosphorus, potassium, magnesium, sulfur, iron, calcium, manganese, zinc, copper, and molybdenum. However, only a small fraction of soil minerals are readily available to microorganisms. In general, the nitrogen, phosphorus, and trace minerals of a soil represent a slowly utilized reservoir rather than a readily available supply.

Soils are classified by particle size, with the three main components being clay, silt, and sand. Size classification is commonly based on sieve analysis using a U.S. Department of Agriculture (USDA) system. Clays pass a 0.002-mm (2-μm) sieve, silts are retained on a 0.002-mm sieve but pass a 0.05-mm sieve, and sands are retained on a 0.05-mm sieve but pass a 2-mm sieve. Particles retained on a 2-mm sieve are classified as gravel or stones.

The unit surface area (area/volume) of the particle types directly impacts the chemical, physical, and biological properties of the soil (Table 2.1). Clay exerts the most influence on the properties of soil, primarily because of the high surface activity. Most clay particles are colloidal in nature, carry a net negative surface charge, and have flat platelike shapes. Clays are good absorbents of water, ions, and gases. Larger particles, such as sand, do not possess the same level of surface

TABLE 2.1
Classification of soil particles

Particle type	Diameter, mm	Surface area, m^2/kg
Sand		
Fine gravel	1.0–2.0	1.1
Coarse sand	0.5–1.0	2.3
Medium sand	0.25–0.5	4.5
Fine sand	0.1–0.25	9.1
Very fine sand	0.05–0.1	22.7
Silt	0.002–0.05	45.4
Clay	< 0.002	1,130

Source: Millar et al., 1958.

activity. The principal effects associated with sand are related to soil pore size and hence the movement of air and water in soil. Sand does not greatly influence the chemical and biological properties of the soil.

The presence of surfaces in soil which strongly adsorb certain classes of compounds may reduce the availability of organic compounds for biodegradation. Another factor influencing the availability of nutrients is the cation-exchange capacity of the soil. Clay minerals and organic materials possessing sites of negative electrical surface charge attract positively charged ions such as NH_4^+, K^+, Na^+, Ca^{2+}, and Mg^{2+}. As a result, ammonium, for example, which is positively charged, is less available for immediate use by plants and microorganisms and is retained longer in the soil than nitrate, its oxidized and negatively charged counterpart.

Soil Organic Matter

The organic fraction of the soil is made up of plant and animal debris, microbial cells, and products of microbial metabolism, and is often referred to simply as humus. The term humus refers to organic material that has undergone enough degradation and transformation to make the parent material unrecognizable (Atlas and Bartha, 1987). Humus is largely composed of polymerized substances: aromatics, polysaccharides, amino acids, uronic acid polymers, and phosphorus-containing compounds (Alexander, 1991). The amount of humus in soil is greatly influenced by agricultural activities. Humus rarely exceeds 10 percent by weight of mineral soils. In highly organic soils such as peat, however, humus can be as high as 90 percent of the soil weight.

Humus is a complex mixture of a large variety of ill-defined compounds with molecular weights ranging between 700 and 300,000. Much of the organic matter in soil, particularly humus, is only slightly soluble in water and somewhat resistant to biodegradation. Based on their solubility characteristics, humic substances are divided into humic acid, fulvic acid, and humin. Experimentally, the three components are differentiated by the fact that humic acid is extractable with alkali but precipitated with acid; fulvic acid is extractable with alkali but soluble in acid, while humin is not extractable with alkali (Gray and Williams, 1971). The chemical structure of the three components is unknown, or perhaps is better described as

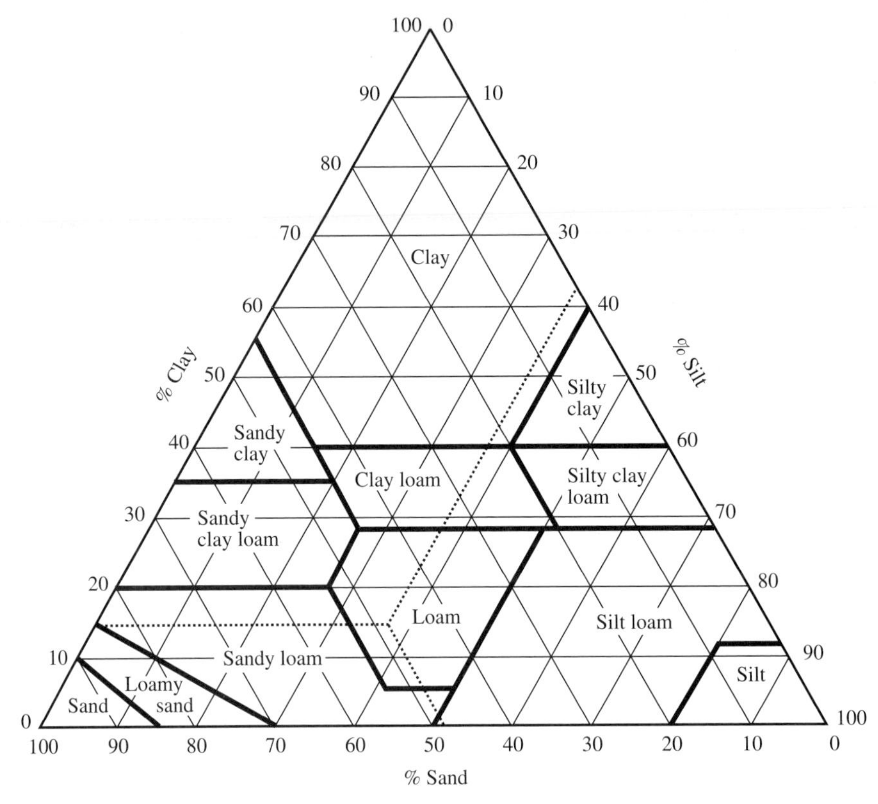

FIGURE 2.1
U.S. Department of Agriculture triangular soil classification chart.

undefined. Rather, it is believed that humic substances are made up of "randomly assembled irregular polymers" (Atlas and Bartha, 1987). Humic material, like clay, has colloidal properties and a net negative charge; hence its importance in forming mineral–organic matter complexes is significant, as will be explained in the following sections.

SOIL TEXTURE

The amounts of sand, silt, and clay in a certain soil determine the textural group to which that soil belongs. A textural triangle (Figure 2.1) is usually used to determine the "texture" classification for a specific soil. To use the triangle, one would start with the number corresponding to the silt content (percentage) of the soil and draw a line parallel to the left side of the triangle. Then starting at the number corresponding to the clay content, one would draw a horizontal line parallel to the base. The area in which the two lines intersect is then the class name to which the soil belongs (see Example 2.1).

The textural class termed loam, seen in Figure 2.1, simply refers to soils that are not dominated by any one of the particle sizes or types. The importance of the clay content in soil is emphasized in the textural triangle by the fact that soils with less than 40 percent clay are classified as sandy clay.

Sand and gravel are generally regarded as coarse particles, compared to silt and clay, which are considered fine particles. Soils dominated by sand tend to have a low water-holding capacity and high permeability; they are generally well aerated and drain well. Soils high in silt tend to have a problem with crusting, resulting in surface runoff. In general, soils dominated by silt and clay tend to be cohesive and to have a high water-holding capacity and low permeability, leading to low infiltration rates and low aeration.

EXAMPLE 2.1. DETERMINATION OF SOIL TEXTURE CLASSIFICATION. A soil analysis has resulted in the following information:

Sample mass	250.0 g
Sand	120.0 g
Silt	92.5 g
Clay	37.5 g

Determine the appropriate soil texture classification using the USDA method.

1. Calculate percentages of sand, silt, and clay.

$$\text{Percent sand} = \frac{120\ \text{g}}{250\ \text{g}}\quad 100 = 48$$

$$\text{Percent silt} = \frac{92.5\ \text{g}}{250\ \text{g}}\quad 100 = 37$$

$$\text{Percent clay} = \frac{37.5\ \text{g}}{250\ \text{g}}\quad 100 = 15$$

2. Use Figure 2.1 to classify soil texture: *loam*

SOIL STRUCTURE AND AGGREGATION

Soil *structure* can be defined as the arrangement and organization of the different particles in the soil. As such, soil structure defines a qualitative rather than a quantitative property of the soil. It relates to the total porosity within a soil volume, the shapes of the individual pores, and the overall pore-size distribution. As a result, soil structure greatly affects the mechanical properties of the soil, mainly the movement of fluids, including infiltration, water retention, and aeration. Soils in which the particles are loose and unattached to each other, such as the unconsolidated deposits of desert dust, are described either as having no structure or as having a single-grained structure. On the other hand, soils in which the particles are tightly packed, such as in dried clay, are often described as having a massive structure. Soils with a structure of somewhere in between are described as being aggregated (Hillel, 1982).

Soil *aggregation* is the stabilization of sand, silt, and clay, through the formation of clay–organic matter complexes, into aggregates. Compared to the mineral particles, aggregates are temporary structural units. Their stability is strongly affected by microbial activity, changes in climate, and land-management practices such as tilling. Clays, because of their large surface area, their colloidal nature, and their net negative charge, are essential for aggregate formation. Microorganisms and some soil organic matter constituents are also negatively charged at neutral pH values. Aggregate formation results from the ionic bonding between clays and organic matter via the multivalent cations present in the soil. In the case of a divalent cation, for example, one of the bonds would attach to a microorganism or organic matter, and the other would attach to clay. Aggregates are further consolidated through physical forces such as drying, freeze-thaw, root growth, and compaction.

SOIL GASES

The amounts of water and air within a soil volume are directly related, since the part of the pore space that is not filled with water would be filled with gas. The major gases that make up the soil air are essentially the same as those in the earth's atmosphere: nitrogen, oxygen, and carbon dioxide. The relative concentrations of the gases, however, especially oxygen and carbon dioxide, depend on how aerated the soil is, and on the microbial activity within the profile. In our atmosphere oxygen makes up about 20 percent of the air content, while carbon dioxide occupies only 0.03 percent. However, in well-aerated soils the oxygen concentration may range between 18 and 20 percent and the carbon dioxide concentration may be as high as 1 to 2 percent. In less aerated soils, such as a clay soil with a relatively high water content and high microbial activity (respiration), carbon dioxide may occupy as much as 10 percent of the air volume (Paul and Clark, 1989).

The diffusion of gases within a soil profile, as described by Fick's law, is directly related to the concentration gradient within that profile:

$$q = -D\frac{dC}{dz} \tag{2.1}$$

where q = diffusive flux, $g/cm^2 \cdot s$
D = diffusion constant, cm^2/s
C = concentration of the gas, g/cm^3
z = depth, cm

Gases can move either in the air phase, that is, through the pores, assuming that they are interconnected and open to the atmosphere, or in the liquid phase in the dissolved form. The solubility of gases in water depends on several factors, including the gas itself, temperature, and the partial pressures of the gases within the pore space. It should be noted, however, that the diffusion of gases in the water phase is about 10,000 times slower than in the gas phase (Table 2.2).

TABLE 2.2
Diffusion constants for the major gases in air and water and their solubility in water at 20°C

Gas	Diffusivity, cm^2/s		Henry's law coefficient,* dimensionless
	Air	Water	
CO_2	0.161	0.177×10^{-4}	1.07
O_2	0.205	0.180×10^{-4}	30.7
N_2	0.205	0.164×10^{-4}	60.4

Source: Paul and Clark, 1989.
$^*H = C_g/C_L$

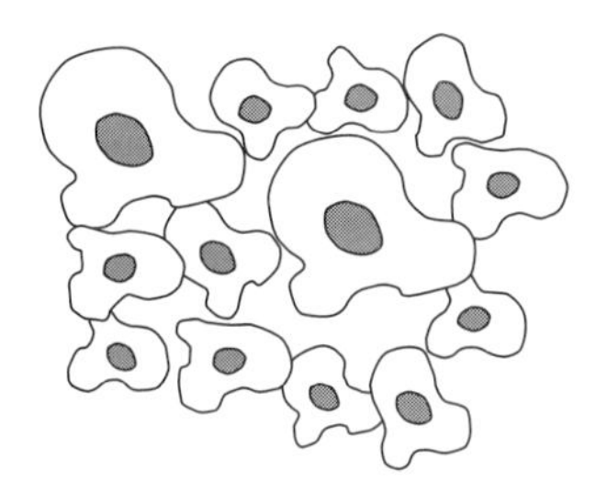

FIGURE 2.2
Soil aggregates with anaerobic microsites in the core and aerobic outer section.

In terms of microbial activity, the overall aeration status of the soil is not as important as conditions within the aggregates. Soils that are generally well aerated may have anaerobic microsites within the aggregate formations. Anaerobic microsites, as shown in Figure 2.2, are a partial explanation for the presence of anaerobic bacteria, such as *Clostridia* in the top soil layers. It has been estimated that soil crumbs larger than 6 mm in diameter have no oxygen at their center. Ironically, it is the aerobic bacteria that create the conditions necessary for anaerobic bacteria to survive. As the aerobic bacteria first colonize the microsites, they consume all of the stored oxygen, thus creating anaerobic conditions and allowing the anaerobic colonies to develop. The transition from aerobic to anaerobic conditions is believed to occur at oxygen concentrations smaller than 1 percent. At the same time, it is believed that to maintain adequate aeration within the soil, the percentage of pore space filled with air should not drop below 10 percent (Paul and Clark, 1989). Improper aeration is generally associated with poor drainage and flooding.

SOIL MOISTURE

Moisture content of the soil strongly influences biological activity. Water is the major component of bacterial protoplasm, and an adequate supply of water is essential for microbial growth and maintenance. Too little moisture in the soil results in dry zones and loss of microbial activity. Too much moisture, however, inhibits gas exchange and oxygen movement into and through the soil and results in the development of anaerobic zones, with the resulting elimination of aerobic bacteria and the ascendance of anaerobes or facultative anaerobes.

Traditionally, soil water has been divided into three types: gravitational, capillary, and osmotic (Gray and Williams, 1971). Gravitational water, as the term itself implies, is the water that drains out under the influence of gravity. It is water that is available to the microorganisms and plant roots and it plays an important role in the transport of contaminants and other material. Capillary water, that which is held by the pores, is also available to the soil microorganisms. Osmotic water is that which is held by the clay particles and humus and, unlike the other two types of water, is less available to the soil microorganisms and plant roots.

Field capacity is an often used term which relates to soil moisture. Veihmeyer and Hendrickson, who first used the term field capacity, defined it as "the amount of water held in soil after excess water has drained away and the rate of downward movement has materially decreased, which usually takes place within 2 to 3 days after a rain or irrigation in pervious soils of uniform structure and texture" (Hillel, 1982). The definition is obviously very vague and lacks reliable quantitative method of measurement. Yet the concept of field capacity has been and still is useful in describing the soil's ability to retain water once internal drainage has stopped. Field capacity of soil typically ranges from 18 to 30 percent by weight and is a function of clay content.

The soil water content, on a mass or on a volume basis, is a function of the soil water pressure potential or, as it is also known, the matric potential. Conceptually, the matric potential is a measure of the tenacity with which water is held in the pores or in the soil matrix. Water is retained in the pores mainly through the capillary effect and through adsorption. Experimentally, the matric potential is a measure of the suction pressure needed to bring a soil volume to a particular water content. The relationship between the matric potential and the soil moisture content is represented graphically with the soil-moisture characteristic curve which is unique for each type of soil. Typical curves for different types of soils are presented in Figure 2.3.

It is noticed from Figure 2.3 that at any particular soil–water pressure potential (suction pressure), a clay soil will retain more water than a loam, and more than a sand soil. In general, the higher the clay content of a soil, the greater its ability to retain water, and the more gradually changing is the slope of its soil-water characteristic curve. That is mainly due to the porosity and the pore-size distribution of the different soils. A sandy soil has a smaller overall porosity compared to a clay soil. However, the pores in a sandy soil are generally larger in size compared to the clay soil. Once suction is applied to a sandy soil, the large pores are emptied quickly, and a small amount of water is left in the soil matrix. In clay soils, water is adsorbed in the small pores and the decrease in water content is more gradual.

The soil–water pressure potential is often reported in the literature in bars (1 bar = 100 kPa) and sometimes in atmospheres (1 atm = 101.3 kPa) or in megapascals (MPa). In the laboratory it is normally measured in centimeters of water suc-

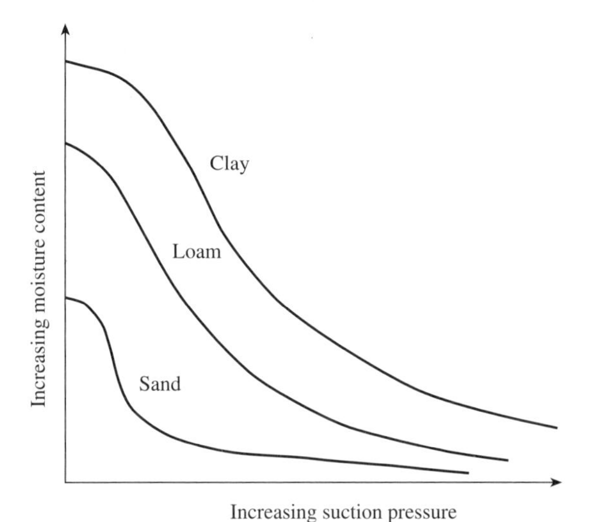

FIGURE 2.3
Typical soil-water characteristic curves for different types of soil.

TABLE 2.3
Suction pressure needed to remove water at different states

	Suction pressure needed to remove all water	
State of water	**m**	**bars**
Gravitational	0–3	0–0.3
Capillary	3–150	0.3–15
Osmotic	150–1500	15–150

Source: Adapted from Gray and Williams, 1971.

tion. The height of a water column is converted to a pressure measurement through the relationship

$$P = \rho g h \tag{2.2}$$

where P = pressure, kg/m· s^2 or pascals (Pa)
ρ = density of water, kg/m^3
g = gravitational constant, 9.81 m/s^2
h = height of water column, m

The three states of soil-water mentioned above, gravitational, capillary, and osmotic, can be removed from the soil matrix under increasing amounts of suction pressure as seen in Table 2.3.

The soil moisture at 1/3 bar is often assumed to be the same as field capacity, while the soil moisture at 15 bars is considered to be the permanent wilting point of plants. Microbial activity, on the other hand, is believed to be optimal at soil moisture contents corresponding to 0.5 bar suction pressure (Paul and Clark, 1989). At lower suction pressures the soil is too moist and may become waterlogged, while at higher pressures the soil is too dry.

MASS AND VOLUME RELATIONSHIPS IN THE SOIL MATRIX

Soil is composed of solids, air, and water. The solid fraction, which is the dominant portion on a mass basis, is made up of minerals, organic matter, and microorganisms, as described earlier in the chapter. Organic matter and microorganisms usually constitute a very small fraction of the solid matter. The void volume in soil is filled with air and water and composes 30 to 60 percent of the total soil volume. Because of the low density of air, dry and wet soil have substantially different bulk densities. The three phases that make up the soil matrix are represented schematically in Figure 2.4. Void volume, that is, the pore space, is expressed as V_f. The mass of air M_a is virtually always assumed to be negligible compared to the mass of water or that of the solids. Solids mass is often referred to as the oven-dry weight since it is obtained by oven drying the soil at 103°C. Mass moisture content is calculated as the difference in soil mass before and after oven drying, and the volume of moisture is calculated from the mass of moisture and the density of water.

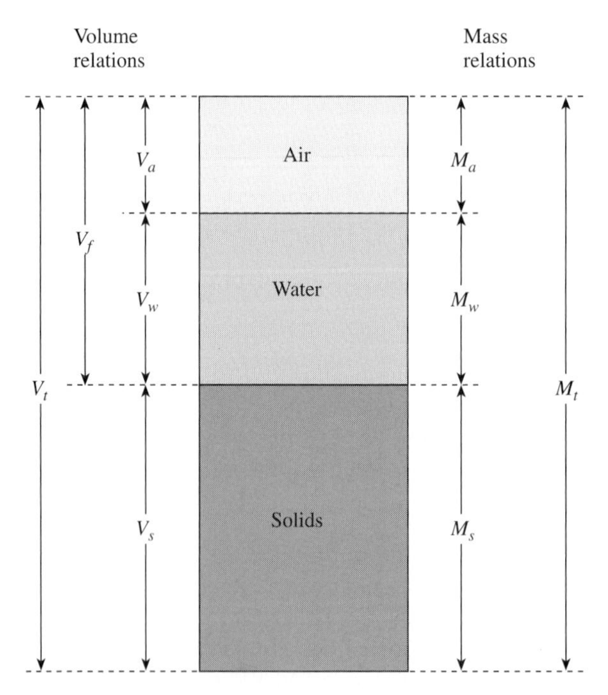

FIGURE 2.4
Mass and volume relationships of the three phases that make up the soil.

Several physical properties of the soil can be expressed with the help of the diagram in Figure 2.4. The most important of these properties are presented below.

Solids density ρ_s

$$\rho_s = \frac{M_s}{V_s} \tag{2.3}$$

In most mineral soil, the average density of the solids is between 2.6 and 2.7 g/cm^3.

Dry bulk density ρ_b

$$\rho_b = \frac{M_s}{V_t} = \frac{M_s}{V_s + V_a + V_w} \tag{2.4}$$

The dry bulk density is the mass of solids divided by the total volume (volume of solids plus volume of pores) and is therefore always smaller than the solids density. Its value ranges between about 1.1 g/cm^3 for clay soils and about 1.6 g/cm^3 for sandy soils (Hillel, 1982). The wet bulk density, which is not often used in the literature, is equal to the total mass ($M_s + M_w$) divided by the total volume.

Porosity ϕ

$$\phi = \frac{V_f}{V_t} = \frac{V_a + V_w}{V_s + V_a + V_w} \tag{2.5}$$

For most soils, the porosity ranges in value between 0.3 and 0.6. Compared to clay soils, the pores in sandy soils are larger in size but the overall porosity is smaller. The porosity of clay soils, however, varies considerably owing to changes resulting from swelling, shrinking, dispersion, compaction, and cracking.

Soil moisture content

Soil wetness, or the soil moisture content, can be expressed on either a mass basis or a volume basis.

Mass basis ω

$$\omega = \frac{M_w}{M_s} \tag{2.6}$$

Volume Basis θ

$$\theta = \frac{V_w}{V_t} = \frac{V_w}{V_s + V_w + V_a} \tag{2.7}$$

Less often used is the degree of saturation f_s, which is defined as the volume of water present in the soil relative to the volume of pores:

$$f_s = \frac{V_w}{V_f} \tag{2.8}$$

The value of f_s varies between zero for dry soil, and 1 for completely saturated soil.

EXAMPLE 2.2. CALCULATION OF SOIL PROPERTIES. A soil core, collected from the field, has a bulk volume of 100 mL, an air volume of 30 cm^3, a wet mass of 145 g, and a dry mass of 125 g. Calculate the total porosity and the bulk density.

$$\text{Total porosity} = \phi = \frac{V_f}{V_t} = \frac{V_a + V_w}{V_t}$$

where

$$V_a = 30 \text{ mL}, \quad V_t = 100 \text{ mL}, \quad V_w = \frac{M_w}{\rho_w}$$

$$M_w = M_t - M_s = 145 - 125 = 20 \text{ g}$$

$$\rho_w = 1 \text{ g/cm}^3$$

therefore,

$$V_w = \frac{20 \text{ g}}{1 \text{ g/mL}} = 20 \text{ mL}$$

and

$$\text{Porosity} = \phi = \frac{30 \text{ mL} + 20 \text{ mL}}{100 \text{ mL}} = 0.5 \text{ mL/mL}$$

$$\text{Bulk density} = \rho_b = \frac{M_s}{V_t} = \frac{125 \text{ g}}{100 \text{ mL}} = 1.25 \text{ g/mL} = 1{,}250 \text{ kg/m}^3$$

PROBLEMS AND DISCUSSION QUESTIONS

2.1. A soil is made up of 24 percent sand, 52 percent silt, and 24 percent clay by weight.
a. How much sand should be added to 500 kg of the same soil to bring the sand content to 35 percent?
b. What are the resulting silt and clay contents on a percent weight basis?

2.2. An experiment is to be conducted with 250 g dry soil. If the soil, in its natural condition, has a moisture content of 18 percent by weight, then how much of the wet soil is needed to account for 250 g dry soil?

2.3. A soil core 10 cm in diameter and 10 cm high is found to have a mass of 1,690 g. After drying in an oven at 105°C, the mass of solids is found to be 1,465 g. Assuming the soil has a particle density of 2.66 g/cm^3, then:
a. What is the moisture content of the soil on a weight basis?
b. What is the porosity of the soil?
c. What is the air-filled porosity?

2.4. A soil sample, 1,000 cm^3, has a porosity of 0.25, a wet bulk density of 2.14 g/cm^3, and a moisture content of 7.9 percent by weight.
a. What is the particle density of the soil?
b. What is the volumetric moisture content of the soil?
c. What volume of water would need to be added to the soil to bring it to saturation?

2.5. Derive an expression for porosity f as a function of dry bulk density ρ_b and particle density ρ_s.

REFERENCES

Alexander, Martin (1991): *Introduction to Soil Microbiology,* Wiley, New York and London.
Atlas, R. M., and R. Bartha (1987): *Microbial Ecology: Fundamentals and Applications,* 2d ed., Benjamin/Cummings Publishing, Menlo Park, CA.
Gray, T. R., and S. T. Williams (1971): *Soil Microorganisms,* Oliver & Boyd, Edinburgh.
Hillel, D. (1982): *Introduction to Soil Physics,* Academic Press, London.
Millar, C. E., L. M. Turk, and H. D. Foth (1958): *Fundamentals of Soil Science,* 3d ed., Wiley, New York.
Paul, E. A., and F. E. Clark (1989): *Soil Microbiology and Biochemistry,* Academic Press, London.

CHAPTER 3

Fate and Transport

Contaminants released into the environment as the result of manufacturing, use of manufactured products, and waste disposal rarely remain at the point of discharge in an untransformed state. Transport through mechanisms of advection, dispersion, and interphase transfer normally takes place. In many cases chemical and biochemical transformations result in significant changes in the nature of the contaminants. In the majority of cases, contaminant mixtures are involved and individual species may be transported at significantly different rates. Additionally some compounds are conservative, that is, highly resistant to transformation, while others are quite chemically or biochemically reactive. Thus, old or weathered contaminants may be quite different in makeup compared to recent discharges. Contamination resulting from leaking underground gasoline storage tanks often provides an excellent example of transport, interphase transfer, and transformation processes. Gasoline contaminant plumes often extend over large distances and move downward to contaminate groundwater. Partitioning of gasoline onto soil organic matter is a function of the individual chemical species in the mixture, as is partitioning to the vapor phase. Most components of gasoline are readily biodegradable, and if nutrients and oxygen are present biodegradation will occur. Weathering is a term used to describe the selective loss or disappearance of components of a contaminant mixture in the soil. Gasoline weathering usually results in relative decreases in the amounts of the more volatile compounds present. With less volatile contaminant mixtures, such as PAHs, weathering may involve permanent sorption of less soluble components into the soil organic matrix.

Successful hazardous-waste management and site remediation requires an understanding of contaminant fate and transport. The required understanding incorporates information from several diverse disciplines including physics, fluid mechanics, earth sciences, chemistry, and biology. This chapter presents many of the fundamental processes affecting the fate and transport of contaminants in the

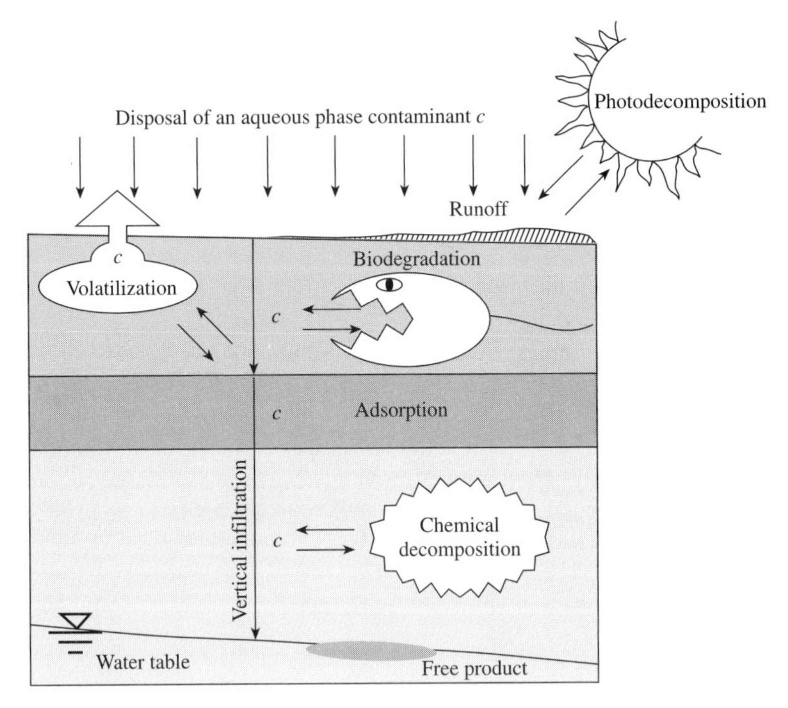

FIGURE 3.1
A schematic diagram of some of the important processes that can affect contaminant fate and transport during vertical infiltration.

environment. Figure 3.1 illustrates how several processes considered here affect the fate of mass as it percolates through the soil column.

HETEROGENEITY

The subsurface is highly variable, or heterogeneous, in structure and texture over relatively small distances, and situations such as those depicted in Figure 3.2 are not uncommon. Fluid flow is strongly affected by the presence of regions having different hydraulic conductivities (or resistance to flow). In one sense flow is accelerated between regions of low permeability or conductivity. Thus as materials flow through the section depicted in Figure 3.2 nearly all of the water would pass through the fine sand and little would pass through the silt or finer (low conductivity) materials. However, contaminants are often found to have accumulated in fine materials. The mechanisms by which the accumulation takes place are not well understood because nonaqueous-phase liquids (NAPLs) are usually hydrophobic and would tend to be excluded from regions with strong surface activity such as pores in clays. Sorption onto the humic materials present in soils may be an explanation, but transport properties observed do not fully support sorption as an explanation. The net result of heterogeneity is that the movement of contaminants is usually very different from the movement of water. In this book the problem will be greatly

FIGURE 3.2
Heterogeneous characteristics of the subsurface.

simplified and sorption will be used as the principal explanation of the observation of differences in fluid and contaminant movement. Those interested in the subject may consult references such as Ginn et al. (1995) and Simmons et al. (1995).

CONSERVATION OF MASS

Description of contaminant migration in soil or in an aquifer is based on application of equations defining conservation of mass. Advanced engineering students are familiar with the use of subunits of systems having defined boundaries called *control volumes* and with the development of the equation of continuity in fluid mechanics. However, consideration of individual chemical species within the fluid, movement of materials between phases, and generation of material through chemical and biochemical reactions result in somewhat more complicated expressions. We will begin with the development of the continuity equation for flow in a porous medium and proceed to expressions for soluble chemical compounds. In each case a word equation [Equation (3.1)] defining conservation of mass in a control volume is converted into mathematical terms.

$$\begin{matrix}\text{Rate of mass} \\ \text{accumulation}\end{matrix} = \begin{matrix}\text{rate of mass} \\ \text{entering}\end{matrix} - \begin{matrix}\text{rate of mass} \\ \text{leaving}\end{matrix} + \begin{matrix}\text{rate of mass} \\ \text{generation}\end{matrix} \tag{3.1}$$

The simplest application of Equation (3.1) is the development of the one-dimensional continuity equation for flow through an open control volume such as is shown in Figure 3.3. Note that similar equations could be developed for flow in the x and y directions and that the three resulting equations would provide a three-dimensional description of flow through the control volume. The objective here is to provide a framework for consideration of movement of contaminants in a porous medium rather than a general discussion of fluid flow.

Consider the case in which flow is only in the vertical direction and the generation term is 0, that is, no chemical or biochemical transformations are occurring, the mathematical form of Equation (3.1) is given by Equation (3.2*a*).

$$\Delta x\, \Delta y\, \Delta z \frac{\partial \rho}{\partial t} = \Delta x\, \Delta y\, \rho v_z - \Delta x\, \Delta y\, \rho v_{z+\Delta z} \tag{3.2a}$$

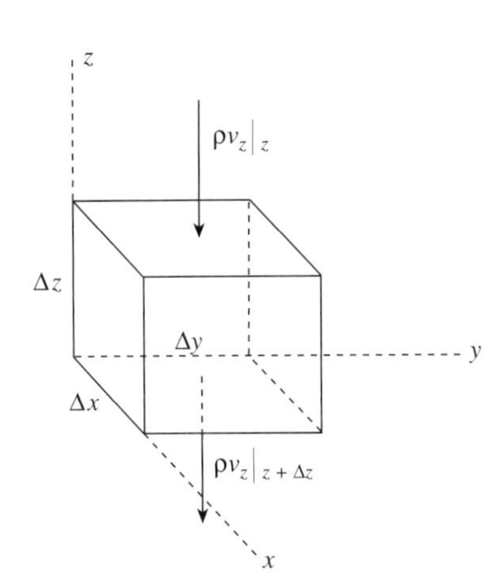

FIGURE 3.3
One-dimensional fluid transport through a control volume with dimensions Δx, Δy, and Δz.

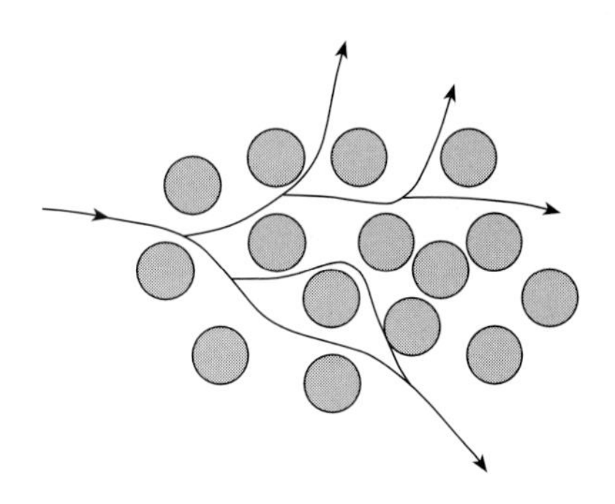

FIGURE 3.4
Local (pore) scale velocity variations mechanically disperse mass in a porous medium.

where ρ = fluid density, kg/m^3
v_z = velocity in z direction, m/s

Taking the limit as $\Delta z \rightarrow 0$ results in Equation (3.2*b*).

$$\frac{\partial \rho}{\partial t} = -\frac{\partial(\rho v_z)}{\partial z} \tag{3.2b}$$

Because liquid density can be assumed constant, the velocity is also constant for one-dimensional flow situations (which are quite rare in the subsurface). For multidimensional flow the divergence of the velocity vector is equal to 0.

$$\nabla \cdot v = \frac{\partial v_x}{\partial x} + \frac{\partial v_y}{\partial y} + \frac{\partial v_z}{\partial z} = 0 \tag{3.3}$$

Continuity in a Porous Medium

In a porous medium, as shown in Figure 3.4, the fluid flow can be described on both a macroscale in which the average velocities are estimated and a microscale in which actual velocities in the pores are considered. In most cases, including this text, average velocities are calculated based on water content of the porous medium and the porosity. The equation of continuity for a porous medium is then

$$\Delta x \, \Delta y \, \Delta z \, \frac{\partial \Theta \rho}{\partial t} = \Delta x \, \Delta y \, \Theta \rho v_z - \Delta x \, \Delta y \, \Theta \rho v_{z+\Delta z} \tag{3.4a}$$

Collecting terms and allowing $\Delta z \rightarrow 0$ gives

$$\frac{\partial(\Theta\rho)}{\partial t} = \frac{\partial(\Theta \rho v)}{\partial z} \tag{3.4b}$$

where Θ = volumetric water content (volume of water per volume of porous medium)

Volumetric water content can vary considerably in soil, as was described in Chapter 2, with typical values for soil ranging from less than 0.1 to greater than 0.3. Because of the difficulty in describing conditions in the soil, mass balances are usually taken over regions in which Θ is constant and the continuity equation can be written as

$$\Theta\frac{\partial\rho}{\partial t} = -\Theta\frac{\partial(\rho v_z)}{\partial z} \tag{3.4c}$$

$$\frac{\partial\rho}{\partial t} = -\frac{\partial(\rho v_z)}{\partial z} \tag{3.4d}$$

and again, for a liquid, the divergence of the velocity vector is equal to 0.

Velocities in porous media are usually described using Darcy's law, an empirical expression first reported by the French hydraulician Henry Darcy in 1856.

$$v = -K_C\frac{\Delta h}{\Delta L} \tag{3.5}$$

where v = superficial velocity along the flow path L, m/s
K_C = hydraulic conductivity, m/s
Δh = head loss, m, over the distance ΔL, m

The hydraulic conductivity is a function of the properties of the porous medium, particularly grain size (or effective grain size), and the properties of the fluid (density and viscosity) (Davis and DeWiest, 1966).

$$K_C = Cd^2\frac{\rho g}{\mu} \tag{3.6}$$

where C = proportionality constant, unitless
d = grain size of porous medium, m
ρ = fluid density, kg/m^3
μ = fluid dynamic viscosity, N · s/m^2

Values of K_C range from approximately 10^{-3} m/s for coarse sands to 10^{-8} m/s for clays. The superficial velocity is based on the total cross-sectional area of flow. Velocities through the pores are higher and in a saturated porous medium average pore velocity v_{pore} can be estimated using Equation (3.7).

$$v_{\text{pore}} = \frac{v}{\phi} \tag{3.7}$$

where ϕ = porosity (volume of voids/volume porous media)

EXAMPLE 3.1. APPLICATION OF DARCY'S LAW. Flow between the two ponds shown in Figure 3.5 is through a fine sand lens having an average cross-sectional area of 30 m^2. Determine the superficial velocity, the average pore velocity, and the volumetric flow rate for the conditions shown.

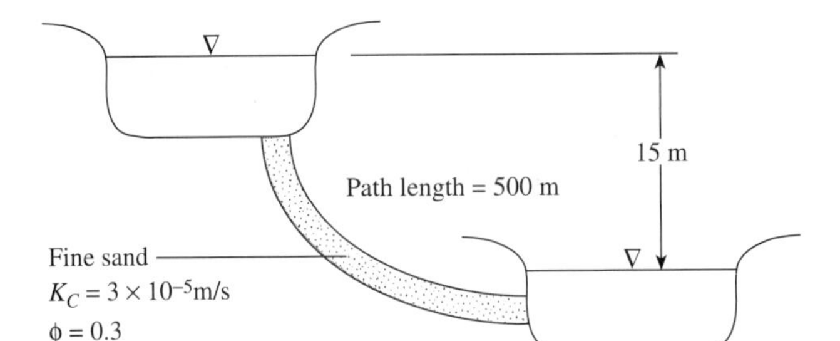

FIGURE 3.5
Definition sketch for Example 3.1.

Solution

1. Determine the superficial velocity using Darcy's law

$$v = -(3 \times 10^{-5} \text{ m/s})\left(-\frac{15}{500}\right) = 9.0 \times 10^{-7} \text{ m/s}$$

$$= 0.0778 \text{ m/day}$$

2. Determine the flow rate

$$Q = vA = (0.0778 \text{ m/day})(30 \text{ m}^2) = 2.33 \text{ m}^3\text{/day}$$

3. Estimate the average pore velocity

$$v_{\text{pore}} = \frac{0.0778}{0.3} = 0.26 \text{ m/day}$$

Conservation of Mass for a Chemical Species

Chemical compounds and ions move with the fluid due to advection and relative to the mean fluid flow due to a process termed *hydrodynamic dispersion.* The advective flux is exactly the same as the convective term in Equations (3.2) and (3.3) except that the fluid density is replaced by the mass concentration of the chemical species. Hydrodynamic dispersion occurs when concentration gradients exist and includes two processes in which a net transport of contaminants takes place, mechanical dispersion and molecular diffusion. In conventional flow situations, such as pipes and open channels, mechanical dispersion results from macroscale movement of fluid (e.g., by turbulent eddies) between regions having different concentrations. However, flow in porous media is virtually always laminar and mechanical dispersion results from differences in velocities along the various flow paths, as indicated in Figure 3.4. Molecular diffusion is the spreading of mass due to the thermal-kinetic energy of the fluid and contaminant molecules. However, the word equation (3.1) remains the same. Separate equations must be written for each

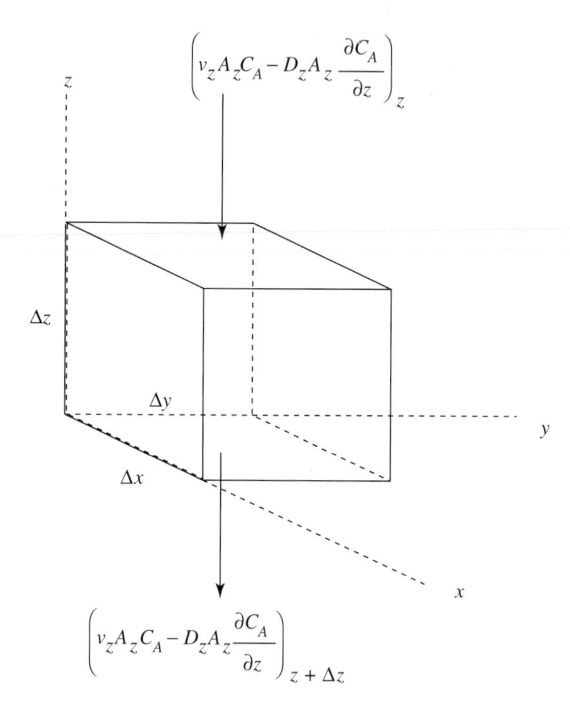

FIGURE 3.6
One-dimensional transport of a contaminant *A* in a fluid.

chemical species being monitored. In some cases the conservation equations are written for surrogate parameters, such as chemical oxygen demand (COD) or total petroleum hydrocarbons (TPH). The concentration term in the species mass balance corresponds to the fluid density term in Equation (3.2) and Figure 3.3. For a fluid system, such as flow in a pipe, a species mass balance on a material *A* in the fluid phase is shown in Figure 3.6 and written in Equation (3.8).

$$A_z \Delta z \frac{\partial C_A}{\partial t} = A_z \left(v_z C_A - D_z \frac{\partial C_A}{\partial z} \right)_z - A_z \left(v_z C_A - D_z \frac{\partial C_A}{\partial z} \right)_{z+\Delta z} + A_z \Delta z (r_A + r_{TA}) \quad (3.8)$$

where A_z = cross-sectional area in the *xy* plain, m^2
C_A = concentration of species *A*, g/m^3 (or possibly $moles/m^3$)
D_z = hydrodynamic dispersion coefficient in the *z* direction, m^2/s
r_A = transformation rate of species *A*, $g/m^3 \cdot s$
r_{TA} = rate of interphase transfer within the control volume, $g/m^3 \cdot s$

The addition of the last term on the right side of Equation (3.8) represents losses (or gains) from reactions or partitioning to the soil and will be discussed in greater detail later in the chapter.

Defining the terms in Equation (3.8) can be quite difficult in soil and groundwater systems. In single-phase systems the mass rates entering and leaving a control volume are the result of advection, the movement of a material with the fluid, designated by the $v_z C_A$ terms in Figure 3.6, and dispersion, the net movement of a material resulting from spatial concentration gradients, designated by the

$-D_z \partial C_A / \partial z$ terms in Figure 3.6. The dispersion term can be broken into a term related to flow variation and a term resulting from molecular diffusion. In fluid systems, such as pipe or channel flow, dispersion results from turbulence and molecular diffusion. Under turbulent-flow conditions, the contribution of turbulent diffusion to the dispersive flux is much greater than the contribution of molecular diffusion, while under laminar flow, molecular diffusion is the dominant cause of mixing.

Advection and Hydrodynamic Dispersion in Porous Media

Advection is the movement of particles or a contaminant by fluid velocity. In soils and aquifers, flow is nearly always laminar and, generally, it is either impossible or impractical to determine all aspects of a particular velocity field. Therefore, advection usually refers to transport due to the mean velocity field. The mass flux of a contaminant species in a porous medium due to advection in the z direction can be written as

$$\text{Advective flux} = \Theta v_z C_A \tag{3.9}$$

where C_A = aqueous-phase mass contaminant concentration of species A, g/m^3

When a porous medium is saturated, Θ is equal to the porosity ϕ.

Mechanical dispersion is the spreading of mass due to variations in velocity at a scale smaller than the size of the control volume (the averaging volume for the mean advective velocity), as suggested in Figure 3.4, and covers our ignorance of the true velocity field within the control volume. In soils and aquifers, flow is nearly always laminar and the flow-variation portion of the dispersion term is used to describe effects of shear in the velocity field and the tortuous path fluid takes within the porous medium. The contribution of "tortuosity" to dispersion is generally much greater than the contribution due to molecular diffusion. Mass flux due to hydrodynamic dispersion in a porous medium can be written as

$$\text{Dispersive flux} = -\Theta D_z \frac{\partial C_A}{\partial z} \tag{3.10}$$

where D_z = coefficient of hydrodynamic dispersion in the z direction, m^2/s

The coefficient of hydrodynamic dispersion is usually expressed as the sum of two terms, the mechanical dispersion $\alpha_z v_z$ and the molecular diffusivity D_m (Bear and Bachmat, 1991):

$$D_z = \alpha_z v_z + D_m \tag{3.11}$$

where α_z = dispersivity, an empirical coefficient accounting for tortuosity and mixing in the z direction, m

The generation-rate term is used to account for transformations of the compound by reaction (r_A) and the interphase transport (r_{TA}) of compound A within the control volume. Reaction-rate functions will be discussed in detail later in this chapter. Interphase transport results from a compound's being in more than one phase (e.g., present as pure product, sorbed onto the solid phase, and/or partitioned into the gas phase when the pores are not saturated). Flow and reaction results in loss of equilibrium conditions between phases, and interphase transfer is the result. The accumulation term of Equation (3.1) accounts for the change in mass within the control volume and is normally applied to only one phase. Thus for the vadose zone three conservation equations would be written, one each for the gas, liquid, and solid phases.

Transport of a Conservative Constituent in Porous Media

A conservative constituent is one for which the generation term is zero. The principles presented thus far form the basis for a model of a conservative constituent undergoing transport by advection and hydrodynamic dispersion but neither volatilizing into the gas phase nor adsorbing onto the solid phase. Based on Figure 3.7, a materials balance for a nonvolatile and nonadsorbing conservative substance A can be written as

$$A_z \Delta z \frac{\partial \Theta C_A}{\partial t} = A_z \left(v_z \Theta C_A - \Theta D_z \frac{\partial C_A}{\partial z} \right)_z - A_z \left(v_z \Theta C_A - \Theta D_z \frac{\partial C_A}{\partial z} \right)_{z+\Delta z} \tag{3.12}$$

Collecting terms and taking the limit as $\Delta z \to 0$ yields

$$\frac{\partial \Theta C_A}{\partial t} = \frac{\partial}{\partial z} \left(\Theta D_z \frac{\partial C_A}{\partial z} \right) - \frac{\partial \Theta v_z C_A}{\partial z} \tag{3.13}$$

Equation (3.13) is the one-dimensional advection-dispersion equation for the transport of a conservative constituent and applies to transport in unsaturated and

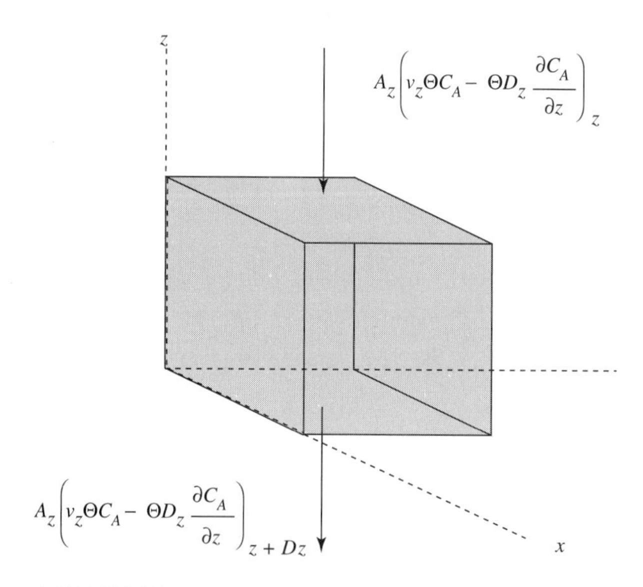

FIGURE 3.7
Control volume for vertical mass transport model in a porous medium.

saturated porous media. Solutions to Equation (3.13) will depend on boundary and initial conditions. In addition, the mean fluid velocity must be known. Movement of contaminants in soils and aquifers is always a three-dimensional problem, and equations corresponding to (3.13) for the x and y directions must be written also. Quite often flow in one, or perhaps two, directions dominates contaminant transport and a relatively simple one-dimensional model is satisfactory.

Interphase Transfer Processes

Processes other than advection and hydrodynamic dispersion that result in movement of materials into and out of a control volume are referred to here as transfer processes. The transfer processes that are of the most importance are adsorption, dissolution, absorption, and volatilization.

Adsorption

Molecules of a compound in solution become adsorbed onto the surface of a solid through chemical reaction (chemisorption) or physical (e.g., van der Waals) forces. Chemical bonding is stronger and less reversible than physical bonding. In most cases chemisorption is essentially permanent, while physical adsorption is readily reversible. Chemisorption limits transport and immobilizes contaminants. Discussion here will focus on physical adsorption. The rates of adsorption and desorption are dependent on the concentration of the contaminant in solution. If the concentration of the dissolved contaminant increases, adsorption increases; if the concentration decreases, desorption increases. Adsorption is controlled by several factors, the most important of which are the properties of the contaminant compound (e.g., molecular structure, charge, polarity, and water solubility) and the properties of the soil (e.g., clay and organic content, pH, moisture content, and temperature). An increase in temperature usually results in a decrease in contaminant adsorption onto the soil, as does an increase in moisture content (Dupont et al., 1988). Adsorption competes with biodegradation and volatilization in determining the fate of VOCs in the soil (U.S. EPA, 1990) and thus influences both volatilization and biodegradation. As for biodegradation, the strong adsorption of organics onto subsurface minerals, or their penetration into mineral structures that are too small for the microorganisms to reach, will render the organics inaccessible to the microbial population or their degrading enzymes (McCarty, 1991).

The most commonly encountered model for equilibrium adsorption of a compound from solution is the empirically derived Freundlich isotherm:

$$\frac{x_A}{m} = K_F C_{Ae}^{1/n} \tag{3.14}$$

where x_A = mass of A sorbed onto the solid phase, g
m = mass of solid phase, g
K_F = Freundlich adsorption coefficient, variable units
C_{Ae} = equilibrium concentration of A, g/m^3
n = empirical coefficient or fitting parameter, unitless

Langmuir developed a theoretical adsorption model based on the assumption that the adsorption surface is saturated when a monolayer has been adsorbed. The Langmuir isotherm can be written as

$$\frac{x_A}{m} = \frac{aC_{Ae}}{b + C_{Ae}} \tag{3.15}$$

where a = factor relating mass-sorbed per mass-sorbent to fractional surface area covered, unit less
b = saturation coefficient, g/m^3

In the soils literature, the distribution of a contaminant between the solid and liquid phase is given by the soil distribution coefficient defined as

$$K_{SD} = \frac{s}{C_{Ae}} \tag{3.16}$$

where K_{SD} = soil distribution coefficient, L/mg or m^3/g
s = mass of solute sorbed per unit dry mass of soil (x_A/m), unitless

The soil distribution is equal to the Freundlich adsorption coefficient when $n = 1$. Both the Freundlich and Langmuir isotherms approximate linear behavior for dilute solute concentrations.

Soils characteristics often vary a great deal over short distances, and prediction of K_{SD} value variation is a significant problem. Sorption of organic compounds is a strong function of the soil organic content and the hydrophobicity of the compounds. A relative measure of hydrophobicity is provided by the octanol water partition coefficient K_{ow}, the ratio of mass concentration of a compound in octanol to the concentration in water. Values of K_{ow} are available in the literature for a large number of organic contaminants, and the relation below can be used to estimate K_{SD} in the absence of other data (note that the units are associated with the coefficient exponent).

$$K_{SD} = 6.3 \times 10^{-7} f_{oc} K_{ow} \tag{3.17}$$

where f_{oc} is the mass fraction of organic carbon in the soil.

EXAMPLE 3.2. DETERMINATION OF K_{ow}. 10 mg of 1,1,1-trichloroethane (1,1,1-TCA) is added to a closed 500-mL bottle containing 200 mL of water and 300 mL of octanol. The bottle is placed on a shaker operated at low speed for 24 h and the fluids are then allowed to separate. Analysis of the 1,1,1-TCA concentrations in each liquid resulted in values of C_{water} = 105 μg/L and C_{octanol} = 33.3 mg/L. Determine K_{ow} and check the mass balance.

Solution

1. Determine K_{ow}.

$$K_{ow} = \frac{C_{\text{octanol}}}{C_{\text{water}}} = \frac{33.3 \text{ mg/L}}{0.105 \text{ mg/L}} = 317$$

2. Check mass balance.

$$\begin{aligned}10\text{ mg} &= V_{\text{octanol}}C_{\text{octanol}} + V_{\text{water}}C_{\text{water}}\\ &= (0.3\text{ L})(33.3\text{ mg/L}) + (0.2\text{ L})(0.105\text{ mg/L})\\ &= 9.99\text{ mg} + 0.021\text{ mg} = 10.01\text{ mg}\end{aligned}$$

The mass balance is well within the ability of standard gas chromatography methods to measure concentration, i.e., a standard deviation of about 1 to 10 percent.

Adsorption will often slow, or attenuate, the transport of a contaminant plume. Because transport rates in soil and groundwater are usually quite slow, the assumption is usually made that the adsorption/desorption process is instantaneous. In this case, the rate of change of mass within a control volume can be written in terms of K_{SD} as

$$\text{Mass accumulation rate} = \left(\frac{\partial s}{\partial t}\rho_b + \frac{\partial \Theta C_A}{\partial t}\right)A_z\,\Delta z \tag{3.18a}$$

$$= \left(\frac{\partial K_{SD}C_A}{\partial t}\rho_b + \frac{\partial \Theta C_A}{\partial t}\right)A_z\,\Delta z \tag{3.18b}$$

Assuming that Θ is constant,

$$\text{Mass accumulation rate} = \left(1 + \rho_b\frac{K_{SD}}{\Theta}\right)\Theta\frac{\partial C_A}{\partial t}A_z\,\Delta z = \Theta R\frac{\partial C_A}{\partial t}A_z\,\Delta z \tag{3.18c}$$

where ρ_b = dry bulk density of soil, g/m^3
R = retardation coefficient defined by Equation (3.18*c*)

Replacing the mass accumulation term for a conservative contaminant to account for sorption in Equation (3.13) by Equation (3.18*c*) and simplifying yields (Bear and Bachmat, 1991):

$$\Theta R\frac{\partial C_A}{\partial t} = \frac{\partial}{\partial z}\left(\Theta D_z\frac{\partial C_A}{\partial z}\right) - \frac{\partial}{\partial z}(\Theta v_z C_A) \tag{3.19a}$$

$$R\frac{\partial C_A}{\partial t} = \frac{\partial}{\partial z}\left(D_z\frac{\partial C_A}{\partial z}\right) - \frac{\partial}{\partial z}(v_z C_A) \tag{3.19b}$$

Equation (3.19*b*) is a one-dimensional equation for the mass transport of a contaminant undergoing advection, hydrodynamic dispersion, and instantaneous equilibrium adsorption/desorption in a uniform porous medium.

A common approach in estimating movement of a contaminant relative to water is to apply the relationship given in Equation (3.20) for the average contaminant velocity as shown in Figure 3.8. The velocities shown in Figure 3.8 (v_1, v_2, v_3) will, under ideal conditions, be equal to $v_{\text{contaminant}}$ given in Equation (3.20).

$$v_{\text{contaminant}} = \frac{v_{\text{pore}}}{R} \tag{3.20}$$

EXAMPLE 3.3. ESTIMATION OF TRANSPORT TIMES IN AN AQUIFER. Estimate the time for 1,1,1-TCA introduced into the upper-elevation pond of Figure 3.5 and Example 3.1 to reach the lower pond. Base the estimate on the movement of the midpoint of the concentration profile, as suggested in Equation (3.20). The fractional organic content of the sand lens is 0.005 and the dry bulk density of the sand is 2,100 kg/m^3.

Solution

1. Estimate K_{SD}.

$$K_{SD} \approx 6.3 \times 10^{-7} f_{oc} K_{ow}$$

From Example 3.2, $K_{ow} = 317$.

$$K_{SD} \approx 6.3 \times 10^{-7}(0.005)(317) = 1 \times 10^{-6} \text{ L/mg}$$

$$\approx 1 \times 10^{-3} \text{ m}^3\text{/kg}$$

2. Determine the retardation coefficient R.

$$R = 1 + \rho_b \frac{K_{SD}}{\Theta}$$

$$= 1 + (2{,}100 \text{ kg/m}^3)\frac{1 \times 10^{-3} \text{ m}^3\text{/kg}}{0.3} = 8$$

3. Estimate the contaminant velocity.

$$v_{\text{TCA}} = \frac{v_{\text{pore}}}{R} = \frac{0.26 \text{ m/day}}{8} = 0.0325 \text{ m/day}$$

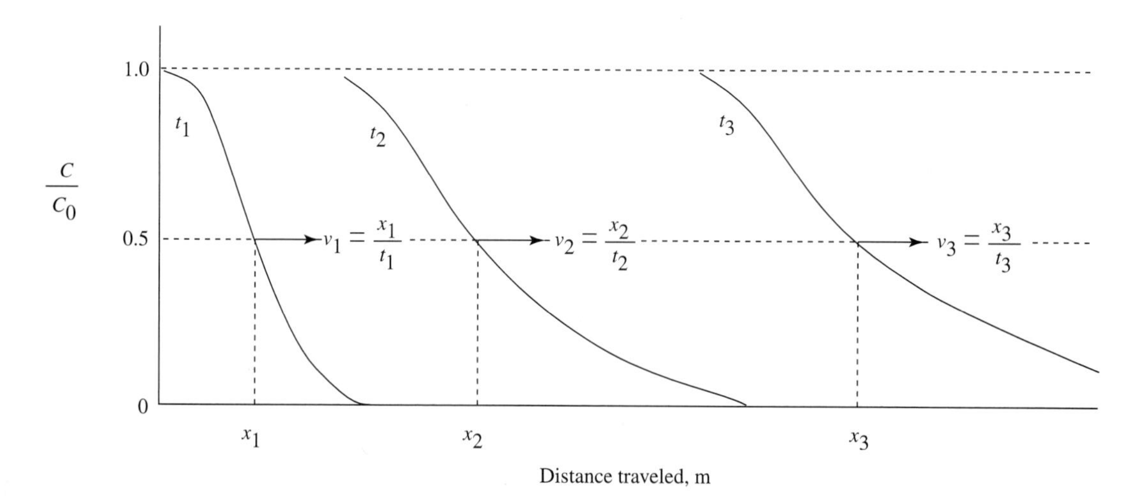

FIGURE 3.8
Time variation of contaminant concentration distribution in an aquifer.

4. Estimate travel time.

$$t = \frac{L}{v_{\text{TCA}}} = \frac{500 \text{ m}}{0.0325 \text{ m/day}}$$

$$= 15{,}385 \text{ days} = 42 \text{ years}$$

The breakthrough time predicted is far greater than could be validated by any currently available mathematical model. Predictions greater than 1 to 2 years should be viewed with great suspicion, although the concept of long breakthrough periods is probably correct. However, channelization and dispersion will certainly result in breakthrough of small concentrations at much shorter time periods than predicted by models.

Dissolution

Pure product in contact with water will dissolve until limited by the solubility of the compound. The interphase transport rate is usually modeled as a function of (1) the difference between the phase equilibrium condition and the actual condition and (2) the interfacial area. When pure product exists, the rate of transfer is a function of the difference between the solubility of the contaminant compound (i.e., the equilibrium condition where pure product exists) and the measured concentration.

$$r_{TA} = K_L a(C_S - C_b) \tag{3.21}$$

where K_L = mass transfer coefficient, m/s
a = interface area per unit volume, m^2/m^3
C_S = equilibrium concentration, mg/L
C_b = measured concentration, mg/L

Determination of the mass transfer coefficient K_L and the specific surface area a is quite difficult in soil. Average values of the mass transfer coefficient can be estimated using correlations developed for chemical reactors (Onda et al., 1965). However, the interfacial area is dependent upon the nature of the nonaqueous-phase material. Globules of pure product may exist in the interstices between medium granules, and the interfacial area will depend on the size of the globules. Similarly, the contaminant may be adsorbed onto the soil surface or onto the soil organics, and the effective interfacial area will be a function of the specific surface area of the soil. For this reason, the mass transfer coefficient K_L and the specific surface area a are usually reported as a single number K_La.

EXAMPLE 3.4. DETERMINATION OF K_La IN A SOIL SAMPLE. Four 15-cm-diameter, 5-cm-deep cylindrical core samples of soil contaminated with tetrachloroethene (PCE) are tested in the apparatus shown in Figure 3.9. Water is run continually through each sample until a steady-state PCE concentration in the effluent is obtained. Using the data given in Table 3.1, estimate hydraulic conductivity and the combined mass transfer coefficient and specific surface area K_La. Note that if the experiment is conducted for too long a period the effluent concentration will begin to decrease as the interfacial area decreases with loss of contaminant. In the velocity range available in porous media the mass transport coefficient should be relatively constant. The solubility of PCE is 160 mg/L (see Appendix E). For the purpose of this example, assume that PCE is in sorption equilibrium with the soil throughout the sample.

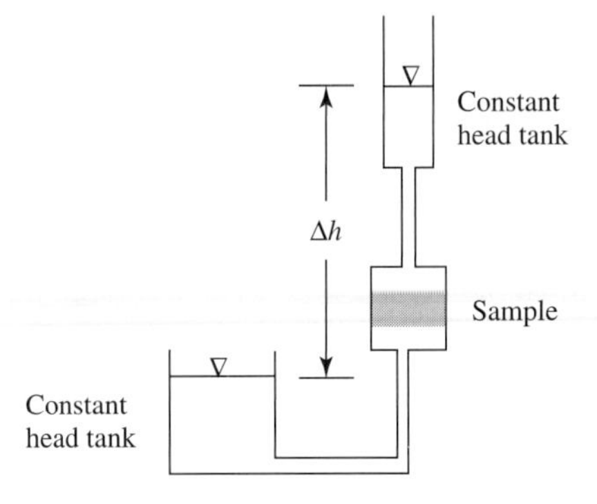

FIGURE 3.9
Experimental system for measuring K_La.

TABLE 3.1
Mass transfer data for Example 3.4

$-\Delta h$, m	Q, m³/day	C_{PCEo}, mg/m³
1	0.305	147
2	0.602	75
3	0.930	49
4	1.19	38

Solution

1. Estimate the hydraulic conductivity K_C.

$$Q = Av$$

$$= -\left(\frac{\pi}{4}d^2\right)K_C\frac{\Delta h}{\Delta L} = -\frac{\pi}{4}(0.15\text{ m})^2 K_C\frac{\Delta h}{\Delta L}$$

$$K_C = -56.6\,Q\frac{\Delta L}{\Delta h} = -56.6\,Q\frac{0.05}{\Delta h} = -2.83\frac{Q}{\Delta h}$$

$-\Delta h$, m	Q, m³/day	v_z, m/day	K_C, m/day
1	0.305	17.3	0.863
2	0.602	34.1	0.852
3	0.93	52.6	0.877
4	1.19	67.3	0.842
		Average	0.859

2. Write the mass balance expression assuming that PCE does not react.

$$A\,\Delta z\frac{\partial C}{\partial t} = QC_z - QC_{z+\Delta z} + K_La(C_s - C)A\,\Delta z$$

3. Assume steady-state conditions develop, rearrange the equation, integrate, and solve for $K_L a$.

$$-\left(\frac{QC_z - QC_{z+\Delta z}}{A\Delta_z}\right) = K_L a(C_S - C)$$

$$v_z \frac{dC}{dz} = K_L a(C_S - C)$$

$$\int_0^C \frac{dC}{C_S - C} = \frac{K_L a}{v_z} \int_0^{-0.05} dz$$

$$\ln \frac{C_S - C}{C_S} = -0.05 \frac{K_L a}{v_z}$$

$$K_L a = -20 v_z \ln \frac{C_S - C}{C_S} = -20 v_z \ln \frac{1.60 \times 10^5 - C}{1.60 \times 10^5}$$

4. Substitute v_z and C values and solve for $K_L a$.

C, mg/m^3	v_z, m/day	$K_L a$, per day
147	17.3	0.318
75	34.1	0.292
49	52.6	0.322
38	67.3	0.320
	Average	0.313

Note that the estimated $K_L a$ value is based on the assumption of steady-state conditions. At some point in time the interfacial area will begin to decrease due to loss of sorbed PCE or a complete dissolution of pure PCE globules. At that time the observed $K_L a$ values will decrease.

Volatilization and absorption

Many contaminants exist in a vapor as well as a liquid and a solid phase. Volatilization is the process whereby chemicals move from either a liquid or a solid phase to a gas phase (vaporize) as a result of molecular diffusion caused by a chemical potential between the phases. Absorption is the process whereby chemicals move from the gas phase to the liquid phase, again the result of molecular diffusion caused by a chemical potential between the gas and liquid phases. The mathematical descriptions of volatilization and absorption are similar.

A mathematical description of the rate of volatilization is necessary to incorporate volatilization in the time-dependent mass balance given by Equation (3.1). The most commonly used volatilization description is the two-film model (U.S. EPA, 1990), which is the same approach as used in describing dissolution, above. In the two-film model it is assumed that both the air and the water phases are well mixed except near the interface, where a stagnant film exists. When the bulk phases are not in equilibrium (as defined by Henry's law) a net flux will occur from one phase to the other and a concentration gradient forms in each of the stagnant films. At the interface, the ratio of the concentration of the chemical in the air to the water is assumed to equal the Henry's law coefficient (U.S. EPA, 1990):

$$H = \frac{C_G}{C_L} \tag{3.22a}$$

Henry's law is stated in several formats, each of which has different units for the coefficient

$$m = HRT = \frac{C_G}{C_L} RT \tag{3.22b}$$

$$m = e^{(A-B/T)} \tag{3.22c}$$

$$x = K_H p \tag{3.22d}$$

where H = dimensionless Henry's law coefficient
m = Henry's law coefficient, atm-m^3/mol
R = universal gas constant (8.2057×10^{-5} atm-m^3/mol-K)
T = temperature, K
C_G = gas-phase contaminant concentration, g/m^3
C_L = liquid-phase contaminant concentration, g/m^3
A,B = empirical constants, which are compound specific
K_H = Henry's law coefficient, atm^{-1}
p = partial pressure, atm
x = mole fraction

The value of the Henry's law coefficient is indicative of the tendency of a chemical to volatilize. A chemical can be considered to be volatile if $m > 3 \times 10^{-7}$ atm-m^3/mol ($H > 1.2 \times 10^{-5}$), while for smaller values of m, volatilization is usually an unimportant pathway of contaminant transport (U.S. EPA, 1990; Lyman et al., 1982). Reported values of the Henry's law coefficient often vary considerably among sources, and consultation of more than one reference is advisable. Henry's law coefficients are strongly temperature-sensitive. The Henry's law coefficient is dependent on the type of compound considered, the activity of the compound in each phase, and the temperature. For a specific gas, m can be roughly estimated from (U.S. EPA, 1990)

$$m \approx \frac{P_v}{C_s} \tag{3.23}$$

where P_v = pure component vapor pressure, atm
C_s = solubility of chemical in water, mol/m^3

At soil-water and soil-air interfaces, the volatilization process is more complex. As depicted in Figure 1.1, VOCs sorbed onto the soil surface may transfer into solution in the soil-water or into the vapor phase in the soil-air (U.S. EPA, 1990). Transfer from the soil-air to the ambient atmosphere will occur through diffusion under ordinary circumstances or by convection if ventilation is imposed on the soil. Within the soil matrix, Henry's law can be used to estimate partitioning between the soil-water and the soil-air interphases.

The rate of volatilization of a contaminant from soil and soil-water is a function of the contaminant concentration, and the vapor pressure and aqueous solubility of the contaminant. Volatilization rates increase dramatically with increasing temperature and, near the soil surface, air turbulence. Other factors that influence the rate of volatilization include soil moisture content, soil temperature and porosity, the organic and the clay content in the soil, soil handling, and the bioremediation technique used.

The rate of mass transfer across the air-water interface is proportional to the difference between the existing aqueous-phase concentration C_A and the equilibrium aqueous-phase concentration C_{Ae} as defined by Henry's law.

$$r_{TA} = K_L a(C_A - C_{Ae}) \tag{3.24}$$

where r_{TA} = mass transfer rate, $g/m^2 \cdot s$

Equation (3.24) can be combined with Equation (3.19) to yield

$$\frac{\partial R\Theta C_A}{\partial t} = \frac{\partial}{\partial z}\left(\Theta D_z \frac{\partial C_A}{\partial z}\right) - \frac{\partial}{\partial z}(\Theta v_z C_A) + K_L a(C_A - C_{Ae}) \tag{3.25}$$

In flow through porous media, the specific surface area will generally be a function of the structure of the particular medium and the degree of saturation Θ. Henry's law can be used to estimate the equilibrium aqueous-phase concentration C_{Ae} as given above. If the compound of interest is essentially absent in the air, then $C_{Ae} = 0$. If the concentration in the air is greater than in the water, the compound will be absorbed by the water.

Phase Distribution of Contaminants

The distribution of contaminants between phases is an important factor in the selection of the type of bioremediation process and the prediction of the rate at which bioremediation will occur. Henry's law, the soil distribution coefficient, and the solubility can be used to determine the distribution of contaminants among the phases to some extent. The presence of pure product or physical discontinuities in wetting may result in isolation of portions of the medium and spatial differences in equilibrium conditions.

EXAMPLE 3.5. DISTRIBUTION OF CONTAMINANTS AMONG WATER, AIR, AND SOIL. A soil is found to contain 30 g of toluene/kg on a wet-weight basis. The dry bulk density of the soil is 2,070 kg/m^3, the porosity is 0.35, the volumetric water content is 0.20, and the organic fraction is 0.03. Properties of toluene are given below.

Toluene: Chemical formula: C_7H_8 — Molecular weight: 92.15
Water solubility at 20°C: 515 mg/L — H at 20°C: 0.235
K_{ow}: 537

Estimate the distribution of toluene in the water, air, and on soil assuming no pure product exists.

Solution

1. Determine the volumes of air and water per m³ of soil.

$$V_{water} = 0.20 \text{ m}^3 \text{ water/m}^3 \text{ soil}$$

$$V_{air} = (0.35 - 0.20) = 0.15 \text{ m}^3 \text{ air/m}^3 \text{ soil}$$

2. Determine the wet density of the soil ρ_{wet}.

$$\rho_{wet} = \frac{V_{soil}\rho_b + V_{water}\rho_w}{V_{soil}}$$

$$= \frac{(1 \text{ m}^3)(2{,}070 \text{ kg/m}^3) + (0.2 \text{ m}^3)(1{,}000 \text{ kg/m}^3)}{1 \text{ m}^3}$$

3. Determine K_{SD} for toluene in the soil.

$$K_{SD} = 6.3 \times 10^{-7}(0.03)(537)$$

$$= 1 \times 10^{-5} \text{ L/mg} = 0.01 \text{ m}^3\text{/kg}$$

4. Write a mass balance on toluene for equilibrium conditions and a 1 m³ soil volume.

$$M_{tol} = M_{soil} + M_{water} + M_{air}$$

$$30 \text{ g} = s\,\rho_b V_{soil} + V_{water}\,C_{water} + V_{air}C_{air}$$

5. Solve for the individual components.

$$s = C_{water}K_{SB}$$

$$C_{air} = HC_{water}$$

$$C_{water} = \frac{30 \text{ g}}{K_{SD}\rho_b V_{soil} + V_{water} + V_{air}H}$$

$$C_{water} = \frac{30 \text{ g}}{(0.01 \text{ m}^3\text{/kg})(2{,}070 \text{ kg/m}^3)(1 \text{ m}^3) + 0.2 \text{ m}^3 + 0.235(0.15 \text{ m}^3)}$$

$$= \frac{30 \text{ g}}{20.7 \text{ m}^3 + 0.2 \text{ m}^3 + 0.0353 \text{ m}^3}$$

$$= 1.43 \text{ g/m}^3$$

$$M_{water} = (1.43 \text{ g/m}^3)(0.2 \text{ m}^3) = 0.286 \text{ g}$$

$$s = (0.01 \text{ m}^3\text{/kg})(1.43 \text{ g/m}^3) = 0.0143 \text{ g toluene/kg}$$

$$M_{soil} = (0.0143 \text{ g toluene/kg})(2{,}070 \text{ kg/m}^3)(1 \text{ m}^3)$$

$$= 29.6 \text{ g}$$

$$C_{air} = (0.235)(1.43 \text{ g/m}^3) = 0.336 \text{ g/m}^3$$

$$M_{air} = (0.336 \text{ g/m}^3)(0.15 \text{ m}^3) = 0.05 \text{ g}$$

$$M_{air} + M_{water} + M_{soil} = 0.05 + 0.29 + 29.6 = 29.94 \text{ g}$$

The difference of 0.06 g is due to roundoff error.

A constituent that is transformed (degraded or generated) within a control volume is said to be nonconservative. Transformation can occur as a result of biological, chemical, and physical processes. In what follows, the most important transformation processes in soil and groundwater are discussed, including biodegradation, natural decay, and chemical degradation (photochemical, oxidation-reduction, and hydrolysis reactions).

Biodegradation

In biodegradation, microorganisms transform contaminants through metabolic reactions. Both organic and inorganic compounds are transformed by soil microorganisms, and bioremediation processes include oxidation-reduction reactions, sorption and ion-exchange processes, and chelation and complexation reactions that result in immobilization of metals.

Biotransformation of organic compounds

The bioremediation processes of greatest interest involve biooxidation of organic contaminants (shown schematically in Figure 3.10). Biodegradable

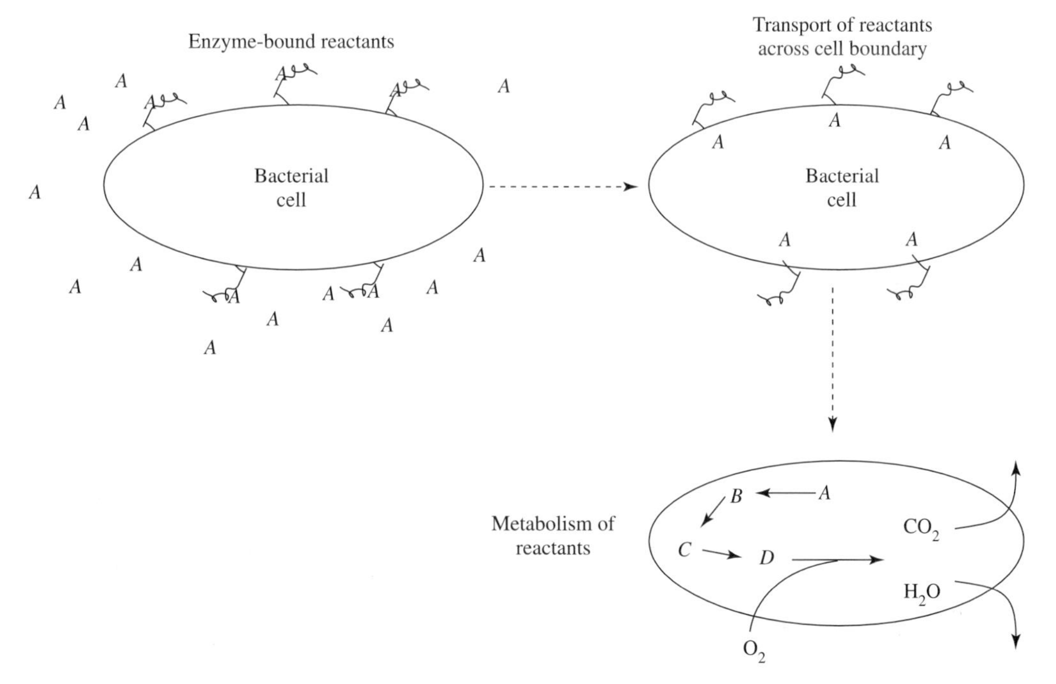

FIGURE 3.10
Schematic diagram of biodegradation process. An organic reactant *A* is bound to an extracellular enzyme and transported within the cell. The biooxidation process involves a series of reactions in which electrons are stripped from the compound and energy released is used to synthesize new cellular material (growth), repair damaged material (maintenance), and transport nutrients against the concentration gradient across the cell boundary.

compounds are initially bound to extracellular enzymes and transported across the cell membrane. A series of transformation reactions then occurs in which electrons are stripped from the compound and the carbon skeleton is oxidized. Energy released in the reactions is used for synthesis of new cell material, for repair of damaged material, transport of compounds into the cell, and, in some cases, movement. When the organic contaminants have been converted to CO_2 and H_2O, mineralization is said to have occurred. Complete mineralization never occurs because a portion of the organic material is converted to cells and a significant fraction of the cell mass is effectively nonbiodegradable. However, conversion of toxic and hazardous materials to a combination of CO_2, H_2O, and new bacterial cells eliminates most of the problems requiring remediation.

Biodegradation does not always result in mineralization. The change in the molecular structure of a contaminant during bioremediation may result in the production of reaction products different from the parent compound but still toxic or hazardous. Nonmineralized products are usually more oxidized and less volatile than the parent compound. However, the products may tend to partition onto soil more than the parent compound, making remediation more difficult.

Biotransformation of inorganic ions and compounds

The most common reactions involving transformation of inorganic compounds are those of the nitrogen cycle (Figure 3.11), with the reactions of the sulfur cycle also being significant (Tchobanoglous and Schroeder, 1985). Nitrogen transformations of particular significance in bioremediation are nitrification and denitrification. Nitrification is carried out by organisms that use either NH_3 or NO_2^- as energy sources and CO_2 as a carbon source. Although the two steps are distinct and different groups of bacteria are involved, the process of nitrification (transformation of NH_3 to NO_3^-, or oxidation of nitrogen from N^{3-} to N^{5+}) can usually be satisfactorily lumped together in modeling of soil processes. Denitrification is carried out by facultative anaerobic bacteria which are quite common in soil and is very similar to the use of oxygen in aerobic respiration. The bacteria use organic material as a carbon and energy source and, in the absence of O_2, use NO_3^- as an electron acceptor. Many species of soil bacteria are also capable of using NO_2^- as an electron acceptor, and in most cases of denitrification the nitrogen is reduced from the +5 oxidation state of NO_3^- to the 0 oxidation state of N_2.

Sulfur metabolism is similar to nitrogen metabolism in that reduced sulfide compounds (e.g., H_2S) are used as an energy source for a select group of aerobic bacteria and a small number of bacterial species use sulfate and sulfite as electron acceptors in the absence of oxygen. The sulfide-oxidizing bacteria use CO_2 as a carbon source while the sulfate-reducing bacteria use organic compounds as energy and carbon sources. Iron metabolism is somewhat analogous to sulfur and nitrogen metabolism, and a number of other metals can be used as electron acceptors under limited conditions (Brock et al., 1984).

Biodegradation rates

Contaminant biodegradation is often described by a Monod rate model. This model is based on an equation which ensures that the degradation rate will not exceed a maximum value (Paul and Clark, 1989). The Monod model for the porous media control volume shown in Figure 3.7 can be written as

$$\Theta A_z \Delta z r_A = -\frac{r_{max} C_A}{K_S + C_A} \Theta A_z \Delta z \tag{3.26}$$

where r_A = rate of biodegradation of chemical species A, g/m³ · s
r_{max} = maximum biodegradation rate, g/m³ · s
K_S = saturation coefficient, g/m³

Inclusion of the volumetric water content with the Monod term is necessary because contaminants are biodegraded only in the aqueous phase. The maximum biodegradation rate r_{max} is dependent on the mass of microorganisms present and can be represented as the product of the degradation rate coefficient k (s^{-1}), and the microbial mass concentration X (g/m^3). Thus, a model for microbial growth may also be needed to determine microbial mass concentration. If the contaminant is also used as a carbon source by the microorganism, the growth and degradation models are coupled.

Incorporating Equation (3.26) into Equation (3.25) yields

$$\frac{\partial R \Theta C_A}{\partial t} = \frac{\partial}{\partial z}\left(\Theta D_z \frac{\partial C_A}{\partial z}\right) - \frac{\partial}{\partial z}(\Theta v_z C_A) + K_L a(C_A - C_{Ae}) - \frac{r_{max} C_A}{K_S + C_A} \Theta \tag{3.27}$$

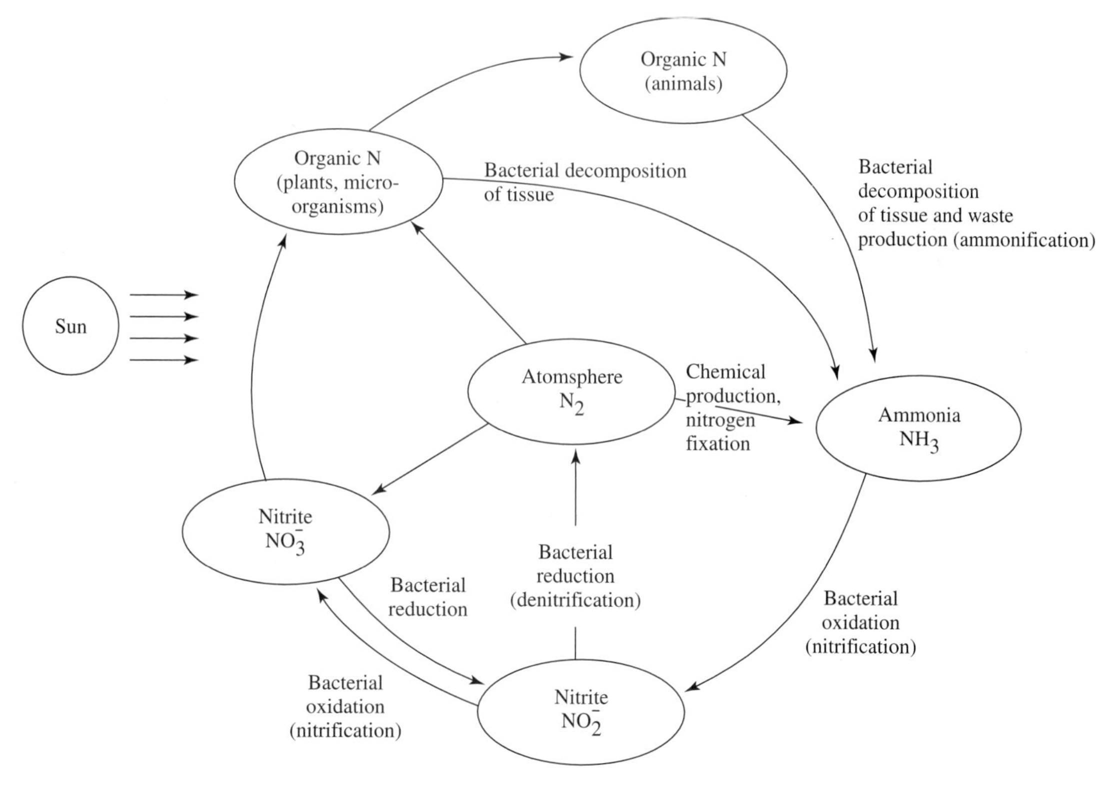

FIGURE 3.11
A simplified representation of the nitrogen cycle in nature.

Equation (3.27) is the one-dimensional equation for the mass transport of a contaminant undergoing advection, hydrodynamic dispersion, volatilization, and instantaneous equilibrium adsorption/desorption and biodegradation.

Natural Decay/Attenuation and Intrinsic Bioremediation

In nature, contaminants will often disappear owing to a number of interrelated factors which are usually too complex to characterize and impossible to enumerate. It is common to refer to the total degradation due to these factors as natural decay, natural attenuation, and/or intrinsic bioremediation. Natural decay rate r_N can be described using a lumped-parameter model, normally assumed to follow first-order kinetics as

$$r_N = -\,k_N R \Theta C_A \tag{3.28}$$

where r_N = natural decay rate, g/m$^3 \cdot$ s
k_N = kinetic rate coefficient, s^{-1}

The retardation coefficient precedes the natural decay because solid- and aqueous-phase degradations are lumped owing to lack of knowledge about the processes involved. A factor in natural decay may be an increasing fraction of material that is impossible to recover from the soil organic matter in laboratory analysis (Young, 1995). This rate coefficient is usually determined from experimental observations. Incorporating natural decay into the one-dimensional transport and fate equation yields

$$\frac{\partial R\Theta C_A}{\partial t} = \frac{\partial}{\partial z}\left(\Theta D_z \frac{\partial C_A}{\partial z}\right) - \frac{\partial}{\partial z}(\Theta v_z C_A) + K_L a(C_A - C_{Ae}) - \frac{r_{\max} C_A}{K_S + C_A}\,\Theta - k_N R\Theta C_A \tag{3.29}$$

Chemical Decomposition

Many chemical reactions lead to the destruction or decomposition of specific contaminants in the environment. Whether or not a specific reaction will occur can be determined from well-known thermodynamic-equilibrium relationships. However, rates of reaction cannot be determined from these relationships. Rates of reaction are given by kinetic relationships which are less understood. A thorough treatment of chemical decomposition is beyond the scope of this chapter. A brief description of three types of reactions of primary concern is presented: photodegradation, hydrolysis, and oxidation-reduction.

Photodegradation

Solar radiation is known to cause and accelerate a number of chemical reactions. Direct photolysis is the process where the adsorption of light by a contaminant causes contaminant transformation. In the aquatic environment, direct photolysis can be important. Since light is attenuated by substances present in water, significant radiation may penetrate only the upper layers of a water body. Photodegradation in subsurface porous media is normally considered to be negligible, as light does not generally penetrate the subsurface to any significant extent.

Hydrolysis

The general definition of hydrolysis is "the reaction of an anion with water to form a weak acid and OH^-, or the reaction of a cation with water to form a weak base and H^+ (Russell, 1980). Of particular interest is the hydrolysis of organic chemicals. The rate of hydrolysis is dependent on temperature, pH, presence of organic solvents, and the ionic strength of the solution. In most bioremediation systems, hydrolysis is considered negligible as a removal pathway for organic chemicals (U.S. EPA, 1990).

Oxidation-reduction

An oxidation-reduction reaction is an electron transfer reaction in which the substance that loses electrons is oxidized, while the one that gains electrons is reduced. Oxidation of organic chemicals is typical of aerobic environments, whereas reduction is typical of anaerobic environments. Like hydrolysis, non-biologically mediated oxidation-reduction reactions are considered to be insignificant as removal pathways of organic chemicals in biological treatment systems (U.S. EPA, 1990).

PROBLEMS AND DISCUSSION QUESTIONS

3.1. A soil containing 3 percent organic matter ($f_{oc} = 0.03$) by weight has been contaminated with a hydrocarbon compound having a $K_{ow} = 10^3$ and a Henry's law constant $H = 4 \times 10^{-12}$. Laboratory experiments have been conducted to measure the K_{SD} value in which 100 μg of the compound was added to 500 g of dry soil. Selected volumes of water were then added to the mixture and the samples were stored in open containers. Data obtained from the experiments is given below.

Intended data

Volume, L	C_e, μg/L
1	11.8
2	10.5
3	9.5
4	8.7
5	8.0

a. Was the use of open containers appropriate in this case? Explain your answer.
b. Determine the appropriate K_{SD} value in units of m^3/g.
c. Does the relationship $K_{SD} = 6.3 \times 10^{-7} f_{oc} K_{ow}$ provide a reasonable estimate of the experimental value?

3.2. In Problem 3.1, the statement was that a single compound was added to the soil. Consider the situation where a mixture of compounds (such as gasoline) is measured by a single lumped concentration parameter such as "total petroleum hydrocarbons" or TPH. If the same experiments as in Problem 3.1 were performed:

a. Would you expect the calculated K_{SD} to be constant?
b. Sketch the probable shape of the K_{SD} versus volume of water added curve, see (*a*).
c. *Briefly* explain the reasoning for your answer and the shape of the curve (two or three sentences should be enough).

3.3. Trichloroethylene (TCE) has been disposed of in open, unlined evaporation ponds for a number of years. The TCE was used to clean engine parts and is contaminated with grease and water. Soil beneath the ponds is of moderate clay (10 percent) and organic content (3 percent). The groundwater table is 50 ft below the bottom of the ponds. Characteristics of TCE are:

$H = 0.3$ $\rho_{TCE} = 1{,}460$ kg/m^3 Solubility = 1,100 mg/L
$K_{ow} = 240$

a. List the possible fates of TCE, assuming that biodegradation *will not* occur.
b. Briefly discuss the probable importance of each fate using the given data.
c. Do you believe drinking-water wells using the aquifer below the ponds and located a distance of 1 mile away are in danger of being contaminated? Explain your answer using the information available and what you have learned during the class.

If you need to estimate any parameter values (e.g., the moisture content of the soil), justify the value you use.

3.4. Consider the effect of temperature on groundwater flow rates using the information in Example 3.1. Assume that the data given in Example 3.1 are for 20°C and determine the superficial velocity at 10 and 30°C using Equation (3.6).

3.5. A solid-waste-disposal site has a 30-cm-thick clay liner ($K_C = 10^{-8}$ m/s at 20°C) and an accumulated liquid depth of 2 m. If the site is 5 ha in area, determine the rate of leakage in m^3/year.

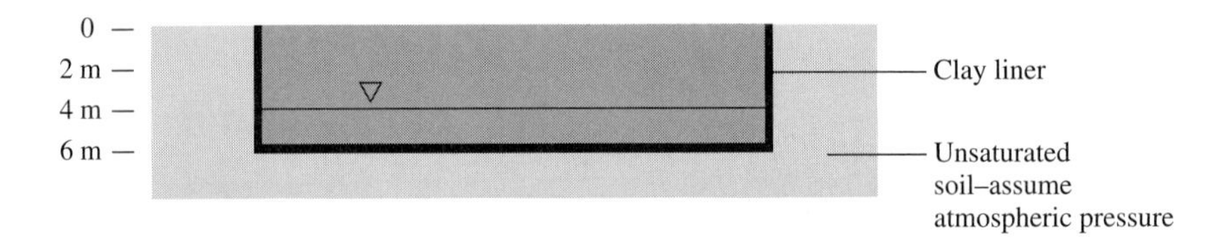

3.6. Water in the waste-disposal site of Problem 3.5 contains phenol and *p*-xylene in significant concentrations. Determine the flux into the soil of these two compounds at 20°C knowing that the dry bulk density, porosity, and organic fraction of the clay are 2,100 kg/m^3, 0.3, and 0.03, respectively. Properties of phenol and *p*-xylene at 20°C are given below.

	MW	*H*	K_{ow}
Phenol (C_6H_6O)	94.14	0.416	28.8
p-Xylene (C_8H_{10})	106.18	0.179	1,820

3.7. A plume of chlorobenzene has been generated from a waste pit as shown in the figure below. Flow from the pit occurs at a point where a permeable layer having a cross section (assume area constant at 10 m^2 and a length $L = 2{,}000$ m). The endpoint of the permeable layer is a small stream, and therefore the chlorobenzene will eventually pollute the stream. Characteristics of the permeable layer are given below.

$f_{oc} = 0.001$ g/g

$K = 10^{-4}$ m/s (assume at 20°C)

$\phi = 0.35$

Solubility of chlorobenzene = 500 mg/L at 20°C

Assume solubility at 10°C = 350 mg/L

$\log K_{ow} = 2.48$ at 20°C (assume constant)

$\rho_s = 2{,}750$ kg/m^3 (note this is the dry solids density, not the bulk density)

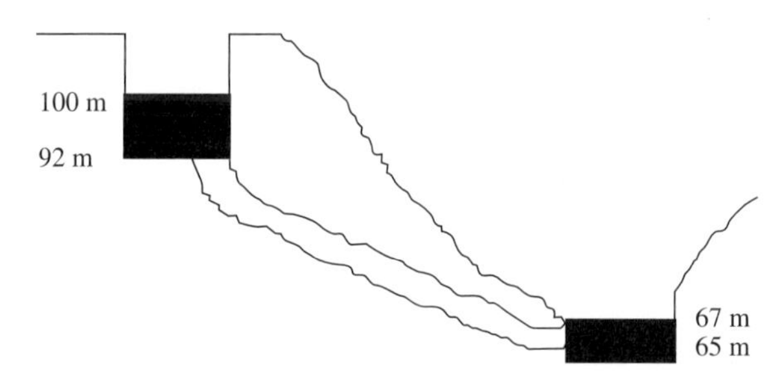

a. Estimate the time of travel of an unretarded contaminate at 10 and 20°C, respectively.

b. Estimate the time required before the chlorobenzene concentration entering the stream is equal to one-half of the saturation concentration, assuming that the initial concentration in the permeable zone was 0 mg/L.

3.8. A completely dry 1-kg soil sample is placed in 10 L of water and mixed until a slurry is formed. 50 g of phenol are added to the slurry during the mixing process. The sealed slurry container has a 6-L head space. Density of the soil is known to be 2,750 kg/m^3. After 3 days of mixing, a sample is removed and filtered. The phenol concentration of the filtrate is determined to be 5 mg/L. Estimate the value of K_{SD}.

3.9. You have been asked to investigate the impact of the presence of a dense nonaqueous-phase liquid (DNAPL) pool found at the bottom of an aquifer. The pool is composed of nearly pure methylene chloride, which has the properties (at 25°C) listed below.

$\rho = 1{,}400$ kg/m^3	Vapor pressure = 0.46 atm	$H = 0.118$
$K_{ow} = 1.3$	Water solubility = 20,000 mg/L	

Give a qualitative estimate of the relative distribution of methylene chloride among the solid, liquid, and gas phases at each of the five sampling points shown. An example answer would be: $S >> C_L > C_g$. Note that in some cases one of the concentrations will be 0. Give a *brief* explanation of your analysis for sample points 1 and 5. For sample point 5 a partially quantitative estimate is possible with appropriate assumptions.

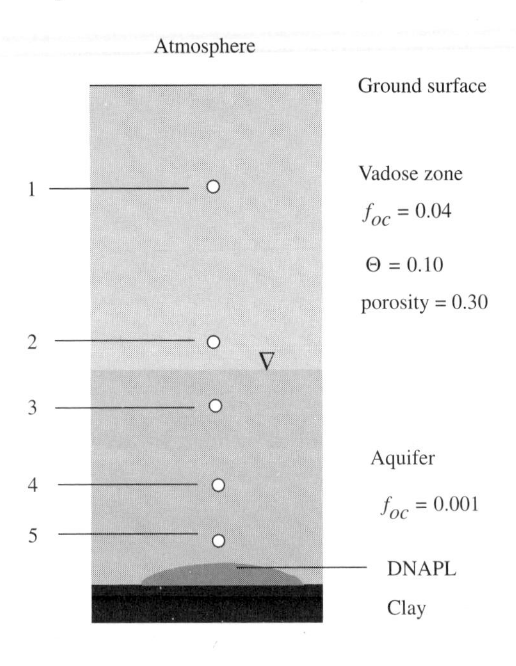

3.10. A leaking underground storage tank has contaminated an aquifer as indicated in the figure below. Characteristics of the contaminant and the aquifer are given below, also. Estimate the rate at which the plume will move.

Aquifer: $\phi = 0.25$ $f_{oc} = 0.01$ $K = 10^{-3}$ m/s $\rho_b = 1{,}900$ kg/m^3
Contaminant: Solubility: 2,500 mg/L $K_{ow} = 250$

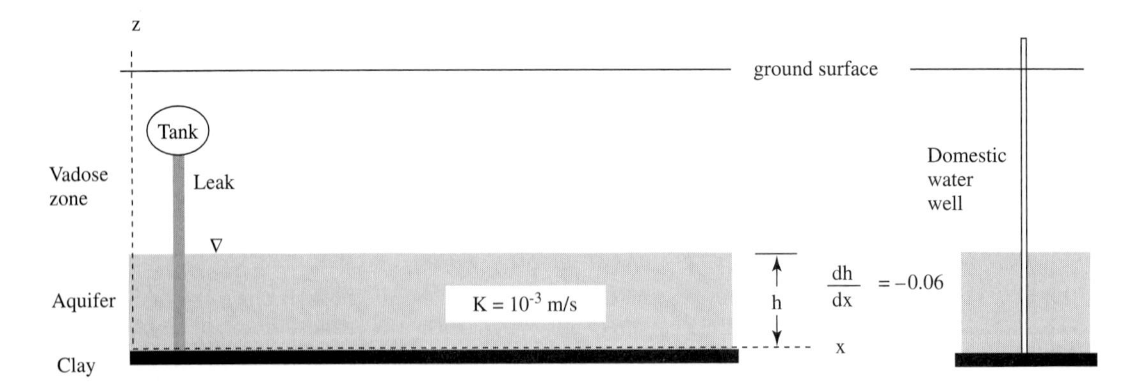

3.11. Xylene (dimethylbenzene) has been found in the vapor phase of soil below a factory which uses the solvent extensively. The contamination was found to be limited to a 150-m^3 block of soil and the average gas-phase xylene concentration was 30 mg/m^3. Given the data below, estimate the mass of xylene contaminating the soil and the soil water.

Soil:	$\phi = 0.25$	$f_{oc} = 0.03$	$\Theta = 0.15$
	$\rho_b = 1{,}900\ kg/m^3$		
Xylene:	$MW = 106.2$	$K_{ow} = 1{,}820$	$P_v = 0.0132$ atm
	$H = 0.30$	Solubility: 198 mg/L	

3.12. A dry-cleaning establishment carries out operations in an unvented room, and the tanks containing the effectively pure fluid are uncovered. Assuming the fluid used is tetrachloroethene (PCE), which has the characteristics given below, determine the vapor-phase concentration in the room. Assume $T = 20°C$.

PCE: C_2Cl_4

$MW = 165.82$

Water solubility at 20°C = 290 mg/L

P_V at 20°C = 5.00 mmHg = 6.6×10^{-3} atm

Henry's law constant coefficients: $A = 12.4$, $B = 4{,}920$

$K_{SD} = 364$ mL/g

$\log K_{ow} = 2.60$

3.13. Gasoline leaks from an underground storage tank into an unsaturated zone having a volumetric moisture content Θ of 0.2 and a downward water velocity of 0.8 m/day. The organic content of the soil f_{oc} is 0.03 and the bulk density ρ_b is 1,600 kg/m^3. Benzene solubility is 1,800 mg/L, and the area bordering the tank is saturated. Estimate the time required for the benzene plume concentration to reach 900 mg/L at the groundwater surface.

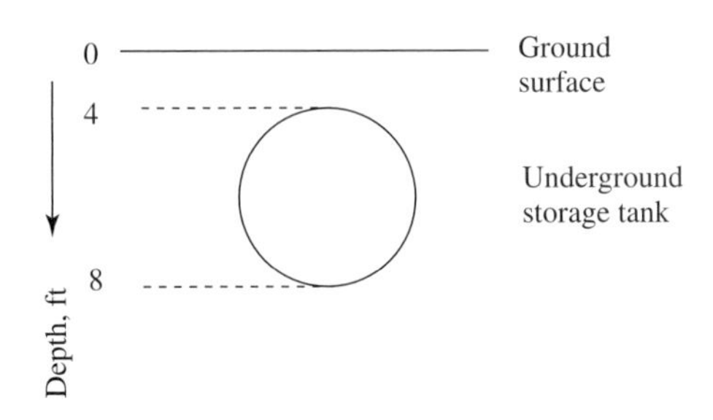

3.14. The most commonly applied biochemical reaction-rate model is the Monod equation:

$$r = -\frac{r_{max} C}{K_S + C}$$

where r = rate of conversion, moles or mass per unit volume · time
r_{max} = maximum rate of conversion, moles or mass per unit volume · time
C = concentration of limiting substance, moles or mass per unit volume
K_S = saturation constant, moles or mass per unit volume

Experiments have been conducted to measure the rate of reaction $3A + B \rightarrow 2C + D$, and the results are given below.

Concentration of *B*, 10^{-3} moles/L	**Rate of disappearance of *B*, 10^{-3} moles/L · min**
10	–5.46
5	–5.00
1	–3.00
0.5	–2.00
0.1	–0.55
0.05	–0.29

a. Determine the value of r_{max} and K_S.
b. Explain the assumption made in deriving these values.
c. At what concentration would you consider the reaction to be saturated? Give a practical answer.

3.15. Many organic compounds found in contaminated soils and groundwaters may be toxic to bacteria at higher concentrations. A number of rate expressions have been used to describe "substrate inhibition," with one of the most widely used given below.

$$r = \frac{r_{max} C}{K_S + C + C^2/K_i}$$

where K_i = inhibition constant

a. Determine the shape of the plot of $-r$ versus C.
b. Identify the key points on the curve and evaluate them in terms of r_{max}, K_S, and K_i.
c. Consider the significance of the dual rate values predicted. Do you believe that both values are real? Explain.

3.16. For the case shown, write a mass balance and set up the differential equations to describe the loss of pure product from the vadose zone under conditions where there is no advection. Assume local phase equilibrium exists and develop a term similar to the retardation coefficient for aquifers.

$C_g \approx 0$

Ground surface

Vadose zone

Pure product pool

Water

REFERENCES

Bear, J., and Y. Bachmat (1991): *Introduction to Modeling of Transport Phenomena in Porous Media,* Kluwer Academic, The Netherlands.

Brock, T. D., D. W. Smith, and M. T. Madigan (1984): *Biology of Microorganisms,* 4th ed., Prentice-Hall, Inc., Englewood Cliffs, NJ.

Davis, S. N., and R. J. De Wiest (1966): *Hydrogeology,* Wiley, New York.

Dupont, R. R., R. C. Sims, J. L. Sims, and D. L. Sorensen (1988): "In Situ Biological Treatment of Hazardous Waste-contaminated Soils," in *Biotreatment Systems,* vol. II, edited by Donald L. Wise, CRC Press, Inc., Boca Raton, FL.

Ehrenfeld, J. R., and J. H. Ong (1985): "Control of Emissions from Hazardous Waste Treatment Facilities," *Proceedings of the 78th annual meeting of the Air Pollution Control Association,* paper 85-70.1.

Ginn, T. R., C. S. Simmons, and B. D. Wood (1995): "Stochastic-Convective Transport with Nonlinear Reaction: 2 Biodegradation and Microbial Growth," *Water Resources Research,* vol. 31, no. 11, pp. 2689–2700.

Lyman, W. J., W. F. Rechl, and O. H. Rosenblatt (1982): *Handbook of Chemical Property Estimation Method, Environmental Behavior of Organic Compounds.* McGraw-Hill, New York, New York.

McCarty, P. L. (1991): Engineering Concepts for in situ Bioremediation, *Journal of Hazardous Materials,* vol. 28, pp. 1–11.

Onda, K., H. Takeuchi, and Y. Okumoto (1965): "Mass Transfer Coefficients between Gas and Liquid Phases in Packed Columns," *Journal of Chemical Engineering of Japan,* vol. 1, no. 1, pp. 58–62.

Paul, E. A., and F. E. Clark (1989): *Soil Microbiology and Biochemistry,* Academic Press, London.

Russell, J. B. (1980): *General Chemistry,* McGraw-Hill, New York.

Simmons, C. S., T. R. Ginn, and B. D. Wood (1995): "Stochastic-Convective Transport with Nonlinear Reaction: 1 Mathematical Framework," *Water Resources Research,* vol. 31, no. 11, pp. 2675–2688.

Tchobanoglous, G., and E. D. Schroeder (1985): *Water Quality Characteristics, Modeling, Modification,* Addison-Wesley, Reading, MA.

U.S. Environmental Protection Agency (1990): Available Models for Estimating Emissions Resulting from Bioremediation Processes: a Review, EPA/600/3-90/031.

Young, T. M. (1996): Phenanthrene Sorption by Natural Organic Matter; Investigation in Aqueous and Supercritical Fluid Systems. Doctoral Dissertation, University of Michigan, Ann Arbor, MI.

CHAPTER 4

Microbial Ecology

The purpose of this chapter is to provide a general introduction to the concepts and methods used in biological treatment of soil and water. Engineering of microbial systems requires consideration of the requirements for microbial growth and determination of the conditions necessary to carry out the desired biochemical reactions.

CLASSIFICATION OF LIVING ORGANISMS

All living organisms can be classified into two major groupings or divisions based on major differences in the cellular structure: eucaryotes and procaryotes (Figure 4.1). A significant difference is the size of the cell, with eucaryotic cells typically being considerably larger than procaryotic cells. However, the subcellular structural differences between eucaryotes and procaryotes are considerably more important. The names eucaryote and procaryote are derived from the Greek where "karyo" means nucleus, and "pro" means before or prior to, while "eu" means good or true. A eucaryotic cell has a well-defined (true) nucleus, with a nuclear membrane protecting the deoxyribonucleic acid (DNA) molecules which make up genetic material. A procaryotic cell, on the other hand, has a nuclear region which is not surrounded by a membrane and which consists of a single DNA molecule. Eucaryotes are divided into two general groups: multicellular organisms in which cells have specific functions, and unicellular organisms in which all cells carry out the same range of functions. Multicellular organisms are classified into two general categories, plants and animals, according to their energy and carbon sources, structure, type of growth, and movement. Unicellular organisms are classified into three general categories, protozoa, fungi, and algae, in a manner analogous to that

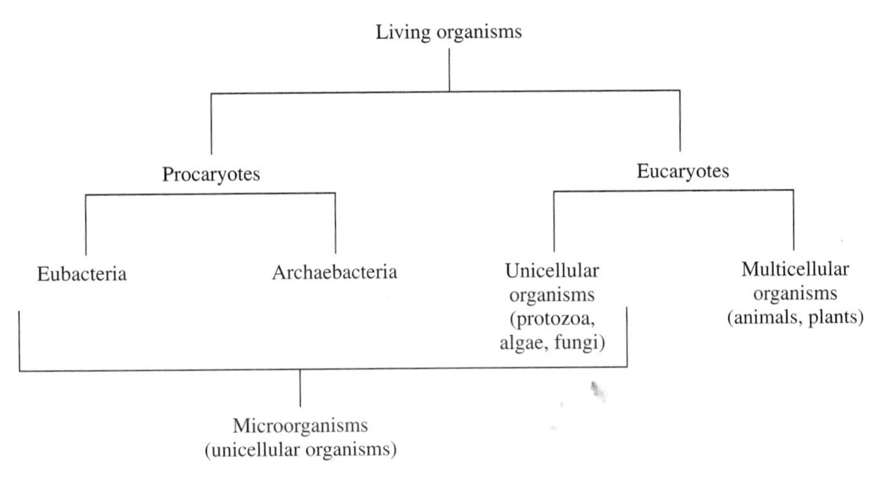

FIGURE 4.1
Classification of groups of living organisms.

for multicellular eucaryotic organisms. A summary of the basic differences between the major groups of organisms, living and nonliving, and their presence in biological treatment processes is listed in Table 4.1.

Within the procaryotic classification, two evolutionary distinct groups of organisms exist: eubacteria and archaebacteria (where archae is Greek for ancient). All procaryotes are microscopic in size, most of the individual cells being smaller than 5 micrometers. Eubacteria and archaebacteria have distinctly different cellular chemistry. The eubacteria include a large number of commonly known bacterial groups, including most of the organisms of importance in biological treatment processes, most of the organisms commonly found in soil and water, and most of the organisms pathogenic to humans and other mammals. The archaebacteria include some anaerobic species, as well as others that live under "extreme" growth conditions. Four subgroups of organisms have been classified as archaebacteria: methanogens, which produce methane under strictly anaerobic conditions; extreme halophilic *Archae,* which thrive in environments having salinities of 2 to 4*M* NaCl, thermophilic *Archae,* which require high temperatures for growth, the optimum for many being 80°C or higher, and thermoplasmas, a thermoacidophilic group which requires a pH of about 2 and a temperature of about 55°C for optimum growth (Brock et al., 1994). For historical reasons, members of both the eubacteria and the archaebacteria are usually referred to simply as bacteria.

Viruses, a group of nonliving particles which have a clear relationship to living organisms, are not shown in Figure 4.1. All viruses are composed of a strand of genetic material: deoxyribonucleic acid (DNA) or ribonucleic acid (RNA) and a protein coat. They do not have a cellular structure and cannot carry out metabolism, reproduction, or other kinds of activity on their own. Viruses are obligate parasites which need to infect living cells to become active and to replicate. Once inside a cell, a virus redirects the metabolic activity of the cell from growth to replication of the virusal genetic code and synthesis of the protein coating. Most viruses are extremely host-specific; that is, only one species or strain of a species

TABLE 4.1
Subdivisions of organisms and principal groups found in biological treatment systems

Group	Cell structure	Properties	Constituent groups	Groups found in biological treatment
Eucaryotes	Eucaryotic	Multicellular, extensive differentiation of cells and tissues	Plants (seed plants, ferns, mosses)	
			Animals (invertebrates, vertebrates)	Nematodes, rotifers
		Unicellular, coenocytic, or mycelial: little or no differentiation of tissue	Protists (algae, fungi, protozoa)	All groups
Eubacteria	Procaryotic	Cell chemistry similar to eucaryotes	Most bacteria	Most gram positive genera, some gram negative genera
Archaebacteria	Procaryotic	Distinctive cell chemistry	Methanogens, halophiles, thermophiles, thermoplasmas	Methanogens
Viruses	None	Nonliving, obligate parasites composed of nucleic acid strand and protein coat or capsid	Animal viruses, plant viruses, bacterial viruses	Bacteriophages

is attacked. This virus-cell association is why viruses are sometimes classified based on the host they infect. For example, a widely studied virus that causes a characteristic pattern of destruction of cell disruption of tobacco leaves is named *tobacco leaf mosaic virus.* The general group of viruses that attack bacteria are known as *bacteriophage.* Viruses have been extensively studied in the medical and agricultural fields as they pertain to causing diseases, but most information is related to specific viruses and viral diseases. Transfer of genetic material between cells is commonly related to viral infection. Such transfer may be important in the induction of metabolic processes for the biodegradation of recalcitrant (difficult to degrade) compounds. However, observations of such genetic enhancement have not been reported in bioremediation systems.

In most cases organisms are identified by their genus and species names, with the genus capitalized and the species uncapitalized and both genus and species in italics or underlined. For example, organisms of the genus *Pseudomonas* are extremely common in both soil and biological treatment processes. Species of *Pseudomonas* often identified are *aeruginosa, cepacia, putida,* and *stutzeri.* Within a species, strains are often identified that behave slightly differently from typical members of the group. For example, some strains of *Pseudomonas putida* may metabolize toluene and others may not. Differences between strains might be considered analogous to characteristics such as height, left- or right-handedness, or ability to distinguish color in humans. Within bacterial groups, new strains are constantly evolving owing to a process called mutation. Mutation is loosely defined as a "mistake" in copying the genetic code, which occurs during cell division. The mutation may be harmless, or it may be detrimental to cell functioning, even resulting in death of the mutant cells. In some cases, however, the mutation may result in a strain that is better capable of surviving in the parent's natural environment. That may happen, for example, if the mutant strain can synthesize an enzyme that is better able to utilize energy from available nutrients, and hence better able to compete in its natural environment.

In most cases, the microbial communities are dominated by mixed or heterogeneous bacterial populations (cultures made up of a number of eubacteria species). A few species generally make up the bulk of the bacterial mass on either a number or mass basis. The reason that predominant species develop is the existence of a competitive advantage over less numerous groups due to the particular environmental conditions or the available nutrient sources. For example, many species of the genus *Pseudomonas* grow well at relatively low temperatures and are able to degrade aromatic compounds through several metabolic routes. Biological processes designed to remove chlorinated aromatic compounds and operating at 10°C would almost always have at least one strain of the genus *Pseudomonas* among the dominant species (Evans and Fuchs, 1988; Reineke and Knackmuss, 1988).

Mixed bacterial populations should be thought of in terms of interacting communities and symbiotic relationships, also. Overall growth and contaminant-removal rates are enhanced by the interaction of the various species making up a population. In unmanipulated mixed populations, species predominate that grow most rapidly, are better adapted to the particular environment (e.g., temperature, pH, or salinity), and are most efficient in energy utilization. However, a type of hierarchy develops in which species depend on each other and in which species occupy ecological niches. For example, in a system treating gasoline-contaminated wastewaters, species that break down aromatic molecules may leave residues that serve as carbon and energy sources for species unable to break the aromatic ring. This close association and interdependence between species of microorganisms is often referred to as a consortium.

MAJOR GROUPS OF MICROORGANISMS

Microorganism is the term often used to describe a free living cell. That definition would include all procaryotes, as well as the unicellular eucaryotes: protozoa, algae, and fungi. Microorganisms are ubiquitous in the environment and are responsible for most of the cycling of carbon, nitrogen, sulfur, phosphorus, and other minerals. Microscopic multicellular organisms and unicellular organisms (microorganisms) play particular roles in bioremediation processes. Bacteria are nearly always the primary degraders, although in some cases fungi are important.

Microscopic plants, for example, serve as surfaces on which smaller organisms grow and, through photosynthesis, act as sources of oxygen. Microscopic animals serve as scavengers that remove floating debris particles.

Bacteria play the largest role in biodegrading organic contaminants in soil and groundwater. Fungi, like bacteria, metabolize organic compounds but do not compete well in most engineered remediation systems. In well-aerated soils both bacteria and fungi will be important, but in poorly aerated soils the bacteria alone are responsible for the biological and chemical changes taking place (Alexander, 1991). Protozoa scavenge particulate materials, including other microorganisms, a role similar to that of animals on a larger scale. Algae are photosynthetic and, like plants, can be used to provide oxygen to microbial systems and to remove inorganic nutrients. However, the light sources required for the growth of algae are rarely feasible in bioremediation systems. A few biological treatment processes that take advantage of the potential contributions of algae and fungal treatment processes are under development for use with some hazardous materials. In the following sections, a more detailed discussion is given of the role each group plays in bioremediation processes.

Bacteria

Bacteria are the most abundant group of organisms present in the soil. Several thousand species have been identified in soils throughout the world, and the presence of several million species is not unlikely. The number of bacteria and the predominant species present is a function of soil characteristics and the specific environment (e.g., temperature and moisture content). However, the range of species present appears relatively constant throughout the world. Bacteria are an extremely diverse group of organisms with widely varying morphological, ecological, and physiological properties and are the primary degraders of natural and xenobiotic organic compounds found in soil. Because of their diversity, bacteria are usually found in heterogeneous communities. Some species will be *primary degraders;* that is, they will initiate the degradation of organic material in the soil. Other species will grow on compounds resulting from partial degradation of complex organics or waste products of primary degraders. Usually the organic compounds available in soil include a broad range of molecular structures and sizes and a number of species will be competitive in their degradation. For example, gasoline is composed of approximately 60 hydrocarbons containing 5 to 12 carbons. Included are aliphatic, branched aliphatic, aromatic, and alicyclic compounds. Based on the observation of the breakdown of these compounds in soil we may conclude that a number of species are involved and that some compounds are considerably more amenable to biodegradation than others.

Bacteria have three general physical appearances, spherical (cocci), rod-shaped (bacilli), and spiral (spirillum). Structurally, bacteria are characterized by a somewhat disorganized outer layer composed mostly of polysaccharides known as the capsule or slime layer, a rigid cell wall, a cell membrane which encapsulates the cytoplasm within which all of the necessary reactions occur, and the nuclear region composed of the genetic component of the cell. A simplified drawing of a bacterial cell is shown in Figure 4.2. Most bacteria are motile. The most commonly observed method of movement is through the waving of the whiplike flagella structure present in many species. The size of a bacterial cell is somewhat dependent on growth conditions. A starved cell may be as small as 0.2 μm in diameter. However, most bacteria grown in laboratories are "fatter." Spherically shaped bacteria may have a diameter of 0.5 to 1.0 μm, whereas rod-shaped bacteria are typically 0.5 by 2 to 3 μm.

Bacteria are classified using physical, chemical, genetic, and metabolic characteristics. Genus and species are assigned on the basis of shape, chemical makeup, and genetic characteristics. The result is that the metabolic potential varies widely within a genus, and sometimes among strains within a species. For example, the ability to degrade toluene is relatively common within the genus *Pseudomonas,* but far from universal. Further, only one strain of the species *Escherichia coli* was responsible for the food-poisoning incidents at fast food restaurants in the United States in 1994 and infected hamburgers in 1997.

Use of and tolerance to oxygen is one of the most general methods of classification. Oxygen is the most commonly used terminal electron acceptor in living organisms. In oxidation-reduction reactions, electrons removed from the compound oxidized are transferred to a compound being reduced. In living organisms a number of steps are involved and the final step results in the stripped electrons being transferred to a compound that leaves the cell. Molecular oxygen is such a compound, and the product is H_2O. Obligate aerobes are bacteria which require oxygen as a terminal electron acceptor and grow only in the presence of oxygen. Facultative aerobes are bacteria which can use alternative terminal electron acceptors and grow in the presence or absence of oxygen. Obligate anaerobes grow only in the absence of oxygen. Some anaerobes are tolerant to oxygen, but oxygen is toxic to many obligate anaerobic species. A more specific, and somewhat more useful, classification in terms of biological treatment is based on compounds bacteria use as energy and carbon sources. This method of classifying bacteria is discussed in

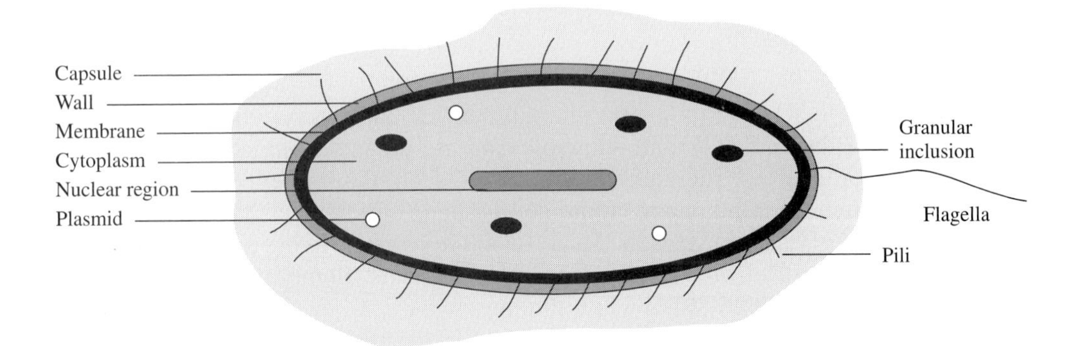

FIGURE 4.2
Schematic representation of the bacterial cell. Note that particular bacterial species are characterized by general shape and the tendency to occur singly, in clumps, or as chains or filaments. Typical cell sizes, excluding the capsule, range from 0.5 to 2 μm. Capsule size tends to be very small during rapid growth.

more detail in Chapter 5. Bacteria are also often classified based on morphology. The major morphological groupings of bacteria are bacilli, or rod-shaped bacteria; the cocci, or spherical cells; and the spirilla, or spirals. Bacteria can also be classified as eutrophs, which grow in the presence of high substrate concentrations, and oligotrophs, which grow at trace concentrations. The most common genera of bacteria in soils are *Pseudomonas, Arthrobacter, Achromobacter, Micrococcus, Vibrio, Acinetobacter, Brevibacterium, Corynebacterium,* and *Flavobacterium.*

Actinomycetes are a transitional group between the more primitive procaryotic bacteria and the eucaryotic fungi. Although taxonomically these organisms are classified as bacteria, the actinomycetes are similar to fungi in that they produce slender extensively branched filaments called hyphae that develop into a mycelium. Hyphae are characteristic of fungal masses which we associate with moldy materials. Many of the actinomycetes also produce spores or chains of spores known as conidia similar to those of fungi. Actinomycetes are present in high numbers in soil, tolerate a wide range of pH and temperature, grow under nutrient-limiting conditions, and are resistant to desiccation. Although their growth rate is slower than that of bacteria, the ability of actinomycetes to thrive under adverse conditions allows them to predominate when selective pressures are great. Actinomycetes have been shown to degrade phenols, aromatics, steroids, chlorinated aromatics, and lignocellulose (U.S. EPA, 1983).

Some bacteria are capable of forming endospores (spores formed within cells). Bacteria go from a vegetative state to an endospore state when growth conditions become too harsh, such as when the soil becomes too dry or when nutrients are limited. The endospore is a temporary condition (which in some cases may last several years), and the cell reverts to a vegetative state as soon as growth conditions improve. Endospores are very resistant to heat and are not easily destroyed by radiation or other chemical factors such as acids and disinfectants. Endospore-forming bacteria are very common in soil, where growth conditions (e.g., moisture content) may be very transient and spore formation is a very useful survival technique.

Fungi

Fungi are higher protists, have cell walls, usually are coenocytic (multinucleate with no cellular subunits), are immotile, and use organic material for both energy and carbon sources. Inorganic contaminants such as nitrogen, phosphorus, and other nutrients are incorporated into cell tissue in stoichiometric amounts, just as in the case of bacterial growth. Some fungi are aquatic, living in fresh water, but most are terrestrial, living in soil or on dead plants. Some of the better-known fungi include molds, yeasts, and mushrooms.

Relative to bacteria, fungi are generally less numerous, grow at considerably lower rates, and do not compete well in most engineered process environments. Additionally, metabolic processes of fungi are generally less diverse than those of bacteria. As a group, fungi tend to be more acid-tolerant than bacteria (many species grow optimally at a pH of 5 or less) and more sensitive to moisture content variability.

A fungus that appears to have considerable potential in the treatment of hazardous organic compounds is *Phanerochaete chrysoporium,* a white rot fungus. This organism produces an extracellular peroxidase enzyme that degrades lignin in the presence of peroxide. The reaction is relatively nonspecific in respect to the type of carbon-carbon bond attacked and has been found to be effective in initiating the degradation of a variety of highly chlorinated, recalcitrant compounds including dioxins (Aust et al., 1988; Hackett et al., 1977). The use of *Phanerochaete chrysoporium* is limited to conditions in which nitrogen is limiting because the peroxidase is not produced otherwise. A possible application of genetic engineering would be in the transfer of the gene responsible for the synthesis of the peroxidase to a bacterial species that grows well in engineered treatment processes.

Algae

Algae, like the fungi, are immotile higher protists with cell walls. Some algae genera are coenocytic but most are not. Many are unicellular, but some are filamentous, or colonial. Some have plantlike structures, with multicellular growth but no real differentiation between cells. Examples are seaweeds and kelps. The majority of algae are aquatic, although there are genuses that grow in soil. Carbon dioxide (CO_2) and/or bicarbonate ion (HCO_3^-) serve as the carbon source for algal growth, and energy is derived from absorption of light by photosynthetic pigments with oxygen being produced as a by-product. The principal wavelength range in which algae absorb light is between 300 and 700 nm. Algae are aerobic organisms and a portion of the oxygen produced is utilized in their respiration (i.e., as terminal electron acceptor). Excess oxygen accumulates in the surrounding water as long as enough light is available.

Algae are not important players in the field of bioremediation. In a few cases algae have been used in the bioremediation of aquatic systems either by bioaccumulation of hydrophobic compounds in their lipids followed by harvesting of the algal biomass or by degradation in the presence of sunlight (Okelley and Deason, 1976; Matsumura and Esaac, 1979). Algae are sometimes used in nutrient-removal systems but are extremely difficult to separate from water and often become troublesome contaminants themselves. Nutrient-rich ponds, sloughs, and lakes often have algal "blooms" over short periods of time that result in unaesthetic green mats on the water surface. The mats may be blown onto beaches, where they decay or are troublesome to people using the waters for recreation. Some algal species produce organic compounds that cause taste and odors in drinking water supplies.

Protozoa

The protozoa include motile and nonmotile, unicellular, higher (eucaryotic) protists that lack cell walls. Many protozoa feed solely by predation, grazing on bacteria or on other organisms such as yeasts, fungal spores, and other protozoa. Some also use dissolved organic substances for food. Thus protozoa are heterotrophic predators, and in a sense are higher on the trophic ladder, or food chain, than bacteria or algae. Protozoa require water to carry out metabolic activity. However, many protozoan species are found in soil where they extract sustenance from particulate material in the water films or sorbed onto soil surfaces. A large number of protozoan species exist and a number are typically seen in the microbial communities found in biological treatment systems.

In biological treatment processes, protozoa may play an important role by feeding on and reducing the numbers of bacteria degrading target contaminants. A rise in bacterial numbers in soil undergoing active biodegradation is often accompanied by a rise in numbers of protozoa. In laboratory studies, a single protozoan has been estimated to consume somewhere between several hundred and several hundred thousand bacterial cells per hour, the numbers being dependent on the type of protozoa and the bacteria (Paul and Clark, 1989). In soil, consumption rates are much slower, probably owing to inaccessibility. Bacteria tend to colonize pores and crevices within aggregates that are too small for protozoa to penetrate.

Grazing by protozoa may be beneficial in some cases. For example, protozoa help control bacterial growth near injection wells in in situ bioremediation processes, where excessive bacterial growth may cause clogging of porous media and decreased hydraulic conductivity (Sinclair et al., 1993). Also, by selectively grazing on bacteria, protozoa may be responsible for altering the balance between different genera of bacteria within a mixed culture. The numbers of protozoa present in soil are generally several orders of magnitude less than those of bacteria, and their growth rates are substantially lower than those of bacteria. Although protozoa are not generally known to biodegrade pollutants, they can ingest such organics adsorbed to bacterial cells or trapped within the extracellular secretions surrounding the cells (Wolfaardt et al., 1994*a*). Generally, the amount of contaminants removed by protozoans is negligible relative to that removed by bacteria or fungi.

Consortia of Microorganisms

It is important to see the microorganisms used in biological treatment as a community or consortium. A consortium can consist of two or more organisms living in close proximity to one another and interacting with each other. A consortium generally implies a positive interaction, where one group benefits from the actions of the other. In the biodegradation of chemical waste, for example, several groups of bacteria may be necessary to completely mineralize one compound. Individual species isolated in pure culture and given the target compound as a sole carbon source may be found to be incapable of mineralizing it (Wolfaardt et al., 1994*b*; Salanitro et al., 1994). In other cases, a consortium of microorganisms may be able to carry out the degradation faster and more efficiently, as compared to a pure culture. Additionally, consortia often include a certain amount of redundancy. Several organisms may be involved in each step and compete for the target compound or breakdown products during mineralization. The species best suited to the particular environment (e.g., temperature, pH, moisture content, and oxygen content) will predominate. However, a change in environmental conditions will result in a different species rising to dominance.

The interactions between groups of microorganisms are not always positive. In some cases toxic products resulting from one group's metabolism may be inhibitory to the activity of the other. In other cases, competition between groups for the same compound may result in predominance of one microbial species over others and a loss in culture stability. Each of the organisms present in a consortium occupies an ecological niche, and the relative numbers of the species will change as conditions change. For example, increases or decreases in temperature will modify the competitive advantages among the species present and result in changes in species predominance. Cyclic operations are more favorable to some groups than to others. The ability to utilize particular materials, such as benzene rings or ammonia, as energy sources is not widespread among microbial species, and the presence of these materials may give some species either a niche allowing them to survive or a competitive advantage over other species.

Distribution of Microorganisms in Soil and Groundwater

Microorganisms, especially bacteria, can be found almost everywhere in soil and groundwater. Microorganisms have been detected at depths as great as 500 to 600 m below the surface (Thomas and Ward, 1992). The numbers of different genera present, and the diversity within each group may vary dramatically with environmental factors such as pH, moisture content, soil structure, and nutrient availability. In unsaturated soil, microorganisms can be found attached to surfaces of aggregates, inside pores and crevices, or growing on plant roots. In groundwater, some microorganisms can be found floating with the moving water, but the larger majority will be attached to soil surfaces. Microorganisms, especially bacteria, tend to adsorb to surfaces because that is where nutrients accumulate, especially in oligotrophic environments.

The soil surface, particularly the rhizosphere, which is defined as the part of the soil under the direct influence of the roots of higher plants (Tate, 1995), is generally the richest in microbial densities and activity. Plant roots provide surface area for microorganisms to grow on, but more importantly, root exudates provide organic and inorganic nutrients, as well as vitamins and growth factors, thus creating a very rich growth environment. As a result, the rhizosphere generally has a larger number of microorganisms, higher growth rates, and higher metabolic diversity as compared with subsurface soils. The rhizosphere effect, which is a measure of activity (whether it is population density, growth rate, or metabolic capability) of microorganisms in the rhizosphere compared to that in subsurface soil, can have a value as small as 1 to as high as 20 or 100. A value of less than 1 would indicate some toxic or inhibitory effect in the rhizosphere.

$$\text{Rhizosphere effect} = \frac{R}{S} = \frac{\text{activity/unit weight rhizophere soil}}{\text{activity/unit weight subsurface}} \tag{4.1}$$

The intense microbial activity in the rhizosphere has been utilized to biodegrade relatively recalcitrant compounds, such as polynuclear aromatic hydrocarbons and chlorinated solvents. Degradation rates in the rhizosphere are often higher, and acclimation periods are shorter in rhizosphere soil as compared to nonrhizosphere soil. However, a long history of exposure to a contaminant may result in no noticeable difference in microbial activity in the two types of soil (Haby and Crowly, 1996). Bioremediation in the rhizosphere is a relatively new field which is still in the experimental stages.

In surface soils where organic nutrients are abundant, bacteria generally dominate. Numbers can range between 10^7 and 10^{10} colony-forming units/g dry soil weight, depending on soil conditions and the techniques used in estimating population densities. The numbers of protozoa, algae, and fungi are typically several orders of magnitude less than those of bacteria. High numbers of protozoa generally correlate with high numbers of bacteria. Numbers of algae decrease with increasing distance from the surface as solar energy decreases, and fungi tend to flourish in relatively acidic, dry conditions.

Moving deeper into the subsurface, organic nutrients become more limited. The numbers of genera generally decrease, although slightly in some cases, and the diversity within each group of microorganisms generally decreases, also. Sinclair and Ghiorse (1989) reported that even at depths of over 200 m considerable numbers of fungi and protozoa, as well as algae, could be found. Numbers of bacteria ranged between 10^6 and 10^7/g dry soil weight throughout most of the soil profile. Within the different subsurface strata microbial population density, diversity, and activity correlated negatively with clay content, low pH, and heavy-metal concentration, but positively with sand content and moisture content.

THE BACTERIAL CELL

Because bacteria are the most abundant microorganisms in soil, groundwater, and bioremediation systems, emphasis will be placed on properties of this group of protists over the properties of other microorganisms. The focus in this section will be on the chemical composition and general structure of bacterial cells. In most engineering problems the fundamental composition and structure of microbial cells is of less importance than overall size (typically 1 to 5 μm for bacterial cells) and elemental chemical composition.

Chemical Composition of Cells

An overall understanding of the typical composition of bacterial cells can be obtained from Table 4.2. Although the composition given is for a particular bacterial species, *Escherichia coli,* grown under specific conditions, the relative makeup of most bacteria is similar. Note that carbon is the major element in cell makeup. Bratback (1985) estimated the carbon content of a single bacterial cell to be 1×10^{-13}g. The bulk of the bacterial cell is made up of water, which is estimated to be 90 percent of total weight. Proteins make up the majority of macromolecules within bacterial cells.

The principal elements of all living cells, on a mass basis, are carbon, oxygen, nitrogen, hydrogen, phosphorus, and sulfur. A number of metals (iron, manganese,

TABLE 4.2
Typical elementary composition of bacterial cells on a dry-weight basis. Note that approximately 90 percent of the mass of living cells is water

Element	Percent of dry weight	General physiological function
Carbon	50	Constituent of organic cell materials
Oxygen	20	Constituent of organic cell materials and cellular water
Nitrogen	14	Constituent of proteins, nucleic acids, coenzymes
Hydrogen	8	Constituent of cellular water and organic cell materials
Phosphorus	3	Constituent of nucleic acids, phospholipids, coenzymes
Sulfur	1	Constituent of proteins and coenzymes
Potassium	1	Major cation in cell processes
Sodium	1	Major cation in cell processes
Calcium	0.5	Major cation in cell processes and enzyme cofactor
Magnesium	0.5	Major cation in cell processes, cofactor in ATP reactions
Chlorine	0.5	Major anion in cell processes
Iron	0.2	Constituent of cytochromes and other proteins, enzyme
Σ trace elements	0.3	Inorganic contituents of special enzymes

Source: Stanier et al., 1986.

potassium, cobalt, calcium, copper, and zinc) are essential to life because they serve as cofactors (or mediators) to the electron transport that takes place in specific enzyme-catalyzed reactions. However, these *trace elements* need to be present only in very small quantities and are generally available in excess in soils and groundwaters.

Empirical cell formulas, or elemental ratios, of the chemical constituents of microbial cells are used to estimate nutrient requirements for growth and convert gravimetric cell mass measurements into theoretical oxygen demand of cell tissue. The most widely used empirical cell formula, $C_5H_7NO_2$ (Porges et al., 1953), omits the essential nutrient phosphorus. Inclusion of phosphorus results in considerably more complex formulas, such as $C_{42}H_{100}N_{11}O_{13}P$ (McCarty, 1965). Empirical cell formulas are representative of growth under specific environmental conditions, and care must be used in general application of the relationships, as can be seen in Example 4.1.

EXAMPLE 4.1. THEORETICAL OXYGEN DEMAND OF BACTERIAL CELLS. Determine the theoretical oxygen demand (THOD) of 1 g of microbial cells using the two empirical cell formulas given above. Assume that organic nitrogen in the cells is not oxidized and remains in the –3 oxidation state.

Solution

1. Write a stoichiometric equation for the oxidation of $C_5H_7NO_2$.

$$C_5H_7NO_2 + 5O_2 \rightarrow 5CO_2 + NH_3 + 2H_2O$$

2. Determine the THOD of 1 g of $C_5H_7NO_2$.

$$\text{Empirical molecular weight} = 113$$

$$1\text{ g} = 0.00885\text{ mole}$$

5 moles of O_2 is required per empirical mole of cells

(0.00885 mole cells)(5 moles O_2/mole cells)(32 g O_2/mole O_2) = 1.42 g

3. Write a stoichiometric equation for the oxidation of $C_{42}H_{100}N_{11}O_{13}P$.

$2C_{42}H_{100}N_{11}O_{13}P + 107\ O_2 = 84CO_2 + 22NH_3 + 64H_2O + 6H^+ + 2PO_4^{-3}$

Empirical molecular weight = 997

1 g = 0.00100 mole

53.5 moles of O_2 are required per empirical mole of cells

(0.001 mole cells)(53.5 moles O_2/mole cells)(32 g O_2/mole O_2) = 1.71 g

Comment. Note the large difference in oxygen demand that would be predicted for the same mass of cells from application of the two empirical formulas. The empirical formula used most commonly is $C_5H_7NO_2$; however, the general validity of this formula is not well established.

Cell Structure

The physical structure of the bacterial cell can be characterized by shape (spherical, rod, spiral) and components, including their chemical makeup, size, and the manner in which they grow (individual cells, colonies, filaments). Shape and size are somewhat variable with stage of growth. A species listed as a rod may appear to be spherical under some growth conditions and cell size changes during the growth cycle. Bacteria that are normally found in groups, such as the filamentous species, grow as single cells under particular environmental conditions, also. Thus visual observation does not provide a method of species identification. Most methods of identification are based on bacterial metabolism, resistance to antibiotics, and chemical characteristics of the cell wall rather than morphology.

The nature of the structural components of the cell is of considerable interest in biological treatment. Reference can be made to Figure 4.2 in discussing the most important components relative to contaminant removal: the genetic component, enzymes, storage bodies, cytoplasmic (cell) membrane, cell wall, and capsule or slime layer.

Nucleoid and Plasmids

The genetic component of bacterial cells includes both the single deoxyribonucleic acid (DNA) nuclear strand (the more complex eucaryotic cell has a double strand and the individual subunits are paired) located near the center of the cell and relatively small DNA circlets called plasmids located in the cytoplasm. The nuclear strand is typically greater than 1,000 μm in length and has a molecular weight of the order of 10^9. Because bacterial cells average about 2 μm in length or diameter and the nuclear region is only a small part of the cell, the nuclear strand must be tightly folded. The nuclear strand of DNA is essential to the life of a cell. Without the nuclear DNA the information required for production of necessary enzymes and other structures necessary for growth is missing. Damage to the DNA can result in loss of an essential activity and death. Plasmids are generally not a required component of the cell but seem to provide particular capabilities that make a strain more or less competitive in a given environment. These include resistance to specific antibiotics, toxin production, ability to metabolize unusual compounds or ions, and possibly resistance to attack by bacteriophages.

Cytoplasm

The tightly packed, granular region within the cytoplasmic membrane, but excluding the nucleoid, is the cytoplasm. Principal components of the cytoplasm are the ribosomes (the structures made up of ribonucleic acid (RNA) and proteins where cell components are synthesized), enzymes (proteins) that catalyze necessary chemical reactions, plasmids, organic and inorganic compounds involved in metabolism, and granular inclusions containing storage materials such as glycogen and poly-β-hydroxybutyrate (PHB), polyphosphate (volutin granules), and sulfur. Glycogen granules are detectable only using electron microscopy (Ingraham et al., 1983) and are relatively evenly distributed. Poly-β-hydroxybutyrate, volutin, and sulfur granules can often be seen, with special preparation of the slides, as refractile bodies under light microscopy. Volutin is associated with a primitive form of energy metabolism, and the accumulation of large amounts of volutin in bacteria such as *Acinetobacter* under cyclic anaerobic/aerobic operation is now being used in biological phosphorus-removal processes (Barnard, 1975; Buchan, 1983; Metcalf and Eddy, 1991).

Cell membrane

Surrounding the cytoplasm is a bilayer unit membrane composed primarily of phospholipids and proteins. The cell membrane acts as an osmotic barrier (the phospholipid function) and has specific transport functions (a protein function) in which *transferase* enzymes carry out transport reactions for specific molecules. In aerobic bacterial cells, those that use O_2 as a terminal electron acceptor, the system of electron transport enzymes in which the energy-rich compound adenosine triphosphate (ATP) is produced and O_2 is consumed is attached to the cell membrane. Typical thickness of the cell membrane is about 7 to 8 nm.

Cell wall

Structural strength and rigidity are provided by the cell wall. However, this structure also acts as a molecular sieve that screens out toxic molecules and antibiotics, and contains binding and hydrolytic enzymes that aid in nutrient gathering and transport. Two general types of cell wall are identified by the gram stain process in which an applied dye is found to be permanently fixed (positive) or washed out of the cell (negative). Gram-positive cells are resistant to desiccation and are more commonly found in soil and other oligotrophic environments. Bacteria found in the human body are usually gram-negative.

Capsule (slime layer)

The capsule is an amorphous structure surrounding the cell, made up mostly of water (more than 90 percent) and of extracellular polymer substances (EPSs). Some EPSs are specific to certain strains of bacteria and can be used in species identification. Most EPSs are polysaccharides, although nucleic acids and proteins are also common. An extracellular polymer may remain attached to the cell, forming a covering layer around it, in which case it is called a capsule. The polymer may also detach and part away from the cell, in which case it is called free slime. The capsule is considered part of the cell and helps protect the cell wall from attachment of a bacteriophage. Slime layers seem to concentrate loosely around the cell and decrease in density with increasing distance from the cell. Both capsules and slime layers play an important role in the aggregation of cells in colonies and in biofilms (Characklis et al., 1990). For the most part, EPSs are responsible for cell adhesion to surfaces and for maintaining the integrity of biofilms by holding cells and clusters of cells together.

Pili

Single strands of protein originating in the cell membrane and extending about 10 μm from the cell wall are the pili. These hairlike apparati appear to have a function of binding the cell to specific structures. The pili are probably involved in attachment of bacteria to soil particles.

Flagella

These long (15 to 20 μm) and very thin (10 to 20 nm) filaments move the bacterial cell with a whiplike rotational motion, similar to that of a propeller. Movement of motile bacteria (not all species have such structures) toward favorable environments is a complex response to chemical gradients and "attractant solutes," such as nutrients, in the cell's surroundings. Movement can also be away from unfavorable environments, in response to gradients of oxygen and "repellent solutes," such as toxicants. Movement, in response to chemical gradients, is termed *chemotaxis*.

BACTERIAL GROWTH

Growth is simply defined as an increase in number of microorganisms per unit time. Most bacteria reproduce through the process of binary fission, where two cells are formed from the same parent cell. In a rod-shaped bacteria, for example, the parent cell is seen to grow in size and elongate to about twice the original length, and then form a partition called the septum, which divides the cell into two identical cells. The time it takes for two cells to form, from the same parent cell, is called generation time. Since the generation time is also the time needed to double the cell number, the term doubling time is sometimes used. The generation time varies dramatically depending on the species and on the growth conditions. Generation times can be as short as a few minutes or as long as several hours (Table 4.3).

Based on binary fission alone, growth would be expected to be exponential as cell numbers keep multiplying. Exponential growth can be carried out only up to a certain point, however, before nutrients run out or other environmental factors become limiting. Figure 4.3 depicts a typical growth curve for a bacterial culture inoculated into a fresh medium, in a closed-batch system, with optimum conditions for growth. The growth cycle can be divided into four principal phases: lag, exponential, stationary, and death.

TABLE 4.3
Maximum recorded growth rates for some bacteria, measured at or near their respective optimal temperature, in complex media

Organism	Temperature, °C	Doubling time, h
Vibrio natriegens	37	0.16
Bacillus stearothermophilus	60	0.14
Escherichia coli	40	0.38
Bacillus subtilis	40	0.43
Pseudomonas putida	30	0.75*
Vibrio marinus	15	1.35
Rhodobacter sphaeroides	30	2.2
Mycobacterium tuberculosis	37	≈ 6
Nitrobacter agilis	27	≈ 20*

Source: Stanier et al., 1986.
*Grown in synthetic media.

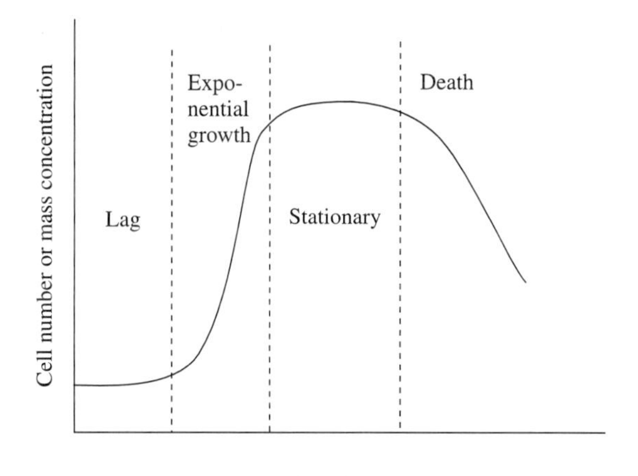

FIGURE 4.3
Typical growth cycle for a bacterial population in a closed-batch system.

Lag Phase

The lag phase is that time period it takes for bacteria to acclimate to the new environment before they start growing. During the lag phase the growth rate is nearly zero. The lag phase may be short, such as when exponentially growing bacteria are transferred into a medium with a nutrient composition similar to the one in which the bacteria had been growing. Inoculation of a sterile medium with cells in the stationary phase or inoculation of a medium having a different nutrient composition may result in a lag phase of 10 to 20 h or longer. The lag phase may be indicative of the time it takes for bacteria to synthesize the necessary enzymes to metabolize the new nutrients. This is often the case when trying to isolate a culture capable of degrading a specific pollutant. The time required for the degraders of a specific contaminant to begin metabolism of the compound is sometimes called the acclimation period. Acclimation periods may be very long, hundreds of days in some cases, depending on the biodegradability of the target chemical, the presence of preferential carbon sources that have to be exhausted first, and the presence of a degrading culture. In some cases, microbial enzymes may be induced only after lengthy exposure to the chemical. Genetic mutations or genetic exchange between indigenous populations may be necessary to develop a degrading culture, with the needed enzyme system. Note that when using the term acclimation period, the emphasis is on the biodegradability of a specific compound and not on growth of the degrading culture.

Exponential Growth

The exponential growth phase generally follows acclimation as nutrients are used to build up new cellular material. Initially cell numbers (or mass) begin to increase measurably, and the short period between zero growth and true exponential increase in numbers with time is sometimes referred to as the *increasing growth stage.* In exponential growth cell numbers increase exponentially, as can be seen by plotting numbers or mass concentrations on semilog paper. Eventually a limitation on growth rate occurs and the growth rate decreases [a period referred to as the *retardation growth stage* (Pitter and Chudoba, 1990)]. Most often the limiting factor is substrate concentration, but in some cases a decrease in oxygen levels (where oxygen is needed for biodegradation), lack of a required nutrient, the accumulation of a toxic product of biodegradation, or a change in pH. In closed systems production of CO_2 may result in a decrease in pH that affects the growth rate.

During exponential growth the change in numbers of bacterial cells is directly proportional to the number of cells present. This relationship can be expressed as

$$\frac{dN}{dt} = k\,N \tag{4.2}$$

where N = number of cells per volume of medium

t = time

k = proportionality constant, often referred to as the specific growth rate t^{-1}

Equation (4.2) can be integrated over a time period between t = 0 and t during which cell numbers increase from N_o to N:

$$\int_{N_o}^{N} \frac{dN}{N} = \int_{0}^{t} k\,dt \tag{4.3}$$

resulting in

$$\ln\left(\frac{N}{N_o}\right) = kt \tag{4.4}$$

or

$$N = N_o e^{kt} \tag{4.5}$$

The specific growth rate during exponential growth is directly related to doubling time (t_d). For one doubling time, $N = 2N_o$, substituting back into Equation (4.4),

$$\ln(2) = k\, t_d$$

or

$$k = \frac{\ln(2)}{t_d}$$

EXAMPLE 4.2.
Given that bacterial cell numbers in a batch reactor measure 34,000/L 4 h after inoculation, and 5.2×10^6/L at 24 h, and assuming a negligible lag phase, determine:

1. The specific growth rate.
2. The cell number at time of inoculation.

Solution

1. Using Equation (4.2) and integrating over the time interval t_1 to t_2 (4 to 24 h):

$$k = \frac{\ln N_2 - \ln N_1}{t_2 - t_1}$$

$$k = \frac{5.03}{20} = 0.25/\text{h}$$

2. The cell number at $t = 0$ can be obtained using Equation (4.5)

$$N_o = \frac{N}{e^{kt}} = \frac{34{,}000}{\exp(0.25 \times 4)}$$

$$N_o = 12{,}500 \text{ cells/L}$$

Stationary Phase

The stationary phase begins when the specific growth rate drops to approximately zero following exponential growth. Duration of the stationary phase is of the order of 12 to 36 h in most cases. Growth does not actually stop, but the net growth is approximately zero. Cells divide using stored organics or components of cells that die and break up (lyse). Overall metabolic activity, as measured by oxygen uptake rate, for example, is very low. Cells may begin to form cysts or enter into other routines necessary to survive during periods of nutrient limitation.

The way in which bacterial species respond in the stationary phase is an important factor in their ability to compete in engineered systems. For example, in batch reaction processes the organic concentration is initially high and decreases to very low values as the reaction proceeds. A period of hours or days may occur during which the culture is separated from the liquid and a new load of organics is added. Species capable of starting up rapidly, growing well at low nutrient concentrations, and maintaining activity (or potential for activity) during the stationary phase have a major competitive advantage. Some bacterial species store substantial quantities of organics as glycogen and poly-β-hydroxybutyrate (PHB) during periods of high concentration while others do not store organics at all. In cyclic processes, species that store organics have a significant advantage and become predominant.

Death

When bacteria stop growing, they eventually die. Death may simply mean inactivation of metabolic activity, or it may be real cell decomposition. Endogenous decay is the term used to describe viable bacteria feeding on the organic matter of dying bacterial cells. As in the stationary phase, death is measured in a net sense. Both growth and death are occurring but the balance results in a decrease in cell numbers and mass. The net decrease in microbial cell numbers due to death can be approximated as an exponential function, similar to that for growth:

$$\frac{dN}{dt} = -bN \tag{4.6}$$

where b is the rate constant of cell number decrease.

FACTORS INFLUENCING GROWTH AND BIODEGRADATION

The density and composition of the microbial community and the rate of transformation of pollutants are influenced by environmental factors, substrate factors, and microbiological factors. Primary environmental factors include moisture, aeration, temperature, pH, and nutrient availability. Properties of the substrate which can affect biotransformation include toxicity, concentration, solubility, volatility, solid-phase partitioning, and chemical structure. Microbiological factors include the presence of microorganisms with pathways for degrading the compound of interest, acclimation of microbial populations, and ecological factors. Some of the most important factors influencing growth and biodegradation are discussed below.

Nutrient Requirements

Microbial metabolism is directed toward reproduction of the organisms, and this requires that chemical constituents of cell components be available for assimilation and synthesis of new cell materials. The elements listed in Table 4.2 are required in the approximate proportions given. Note that the two empirical formulas given above have approximately the same percentages of the principal elements. For example, in the empirical formula $C_5H_7NO_2$ the percentages of carbon, hydrogen, nitrogen, and oxygen are 53, 6, 12, and 28, respectively.

Nutrients are usually assimilated from a limited number of elemental states, and therefore both the amount of nutrient present and the state of the nutrients are important. For example, heterotrophic bacteria require carbon in the organic form. Many bacterial species are able to utilize only a limited range of organic compounds. Only a few species are capable of metabolizing the five-carbon-sugar lactose in the absence of oxygen, and this property is used to determine the presence of *coliform* organisms (species that are common in the gut of warm-blooded animals) in water samples. The most probable number (MPN) test is constructed on the basis of this fact. Most bacteria can assimilate nitrogen in the ammonia (–3), nitrite (+3), and nitrate (+5) oxidation states. Sulfur assimilation is usually only from the sulfate (+6) oxidation state. Metals are assimilated from the ionic state in almost all cases.

Often one or two nutrients in the environment limit microbial growth. The *limiting-nutrient* concept is extremely useful in predicting the impacts of pollutants on receiving waters and in designing and operating biological treatment processes. The Great Lakes are phosphorus-limited and the increases in phosphorus inflow rates that resulted from the introduction of biodegradable detergents in the late 1960s greatly increased the rates of eutrophication. The removal of phosphorus from wastewater discharges to tributaries of the Great Lakes was an application of the limiting-nutrient concept. Many industrial wastewaters are unbalanced with respect to nutrients and stoichiometric additions of the growth-limiting nutrients (usually nitrogen and/or phosphorus) must be made.

Soil pH

Soil pH can significantly affect microbial activity. The growth of most microorganisms is usually greatest within a pH range of 6 to 8, although as noted above some fungi have optimal growth regions at pH levels of less than 5. Highly acid or alkaline conditions generally inhibit microbial activity, and most bacteria favor neutral conditions. There are, however, bacteria that are well adapted to acidic or basic conditions. For example, the sulfur-oxidizing bacteria, an obligate aerobic chemoautotrophic genus that produce sulfurous acid through oxidation of H_2S, function well at pH values of 1.

The soil pH also affects the solubility of phosphorus, an important nutrient for microbes, and the transport of hazardous metals in soil. Phosphorus solubility is maximized at a pH level of 6.5, and metal transport is minimized at a pH level greater than 6 (Sims et al., 1990). Most soils are acidic in nature. To increase the pH, calcium or calcium/magnesium-containing compounds can be added to the soil. This process is known as liming, and examples of the compounds used are calcium oxide (lime), calcium hydroxide, calcium carbonate, magnesium carbonate, and calcium silicate slags. Should the soil pH be high, because of a high carbonate concentration or because of the presence of hazardous wastes that are high in pH, then "acidification" may be necessary. Acidification, or the reduction of the soil pH, can be achieved by adding elemental sulfur or sulfur-containing compounds such as sulfuric acid, liquid ammonium polysulfide, and aluminum and iron sulfates (Dupont et al., 1988).

Temperature

The soil temperature greatly affects microbial activity and biodegradation rates. Bacterial species generally grow well in relatively narrow temperature ranges. Bacteria classified as mesophiles grow from about 15 to about 45°C, have optimal growth in the range of 25 to 35°C, and comprise the bulk of soil bacteria (Sims et al., 1990; U.S. EPA, 1985). Psychrophiles develop best at temperatures below 20°C. Thermophiles grow best at temperatures between 45 and 65°C. Within each division, different groups of bacteria have different optimum temperature ranges at which they grow.

A general rule of thumb is that for every 10°C increase in temperature, the rate of biotransformation increases about twofold (U.S. EPA, 1985). This is true, however, only up to some optimal temperature above which there is a decrease in reaction rate. Increases in biotransformation rates with increases in temperature are attributed to increased microbial activity, increased contaminant solubility, and decreased contaminant soil adsorption. In general, at temperatures above 40°C, biodegradation decreases because of enzyme and protein denaturation, and at temperatures approaching 0°C biodegradation essentially stops (Sims et al., 1990). As a general rule, bacteria are more tolerant of low-temperature extremes since they can capsulate and recover once the environmental conditions improve. At very high temperatures, however, the majority of the bacterial population dies.

In a field site, seasonal climatic changes will determine the soil temperature, and consequently the rate of biodegradation. Mulches are sometimes used to modify the soil temperature. Examples of mulch materials are compost, manure, wood chips and bark, sawdust, asphalt emulsion, and gravel or crushed stones (Dupont et al., 1988). Irrigation is also used to regulate soil temperature. Moisture decreases the thermal conductivity of the soil matrix and reduces daily variations in soil temperature. The principle is used in agriculture where sprinkle irrigation protects against frost formation in the winter and cools soil in the summer. In some cases, a cover is used over the site to control the emissions of volatile compounds, causing the soil temperature to increase.

Moisture Content

Moisture content of the soil strongly influences biological activity. Water is the major component of bacterial protoplasm, and an adequate supply of water is essential for microbial growth and maintenance. Water also serves as the transport medium through which organic compounds and nutrients are moved into the cells and through which metabolic waste products are moved away from the cell. The soil-water content influences aeration (oxygen transport), the solubility of the soil constituents, and the pH. Too little moisture in the soil results in dry zones and loss of microbial activity. Too much moisture, however, inhibits gas exchange and results in the development of anaerobic zones with the resulting elimination of aerobic bacteria and the ascendance of anaerobes or facultative anaerobes. Aeration and moisture are directly related because the pore space in soil not filled by water is filled with gas. The soil atmosphere generally contains more carbon dioxide and less oxygen than the atmosphere above the ground as a result of the respiration of microorganisms and plant roots, and coupled with the difficulty of gas movement into small pores. Most aerobic bacteria operate optimally at moisture levels of 50 to 75 percent of the soil's retention capacity (field capacity) (U.S. EPA, 1985).

Substrate Factors

The structure and makeup of organic compounds has a major impact on their availability to microorganisms and their biodegradability. Some of the more difficult to degrade, or recalcitrant, chemicals found in soil and groundwater include synthetic polymers, chlorinated and aromatic compounds, and pesticides such as DDT and chlordane. Compounds which are too large to penetrate the microbial cell and which cannot be modified by extracellular enzymes, such as polyvinyl chloride and polyethylene, cannot be degraded. Compounds which have very low water solubilities cannot be transported into the cell, and hence are not easily biodegradable. Molecules with certain structural properties may sterically hinder enzymatic attack. Structural factors which inhibit the degradation of compounds include the presence of amine, methoxy, sulfonate, and nitro groups; extensive halogenation; very high molecular weights or long chain lengths; benzenes substituted in the meta position; ether linkages; and branched carbon chains. The relationship between molecular structure and biodegradability is discussed in more detail in Chapter 6.

The concentration of chemicals in the environment can greatly affect the rate of their biodegradation. Compounds may be present in concentrations lower than the threshold concentration which will support growth or maintenance of the microbial population. Examples are 2,4-D and dichlorophenol: these compounds are readily degraded at concentrations in the 1 to 100 ppm range but may persist for years when present at concentrations in the ppb range (Alexander, 1981). An explanation of the phenomenon is that at very low concentrations the compounds do not provide sufficient energy for microbial growth, or perhaps induction of the necessary enzyme systems in situations where other energy sources are available. At the other end of the concentration range, compounds present in moderate to high concentration may be toxic to indigenous microorganisms in water, soil, sediment, or sewage. Oligotrophic organisms may experience toxicity and inhibition of biodegradation at lower concentrations compared to the microorganisms isolated and maintained under typical laboratory conditions such as those used in pure culture studies.

Some organic compounds do not serve as growth substrates but are degraded through a process termed cometabolism. Compounds degraded via cometabolism require specific cosubstrates which must be available as primary substrates to induce synthesis of the necessary enzymes in the cometabolizing population. If the cosubstrate is present in too high a concentration, however, competition for the enzyme inhibits metabolism of the nongrowth compound. Some schemes where the cosubstrates are pulsed into the bioremediation site are being explored in an attempt to get around this problem (McCarty, 1988). The presence of a second substrate which is not a cosubstrate can inhibit the degradation of a pollutant due to diauxic effects. In this case, metabolic control operates in a way that enables organisms to select the substrate which allows them to grow at the highest rate. The less desirable substrate, usually the pollutant, will be degraded only when the concentration of the more readily degraded substrate is limiting.

Biological transformations of organic compounds are catalyzed by the action of enzymes. The biodegradation of a specific compound is often a stepwise process that involves many enzymes and many organisms. Enzymes are specific in terms of the compounds attacked and reactions catalyzed. More than one enzyme is typically required to break down organic substances. Most frequently, the organisms which have the enzymes to break down the pollutants are already present in the soil. For example, this is generally the case for petroleum hydrocarbons. However, degradation of the pollutant often does not occur because of environmental limitations such as oxygen, nutrients, moisture, or pH.

MODELING GROWTH AND BIODEGRADATION

As was discussed above, the specific growth rate of a bacterial population varies depending on the growth conditions. When environmental conditions are maintained at an optimum, the factor that influences growth the most is *substrate* availability. Substrate is a term applied to the growth-limiting nutrient. In most cases the carbon source is the substrate, but an inorganic nutrient, an amino acid, a vitamin, or anything else needed for cell synthesis can be referred to as the substrate in particular cases. Often the carbon source is a mixture of compounds and a lumped parameter such as the biochemical oxygen demand (BOD), chemical oxygen demand (COD), total organic carbon (TOC), or total petroleum hydrocarbons (TPH) is used as a concentration parameter. In soil systems, concentrations may be reported as mass substrate per unit soil mass while in liquid systems concentrations are nearly always given on a mass per volume basis.

The rate of growth will be a function of environmental conditions, substrate concentration, oxygen concentration, and cell concentration. Usually environmental factors are accounted for in a growth-rate coefficient and functional relationships are used to describe the effect of substrate, oxygen, and cell concentrations. Saturation-type reaction-rate expressions such as Equation (4.7) are usually found to describe growth-rate data quite well.

$$r_g = \left(\frac{\mu_{max} C}{K_S + C} - k_d \right) X \tag{4.7}$$

where r_g = growth rate, mg/L · day
μ_{max} = maximum specific growth rate, per day
C = substrate concentration, mg/L
K_S = saturation coefficient, mg/L
k_d = maintenance energy coefficient, per day
X = cell mass concentration, mg/L

An assumption is usually made that the growth rate is linearly related to the cell mass or cell mass concentration, as indicated in Equation (4.7). This assumption should be valid unless a case can be made for a change (probably a decrease)

TABLE 4.4
Half-saturation constant values for selected substrates and for different groups of microorganisms

Organism (genus)	Substrate	K_S, mg/L
Escherichia	Glucose	0.008
	Glucose	4.0
	Lactose	20.0
Saccharomyces	Glucose	25.0
Aspergillus	Glucose	5.0
Pseudomonas	Methanol	0.7
	Methanol	0.4
Klebsiella	Carbon dioxide	0.4

Source: Pitter and Chudoba, 1990.

in growth rate with increasing cell concentration. A decrease might result from growth of cells in biofilms or large flocs where internal substrate or oxygen diffusion become limiting. However, in bioremediation systems measurement of the cell concentration is usually extremely difficult and an alternative approach is used. The maintenance energy term is usually dropped and the cell concentration is often considered to be approximately constant.

$$r_g \approx \frac{\mu_{max} C}{K_S + C} X \tag{4.8a}$$

$$\approx \frac{u_{max} C}{K_S + C} \tag{4.8b}$$

where $u_{max} = \mu_{max} X$, mg/L · day

Note that when substrate is abundant, as defined by $C >> K_S$, the growth rate becomes approximately equal to u_{max}. Some values for K_S for different substrates and selected groups of microorganisms are listed in Table 4.4.

The lower the K_S value, the higher the microorganism's affinity for the substrate. In comparing the curves in Figure 4.4, it can be seen that for a low K_S value, the change in specific growth rate happens over a relatively small range of substrate concentrations. Hence, for a large range of S (or C) values, the specific growth rate is close to maximum. The growth rate drops to zero when the substrate concentration is very dilute. The value of μ_{max} is dependent on several different factors including temperature and pH, composition of the nutrient medium, and the bacterial species itself.

Substrate Disappearance

In biological treatment processes, the substrate that bacteria use as a carbon or energy source is usually the pollutant that the process is designed to eliminate. Hence, modeling substrate disappearance is important in process design and in predicting process performance. However, in most cases, especially in field-scale

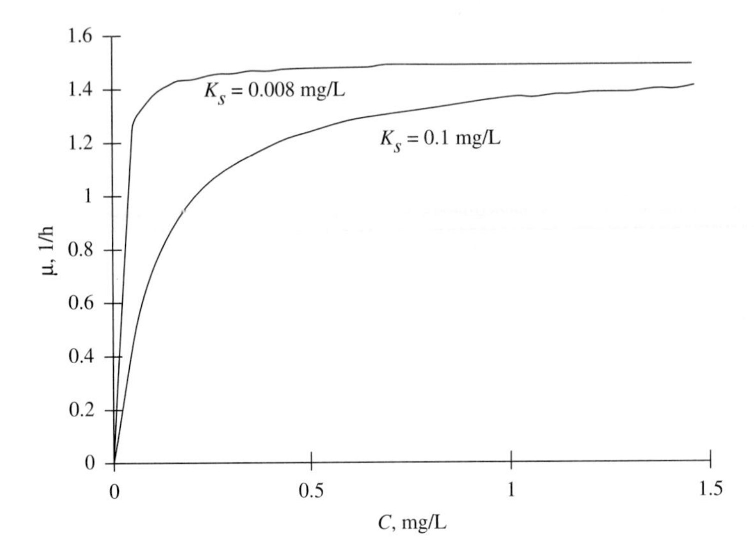

FIGURE 4.4
Change in specific growth rate for different values of K_S, and a μ_{max} value of 1.5/h.

applications, substrate disappearance is dependent on many variables and modeling becomes problematic. Some of these variables include predation by other microorganisms, bioavailability of the substrate, the presence of preferential carbon sources, and the need for an acclimation period. Simple models have been formulated for the case where substrate disappearance is only a function of substrate concentration and microbial biomass, and these are discussed in this section.

In deriving a general equation for substrate disappearance, modified Monod kinetics [Equation (4.8)] are assumed to be adequate for describing the growth dynamics of a bacterial culture, limited only by the concentration of the substrate. An equation relating the mass of cells produced per mass of substrate utilized is also defined:

$$r_g = -Yr_o \tag{4.9}$$

where Y = microbial yield, g cells produced/g substrate utilized
r_o = rate of contaminant removal, mg/L · day

Bioremediation of soils and groundwaters is usually a batch process. For homogeneous batch processes the reaction rate is equal to the accumulation rate (Tchobanoglous and Schroeder, 1985). Because cells are attached to soil particles, the solid-phase concentration of cells B having units of mg cells per g soil is substituted for the liquid-phase concentration X.

$$r_g = \frac{\mu_{max} C}{K_S + C} B \tag{4.10}$$

$$= -Y\frac{dC}{dt} = \frac{\mu_{max} C}{K_S + C} B \tag{4.11}$$

Note that Y can be used to estimate the mass of cells produced based on the mass of organic material removed.

$$Y = -\frac{\Delta B}{\Delta C} \tag{4.12a}$$

$$B = B_o + Y(C_o - C) \tag{4.12b}$$

The Y value for aerobic bacteria is close to 0.5 for a carbon source (such as glucose). In attached growth systems, such as soils, the value of Y based on the carbon sources may be less than 0.1 g/g. Corresponding yields can be determined for inorganic nutrients, such as N or P, and these values are generally considerably greater than 1 (Neidhardt et al., 1990).

Rearranging Equation (4.11) gives

$$\frac{dC}{dt} = -\frac{B}{Y}\frac{\mu_{\max} C}{K_S + C} \tag{4.13a}$$

$$= -\frac{\mu_{\max}}{Y}\frac{C[B_o + Y(C_o - C)]}{K_S + C} \tag{4.13b}$$

Equation (4.13*b*) is the general equation describing substrate disappearance in a system where population density and substrate concentration are the only factors which determine degradation kinetics. The integral form of Equation (4.13*b*) is

$$\frac{\mu_{\max}}{Y}t = \frac{1}{Y}\ln\frac{B_o + YC_o - YC}{B_o} - \frac{K_S}{B_o + YC_o}\ln\left(\frac{CB_o}{C_o(B_o + YC_o - YC)}\right) \tag{4.14}$$

Equation (4.13*b*), and hence its integral form, can be approximated into simpler mathematical forms within regions of extreme ratios of B_o to C_o or of C to K_S. For example, for the case where $B_o >> YC_o$ and $C >> K_S$, Equation (4.13*b*) reduces to a zero-order expression:

$$\frac{dC}{dt} = -\frac{\mu_{\max}}{Y}B_o \tag{4.15a}$$

$$C = C_o - \frac{\mu_{\max}}{Y}B_o t \tag{4.15b}$$

In soil systems three simplifications of Equation (4.13) are in common use, the zero-order expression developed above, a no-growth model (B = constant), and a first-order model ($K_S >> C$) (Simkins and Alexander, 1984; Alexander, 1994). The three modified models are summarized in Table 4.5.

The first-order rate model is the most commonly used of the three models given in Table 4.5 in describing substrate disappearance in field applications. Lack of data and ease of application, rather than accuracy, are the reasons for wide acceptance of the first-order model. One should be careful to remember, however, that the first-order rate model is based on many assumptions including a single substrate, a cell density larger than initial substrate concentration, a relatively low

TABLE 4.5
Models for substrate biodegradation by nongrowing microorganisms, under varying conditions

Model	Necessary condition	Differential form	Integral form
Zero-order	$C_o >> K_S$ $B_o >> YC_o$	$\frac{dC}{dt} = -\frac{\mu_{max}}{Y} B_o$	$C = C_o - \frac{\mu_{max} B_o}{Y} t$
Monod—no-growth	$B =$ constant	$\frac{dC}{dt} = -\frac{\mu_{max} B_o C}{Y(K_s + C)}$	$K_S \ln \frac{C}{C_o} + C - C_o = -\frac{B_o}{Y} \mu_{max} t$
First-order—no-growth	$C_o << K_S$ $B_o =$ constant	$\frac{dC}{dt} = -\frac{\mu_{max} B_o C}{YK_S}$	$C = C_o \exp\left(-\frac{\mu_{max} B_o}{YK_S} t\right)$

substrate concentration, and no acclimation period. As a general rule, biodegradation models should not be applied without consideration for the assumptions involved in formulating them.

EXAMPLE 4.3. KINETIC ANALYSIS OF BIOREMEDIATION DATA. A batch experiment was conducted to follow the biodegradation of the polynuclear aromatic hydrocarbon phenanthrene in a liquid culture. An inoculum of exponentially growing bacteria was introduced, and the following results were obtained.

1. Assuming first-order rate kinetics apply, find the biodegradation rate constant k.
2. Based on the results of this experiment, what is the half-life of phenanthrene?

Time, days	C mg/L
0	500
2	450
5	375
7	350
10	298
15	215

Solution

1. For first-order rate kinetics: $C = C_o e^{-kt}$. A linearized form of the equation is

$$\ln \frac{C_o}{C} = \frac{\mu_{max} B_o}{Y} t = kt$$

A plot of this function results in k, the slope of the curve, where $k = 0.055$/day.

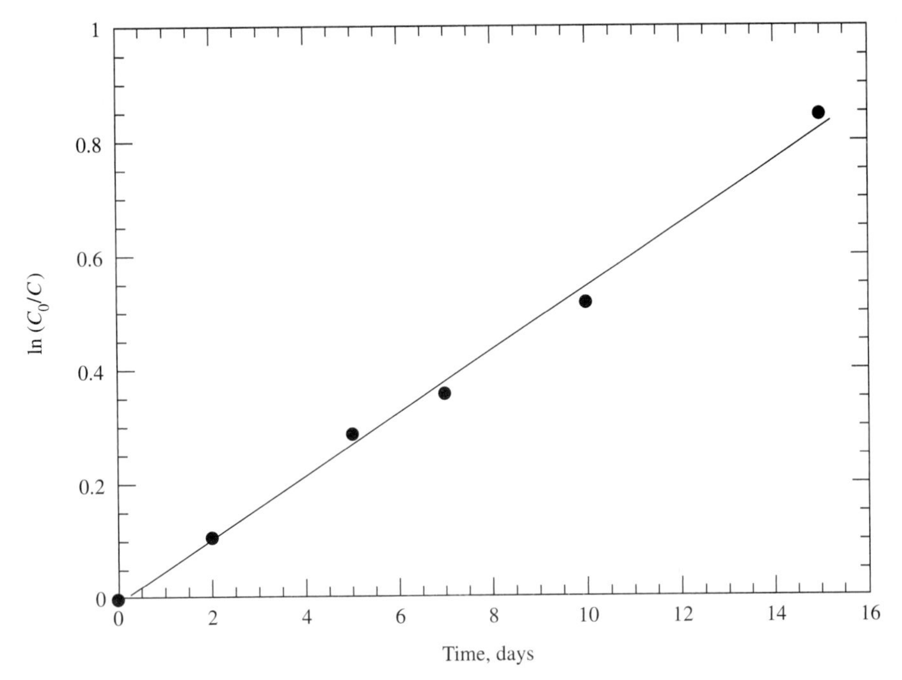

Note that as contaminant concentrations decrease the validity of the first-order model may decrease, also. All kinetic models are empirical approximations and are valid only in limited concentration regions.

2. The half-life of phenanthrene is the time needed to biodegrade 50 percent of the contaminant initially present.

$$t = -\frac{1}{k}\ln\frac{C}{C_o}$$

$$t = -\frac{\ln(0.5)}{0.055} = 12.6 \text{ days}$$

Methods Used to Measure Microbial Populations and Activity

The rate of bioremediation is dependent on both the microbial population density and the activity of the microbial population with respect to degradation of the target contaminants. Because of the variety of microbial species present in soil and groundwater and the difficulty in distinguishing between microbial particles and inert particles using physical methods, analysis of population density and activity is difficult. Available methods must be considered approximate at best. A number of techniques in use for measurement of microbial density and activity are described in this section.

Measuring Growth

Growth is typically measured as the change in cell number or in mass of cells. Cell numbers are measured in two ways: culture media and microscopy. In the first method, an agar medium containing the necessary nutrients for growth is usually used. A small volume of a diluted culture may then be spread over the agar surface (spread-plate method), or it may be mixed in with the agar before it is poured in a petri dish (pour-plate method). The plates are then incubated and each colony that forms on the surface is assumed to correspond to one viable cell in the original culture. Cells are measured as colony-forming units (CFU). This method of measuring growth is one of the least expensive and easiest to use. The assumption that one colony corresponds to one bacterial cell may underestimate population density. To minimize the underestimation, a number of dilutions can be used. The plate-count method allows for growth of only those cells capable of metabolizing the nutrients in the agar media. That is, selective enrichment for certain groups of bacteria occurs. Failure to form colonies is not necessarily an indication of the absence of a certain group of bacteria.

Microscopy is sometimes called a direct-count method. Because distinguishing between viable (living) and dead cells is difficult, overestimation of cell density is a problem. Confusion between microorganisms and other particulate matter is also a problem. Microscopy is generally not recommended for suspensions with a cell density less than 10^6 cells/milliliter (Brock et al., 1994).

Epifluorescence microscopy is a method used to better distinguish between cells and other particulate matter. Bacteria are stained with fluorescent dyes which concentrate in nucleic acids, thus making identification of living bacteria easier using specialized microscopes. The most commonly used dye in epifluorescence microscopy is acridine orange. Although this method of microscopy is more effective than direct microscopy, the cost is much higher, and the techniques are more problematic.

PROBLEMS AND DISCUSSION QUESTIONS

4.1. A soil is found to have a total heterotrophic bacterial population of 9×10^8 colony-forming units (CFU) per gram of dry soil.

a. Assuming that each CFU accounts for one cell and that each cell has a spherical structure with a diameter of 1 μm, then find the volume of cells in 1 g soil.

b. If the soil has a porosity of 0.35 and a particle density of 2.65 g/cm^3, then what percentage of the pore space is occupied by the bacterial cells?

c. Do you think that competition for space is a problem for bacteria?

d. Comment on any weaknesses you may see in the assumptions used above.

4.2. The following table lists results of plate counts measured at different depths in a cultivated soil.

Soil depth, cm	Colony-forming unit / g soil			
	Aerobic bacteria	**Anaerobic bacteria**	**Fungi**	**Algae**
0–5	8.9×10^6	7.8×10^5	1.1×10^4	150
5–20	9.0×10^5	8.8×10^4	6×10^3	5
20–35	3.2×10^4	5.0×10^3	1.4×10^2	0
150–175	4.3×10^2	1×10^5	0	0

a. Why do you think the numbers of microorganisms decrease with depth between 0 and 35 cm?
b. How can aerobic and anaerobic bacteria be found together in the same soil depth?
c. How can you explain the increase in numbers of anaerobic bacteria at depths below 150 cm?
d. Would you expect the numbers of microorganisms to decrease further at a depth below 400 cm? Is there a chance the soil will be sterile?
e. What if there were a leaky underground storage tank at a depth of 2 m?

4.3. Describe four ways in which moisture content can affect microbial activity.

4.4. Mass balances on microbial cells in ideal stirred-tank reactors operated at steady state result in the equations below:

$$r_g = \frac{X}{\theta_H}$$

where r_g = rate of microbial growth, mg/L · day

X = cell mass concentration, mg/L

$$\theta_H = \frac{V}{Q} = \text{hydraulic residence time, days}$$

For the classic Monod rate model:

$$r_g = Y\frac{kCX}{K+C} - k_d X$$

$$R_g = \frac{r_g}{X}$$

A plot of C vs. θ_H results in four asymptotes:
a. Derive the asymptote values.
b. Sketch the curves and determine which are real and which are not real. Explain your answers.

4.5. A well-sealed pond has been used for disposal of wastes from an industrial process utilizing a large number of organic chemicals for a 20-year period ending in 1985. Records of materials that have been placed in the pond are very incomplete but compounds commonly used in the plant include substituted phenols (e.g., 2-chlorophenol and 2,6-dichlorophenol, and pentachlorophenol), unchlorinated solvents (e.g., acetone, toluene), and chlorinated solvents (1,1,1-trichloroethane, trichloroethene, and dichloromethane). Periodic evaluation of the pond contents over the last 10 years has resulted in the data below.

Date	BOD_5, mg/L	COD, mg/L
2/25/85	6,850	15,850
3/15/86	6,200	14,900
2/28/87	5,400	13,640
2/14/88	4,900	12,900
3/3/89	2,300	10,400
3/1/90	2,250	10,250
2/15/91	2,150	10,150
2/12/92	2,090	10,020

A small mixer was installed on 2/28/88.

The mixer was installed in 1988 because the decreasing contaminant concentration values were noted and it was believed that biodegradation would be enhanced.

a. What is the basis for assuming that biodegradation might be occurring?
b. Is biodegradation the only explanation for the decrease in COD and BOD?
c. What explanation might be given for the fact that the BOD_5 remains at 2,090 mg/L after 7 years (i.e., the BOD is 2,090 mg/L after a 5-day test in the laboratory)?
d. What tests might be run to verify your answer to *c*?

4.6. What is the specific growth rate of a bacterial culture undergoing exponential growth, if the generation time is 27 min? If the cell count at time zero is 2×10^4/L, what is the bacterial density at 4 h? Plot a curve showing the change in population density with time.

4.7. During field-scale bioremediation, the disappearance of an organic contaminant is being monitored over time. The concentration of the contaminant is seen to drop from 950 to 500 ppm within 40 days. What would the concentration be at 70 days if:

a. Zero-order rate kinetics model is used.
b. First-order rate kinetics model is used.
c. Based on this data alone, which model would you recommend using to predict process performance?

4.8. Estimate the time required for the concentration of pesticide in a soil to decrease from 1 mg/kg to 0.01 mg/kg (99 percent removal) using *(a)* zero order kinetics (μ_{max}/Y = 0.0015 μg/g.hr) *(b)* first order kinetics ($B_o\mu_{max}/Y$ = 0.0015/hr) and *(c)* Monod no-growth kinetics ($B_o\mu_{max}/Y$ = 0.0015 μg/g.hr, K_S = 0.1 mg/kg).

REFERENCES

Alexander, M. (1981): "Biodegradation of Chemicals of Environmental Concern," *Science,* vol. 211, January 9.

Alexander, M. (1991): Introduction to Soil Microbiology, Wiley, New York and London.

Alexander, M. (1994): Biodegradation and Bioremediation, Academic Press, CA.

Aust, S., T. Fernando, B. Brock, H. Tuisel, and J. Bumpus (1988): "Biological Treatment of Hazardous Wastes by *Phanerochaete Chrysosporium,*" *Proceedings, Conference on Biotechnology Applications in Hazardous Waste Treatment,* G. Lewandowski, B. Baltzis, and P. Armenante (eds.), Longboat Key, Florida, Oct. 30–Nov. 4, Engineering Foundation, New York.

Barnard, J. L. (1975): "Cut P and N without Chemicals," *Water and Water Engineering,* vol. 11, p. 7.

Bratback, G. (1985): "Bacterial Biovolume and Biomass Estimations," *Applied Environmental Microbiology,* vol. 49(6), pp. 1488–1493.

Brock, T. D., M. T. Madigan, J. M. Martinko, and J. Parker (1994): *Biology of Microorganisms,* 7th ed., Prentice-Hall, Englewood Cliffs, NJ.

Buchan, L. (1983): "Possible Biological Mechanism of Phosphorus Removal," *Water Science and Technology,* vol. 15, p. 87.

Characklis, W. G., K. C. Marshall, and G. A. McFeters (1990): "The Microbial Cell," in W. G. Characklis, and K. C. Marshall (eds.), *Biofilms,* Wiley, New York, pp. 131–159.

Dupont, R. R., R. C. Sims, J. L. Sims, and D. L. Sorensen (1988): *In situ* Biological Treatment of Hazardous Waste-contaminated Soils, in *Biotreatment Systems,* vol. II, edited by Donald L. Wise, CRC Press, Inc. Boca Raton, FL.

Evans, W. C., and G. Fuchs (1988): "Anaerobic Degradation of Aromatic Compounds," *Annual Review of Microbiology,* Ornston, L. N., A. Balows, and P. Baumann, (eds.), Annual Reviews, Inc., Palo Alto, CA.

Haby, P. A., and D. E. Crowly (1996): "Biodegradation of 3-Chlorobenzoate as Affected by Rhizodeposition and Selected Carbon Substrates," *Journal of Environmental Quality,* vol. 25, pp. 304–310.

Hackett, W. F., W. J. Connors, T. K. Kirk, and J. G. Zeikus (1977): "Microbial Decomposition of Synthetic 14C-Labeled Lignins in Nature: Lignin Biodegradation in a Variety of Natural Materials," *Applied Environmental Microbiology,* vol. 33, p. 43.

Ingraham, J. L., O. Maaloe, and F. C. Neidhardt (1983): *Growth of the Bacterial Cell,* Sinauer and Assoc., Inc., Sunderland, MA.

Matsumura, F., and E. G. Esaac (1979): "Degradation of Pesticides by Algae and Microorganisms," in *Pesticides and Xenobiotic Metabolism in Aquatic Organisms,* N. A. Q. Kahn (ed.), American Chemical Society.

McCarty, P. L. (1965): "Thermodynamics of Biological Synthesis and Growth," *Proceedings, Second International Conference on Water Pollution Research,* Pergamon Press, New York, p. 169.

McCarty, P. (1988): "Bioengineering Issues Related to In Situ Remediation of Contaminated Soils and Groundwater," in G. S. Owen (ed.), *Environmental Biotechnology,* Plenum Publishing Company, pp. 143–162.

Metcalf and Eddy, Inc. (1991): *Wastewater Engineering,* 3d ed., McGraw-Hill, New York.

Neidhardt, F. C., J. L. Ingraham, and M. Schaechter (1990): *Physiology of the Bacterial Cell: A Molecular Approach,* Sinauer Associates, Inc., Sunderland, MA.

Okelley, J. C., and T. R. Deason (1976): "Degradation of Pesticides by Algae," U.S. Environmental Protection Agency Office of Research and Development, Athens, GA.

Paul, E. A., and F. E. Clark (1989): *Soil Microbiology and Biochemistry,* Academic Press, London.

Pitter, P., and J. Chudoba (1990): *Biodegradability of Organic Substances in the Aquatic Environment,* CRC Press, Boca Raton, FL.

Porges, N., L. Jaiswicz, and S. R. Hoover (1953): "Biological Oxidation of Dairywaste, VII," *Proceedings, 24th Industrial Waste Conference,* Purdue University, West Lafayette, IN.

Reineke, W., and H. J. Knackmuss (1988): "Microbial Degradation of Haloaromatics," *Annual Review of Microbiology,* Ornston, L. N., A. Balows, and P. Baumann (eds.), Annual Reviews, Inc., Palo Alto, CA.

Salanitro, J. P., L. A. Diaz, M. P. Williams, and H. L. Wisniewski (1994): "Isolation of a Bacterial Culture That Degrades Methyl t-Butyl Ether," *Applied and Environmental Microbiology,* vol. 60, no. 7, pp. 2593–2596.

Simkins, S., and M. Alexander (1984): "Models for Mineralization Kinetics with the Variables of Substrate Concentration and Population Density," *Applied and Environmental Microbiology,* vol. 47(6), pp. 1299–1306.

Sims, J. L., R. C. Sims, and J. E. Mathews (1990): "Approach to Bioremediation of Contaminated Soil," *Hazardous Waste and Hazardous Materials,* vol. 7(4), pp. 117–149.

Sinclair, J. L., and W. C. Ghiorse (1989): "Distribution of Aerobic Bacteria, Protozoa, Algae, and Fungi in Deep Subsurface Sediments," *Geomicrobiology Journal,* vol. 7, pp. 15–31.

Sinclair, J. L., D. H. Kampbell, M. L. Cook, and J. T. Wilson (1993): "Protozoa in Subsurface Sediments from Sites Contaminated with Aviation Gasoline or Jet Fuel," *Applied and Environmental Microbiology,* vol. 59, no. 2, pp. 467–472.

Stanier, R. Y., J. L. Ingraham, M. L. Wheelis, and P. R. Painter (1986): *The Microbial World,* Prentice-Hall, Englewood Cliffs, NJ.

Tate, R. L. (1995): *Soil Microbiology,* Wiley, New York.

Tchobanoglous, G., and E. D. Schroeder (1985): *Water Quality Characteristics, Modeling, Modification,* Addison-Wesley, MA.

Thomas, J. M., and C. H. Ward (1992): "Subsurface Microbial Ecology and Bioremediation" *Journal of Hazardous Materials,* vol. 32, pp. 179–194.

U.S. Environmental Protection Agency (1985): *EPA Guide for Identifying Cleanup Alternatives at Hazardous Waste Sites and Spills: Biological Treatment,* EPA 600/3-83/063.

Wolfaardt, G. M., J. R. Lawrence, J. V. Headly, R. D. Robarts, and D. E. Caldwell (1994*a*): "Microbial Exopolymers Provide a Mechanism for Bioaccumulation of Contaminants," *Microbial Ecology,* vol. 27, pp. 279–291.

Wolfaardt, G. M., J. R. Lawrence, R. D. Robarts, and D. E. Caldwell (1994*b*): "The Role of Interactions, Sessile Growth, and the Nutrient Amendments on the Degradative Efficiency of a Microbial Consortium," *Canadian Journal of Microbiology,* vol. 40, pp. 331–340.

CHAPTER 5

Metabolism and Energy Production

Metabolism is the term used to refer to the more than one thousand individual chemical transformations that take place within a cell (Stanier et al., 1986). The ultimate goal of these reactions or transformations is production of new cells. Metabolism is divided into two general processes: anabolism, an energy-requiring process in which microorganisms construct cell material, and catabolism, an energy-releasing process in which microorganisms oxidize compounds. Anabolism, the process of building new cell material, is also called biosynthesis. Both anabolism and catabolism are organized as a series of small steps. For example, the complete oxidation of the six-carbon-sugar, glucose, requires over 20 reactions. Anabolism and catabolism are closely interlinked in that partially oxidized organic compounds along a catabolic pathway may serve as initial steps in the synthesis of a cellular component. Carbon, which makes up almost 50 percent of the dry weight of cells, is the major nutrient required. Hence microorganisms are said to require a carbon source for metabolism. Since metabolism involves energy transfer, microorganisms also need an energy source to carry out metabolic reactions. Most of the organisms of interest in bioremediation are heterotrophic and use organic compounds as both carbon and energy sources.

Microbial metabolism is very diverse in terms of the compounds used as carbon and energy sources, the restrictions or limiting characteristics of particular species, and the environmental requirements for groups to thrive. Classification of microorganisms according to metabolism is very useful. The most common methods of classification are based on carbon source, energy source, and terminal electron acceptors used. With respect to carbon source, the divisions are organic compounds and carbon dioxide (CO_2). When cellular carbon is derived from transformation of organic compounds, the metabolism is termed *heterotrophic,* and when CO_2 is used as the source of cellular carbon, the metabolism is termed *autotrophic.* Energy required for growth is derived from oxidation of chemicals

TABLE 5.1
Classification of microorganisms based on metabolic needs

Classification	Carbon source	Energy source
Based on carbon source:		
Autotroph	CO_2	
Heterotroph	Organic compounds	
Based on energy source:		
Chemotroph		Chemical compounds
Chemolithotroph		Inorganic compounds
Chemoorganotroph		Organic compounds
Phototroph		Light
Combined terms:		
Chemoautotroph	CO_2	Chemical compounds
Photoautotroph	CO_2	Light
Chemoheterotroph	Organic compounds	Chemical compounds
Photoheterotroph	Organic compounds	Light
Other terms:		
Eutroph	Use high concentrations for carbon and energy	
Methylotroph	Use 1-carbon compounds as energy and carbon source	
Oligotroph	Use low concentrations for carbon and energy	
Zymogenous	Grow rapidly when carbon and energy source added	
Autochthonous	Grow slowly and steadily without external inputs	
Saprophyte	Live off of dead organic matter	

(*chemotrophic*) or light (*phototrophic*). Chemical sources of energy include organic compounds (*chemoheterotrophic*) and inorganic compounds (e.g., NH_3, NO_2^-, S, H_2S) (*chemoautotrophic*). The suffix "*-troph*" is derived from Greek and means "to feed." Based on their carbon and energy sources, microorganisms are placed in different metabolic classes (Table 5.1). In the combined terms, the first syllable usually refers to the energy source while the second refers to the carbon source, so that a photoautotroph, for example, is a microorganism that uses light for an energy source and CO_2 for a carbon source. A chemoheterotroph, on the other hand, is a microorganism that uses chemical compounds, as opposed to light, for energy source, and organic compounds for carbon source. With the exception of the groups that oxidize NH_3 and NO_2^-, chemoautotrophic metabolism is generally a characteristic of bacteria that readily oxidize organic compounds if they are available. The metabolic diversity among microorganisms, particularly within different bacterial groups, is discussed in more detail in following sections.

ENERGY

Chemotrophs obtain energy principally through chemical reactions. When an organic or inorganic chemical is oxidized, energy is released. Phototrophs convert light energy into chemical energy to carry out metabolic processes. Nearly all of the reactions carried out in cellular processes require the involvement of special proteins called enzymes that act as catalysts. Hence microorganisms act as reactors in which enzymes act as catalysts mediating oxidation-reduction (and many other types of) reactions. Microbial growth is, to a large extent, dependent on the amount of free energy released from the reaction and on the efficiency with which the energy is captured.

The units of energy most used in biology are the kilocalorie (kcal) and the kilojoule (kJ), where 1 kcal is equal to 4.184 kJ. A kilocalorie is defined as the quantity of heat energy necessary to raise the temperature of 1 kilogram of water by 1°C. When a chemical reaction takes place, a change in energy occurs. The amount of energy released during a chemical reaction can be expressed in two terms:

H = enthalpy = total amount of energy released during a chemical reaction

G = free energy = energy released which is available to do useful work

The difference between enthalpy and free energy is the energy not available to do useful work, which is often lost as heat.

The change in free energy during a reaction can be computed from the standard free-energy change of the reaction as ΔG° and the system composition. The change in free energy can be either a positive or a negative value. For example, for the general reversible reaction:

$$aA + bB \rightleftarrows cC + dD$$

$$\Delta G = \Delta G^\circ + RT \ln\left(\frac{\{C\}^c\{D\}^d}{\{A\}^a\{B\}^b}\right) \tag{5.1}$$

where ΔG = change in free energy at the conditions given
ΔG° = change in free energy at standard conditions
R = gas constant = 1.99 cal/mole · K
= 8.29 joule/mole · K
T = absolute temperature, K
$\{C\}$ = activity of species C; same for A, B, and D
c = stoichiometric coefficient; same for a, b, and d

If ΔG is negative when the reaction proceeds to the right, free energy is released; the reaction will occur spontaneously to the right, and the reaction is said to be exergonic. However, if ΔG is positive, energy is needed to drive the reaction; the reaction will not occur spontaneously as written but rather the reverse reaction will occur, and the reaction is said to be endergonic. Overall, catabolism is exergonic and anabolism is endergonic.

In an exergonic reaction, as the reaction proceeds the concentrations of the products (C and D) build up and the rate of the reverse reaction increases. Equilibrium occurs when the forward reaction rate is balanced by the reverse reaction rate. Because reactions are in fact occurring, equilibrium is a dynamic state, not a static one. In dilute solutions the component activities ($\{A\}$, $\{B\}$, $\{C\}$, . . .) can be approximated by the molar concentrations ($[A]$, $[B]$, $[C]$, . . .). The equilibrium constant K can be calculated by dividing the product of the molar concentration of the products, raised to the power of their stoichiometric coefficients, over the product of the molar concentrations of the reactants, raised to their stoichiometric values.

$$K = \frac{[C]^c[D]^d}{[A]^a[B]^b} \tag{5.2}$$

where K = equilibrium constant
$[C]$ = molar concentration of species C, mole/L

For pure solids, liquids, and elements, the activity is assumed equal to 1. Thus the dissociation constant of water K_W is also the equilibrium constant.

$$K_{H_2O} = \frac{\{H^+\}\{OH^-\}}{\{H_2O\}}$$

Because $\{H_2O\} = 1$, and because the concentrations of H^+ and OH^- are always small, the dissociation constant can be written as

$$K_W = [H^+][OH^-]$$

The concentrations of the reactants and of the products, at equilibrium, are related to the free energy of the reaction. Substituting Equation (5.2) into Equation (5.1) and noting that $\Delta G = 0$ at equilibrium,

$$0 = \Delta G = \Delta G^o + RT \ln K \tag{5.3}$$

Therefore

$$K = \exp\left(\frac{-\Delta G^o}{RT}\right) \tag{5.4}$$

FREE ENERGY OF FORMATION

The free energy of formation (G_f^o) is the energy released or energy required to form a molecule from its elements. By convention, G_f^o of the elements (e.g., O_2, C, N_2) in their standard state is zero. Examples of free energy of formation of some compounds are given in Table 5.2.

Using the free energy of formation, one can calculate the change in free energy for a particular reaction. For the reaction

$$aA + bB \rightarrow cC + dD$$

$$\Delta G_{rx} = cG_f^o(C) + dG_f^o(D) - aG_f^o(A) - bG_f^o(B) \tag{5.5}$$

where ΔG_{rx} = change in free energy for the reaction
c = stoichiometric coefficient
$G_f^o(C)$ = free energy of formation of species C

TABLE 5.2
Free energy of formation values of selected compounds

Compound	G_f^o, kJ/mole
CO_2	–394.4
CH_4	–50.75
$C_7H_6O_2$ (benzoic acid)	–245.6
C_2H_6O (ethanol)	–181.75
$C_6H_{12}O_6$ (glucose)	–917.22
CH_4O (methanol)	–175.39
Fe^{3+}	–4.6
Fe^{2+}	–78.87
H_2	0.0
H^+	0.0 at pH 0, –5.69 per pH unit
O_2	0.0
H_2O	–237.17
NO_3^-	–111.34
NH_4^+	–79.37
N_2O	+104.18
$HS^-_{(aq)}$	+12.58
$H_2S_{(aq)}$	–27.4
SO_4^{2-}	–744.6

Source: Adapted from Brock et al., 1994.

That is, the change in free energy for the reaction is equal to the sum of the free energy of formation of the products minus the sum of the free energy of formation of the reactants.

EXAMPLE 5.1. CHANGE IN FREE ENERGY. One mole of methane (CH_4), and two moles of oxygen are in a closed container. Determine if the reaction written below will proceed as written. What is the equilibrium constant for the reaction?

$$CH_4 + 2O_2 \rightarrow CO_2 + 2H_2O$$

Solution

1. Determine the change in free energy for the reaction.

Compound	G_f^o, kJ/mole
CH_4	–50.75
O_2	0.0
CO_2	–394.4
H_2O	–237.17

Using Equation (5.5):

$\Delta G^o = (1)(-394.4) + (2)(-237.17) - (1)(-50.75) - (2)(0) = -817.99$ kJ/mole

ΔG^o is a large negative value; the reaction is exergonic; it will proceed as written

2. Determine the equilibrium constant, using Equation (5.3):

$$\Delta G = \Delta G^{o} + RT \ln K$$

where $\Delta G = 0$ at equilibrium

$$K = \exp\left(\frac{-\Delta G^{o}}{RT}\right)$$

$$K = \exp\left(\frac{817.99}{(0.0083)(298)}\right)$$

$$K = \exp(330.7)$$

K is very large, which means that the equilibrium is far to the right with much more of the products formed than the reactants remaining.

ACTIVATION ENERGY AND ENZYMES

For molecules to react with one another, an energy input is first required to break bonds within the molecules and bring them to a reactive state. This energy is referred to as the activation energy. A catalyst is a substance that lowers the activation energy of a reaction and increases its rate. Catalysts remain unchanged as a result of the reaction catalyzed. In living organisms, special proteins called enzymes are catalysts for the reactions through which compounds are oxidized and new compounds are synthesized. Most reactions in living organisms require enzymes to proceed at an appreciable rate. Enzymes are highly specific. That is, each enzyme catalyzes only a single type of chemical reaction or a set of closely related reactions.

Enzymes are proteins, made up of amino acids linked together via peptide bonds into chains called polypeptides (Figures 5.1 and 5.2). The polypeptides twist and fold, resulting in a chemically stable, three-dimensional, complex structure which gives the protein its specific characteristics and determines how it will function in the cell. The folding and twisting of the polypeptide subunits into an enzyme molecule results in exposed regions or grooves called *active sites* where other molecules can bind to the enzyme. In a typical reaction, the enzyme binds to the reactant and once the reaction is complete, the product is released and the enzyme returns to its original state. In binding the substrate, the enzyme helps align the reactive groups as well as place strain on specific bonds within the substrate, thus reducing the activation energy needed to drive the reaction. Enzymes can increase the rate of reaction by factors up to 10^{20} times the rate that would occur spontaneously (without the enzyme).

Enzymes are named either for the substrate that they bind or for the chemical reaction that they catalyze, by the addition of the suffix "-ase." For example, cellulase is the enzyme which catalyzes the cleavage of glucose molecules from the polysaccharide cellulose and oxygenase is the enzyme that catalyzes oxygen fixation reactions. The folding patterns of polypeptides that make enzymes so chemically stable

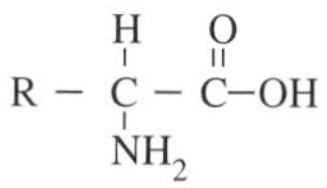

```
     H  O
     |  ||
R – C – C–OH
     |
    NH2
```

FIGURE 5.1
An α-amino acid; the *R* signifies a side chain composed of a skeleton of carbons and various other attached groups.

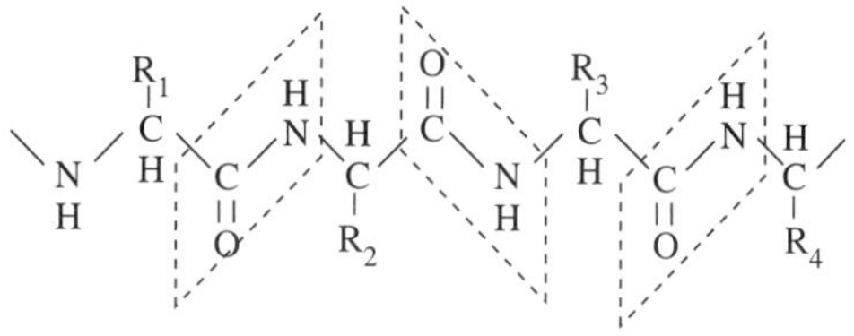

FIGURE 5.2
Polypeptide composed of α-amino acids with side chains designated R_n connected by a peptide bond.

and so specific are also the reason why enzymes are sensitive to temperature and pH variations. Under extreme conditions (e.g., high temperatures), the polypeptide chains unfold, a process termed denaturation. Denaturation destroys the structure of the molecule and breaks the hydrogen bonds that link the different polypeptide chains together, thus rendering the enzyme inactive.

A coenzyme is a low-molecular-weight nonprotein molecule which participates in an enzymatic reaction by transferring electrons or functional groups. Coenzymes bind loosely to enzymes during reactions, and a single coenzyme can be associated with different enzymes at different times. Examples are the coenzymes NAD^+/NADH and $NADP^+$/NADPH and acetyl CoA, which are discussed later in the chapter.

OXIDATION-REDUCTION REACTIONS

Living organisms utilize chemical energy through oxidation-reduction (redox) reactions. A redox reaction is a coupled reaction that involves the transfer of electrons from one molecule to another. In some cases whole hydrogen atoms are also transferred (see coenzymes NAD^+ below). Oxidation is the process in which an atom or a molecule loses electrons and reduction is the process in which an atom or molecule gains electrons. An *electron donor* becomes oxidized after releasing electrons while an *electron acceptor* becomes reduced after receiving electrons. The two molecules involved are called a redox pair. Electron acceptors are also termed oxidizing agents while the electron donors are termed reducing agents. An oxidation and a reduction need to occur simultaneously for a reaction to be complete.

The tendency of a substance to donate electrons or to accept electrons is expressed as the reduction potential (E_o), measured in volts (V). The reduction potentials are expressed for half reactions which are, by convention, written as reductions. For example, the half reaction for the reduction of oxygen is

$$\frac{1}{2}O_2 + 2\,H + 2e^- \rightarrow H_2O \tag{5.6}$$

in which the redox pair is $1/2O_2/H_2O$, and $E_o = +0.816$ V. The half reaction for the reduction of hydrogen is

$$2\,H^+ + 2\,e^- \rightarrow H_2 \tag{5.7}$$

in which the redox pair is $2H^+/H_2$ and $E_o = -0.421$ V.

Note that the redox pairs (O/R) are expressed such that the oxidizing agent (electron acceptor) is written on the left, while the reducing agent (the electron donor) is written on the right (e.g., $1/2O_2/H_2O$). Oxidation-reduction reactions involve two redox pairs, one serving as the electron acceptor and one as the electron donor. So even though both redox half reactions are written as reductions, in a complete oxidation-reduction reaction one of the half reactions must be an oxidation and will therefore proceed in the reverse direction. For example, the redox pairs involved in the formation of water are:

1. Acceptor pair $\quad \frac{1}{2}O_2 + 2\,H^+ + 2e^- \rightarrow H_2O \quad E_o = 0.814$ V
2. Donor pair $\quad H_2 \rightarrow 2\,H^+ + 2e^- \quad E_o = -0.421$ V
3. Combined $\quad \frac{1}{2}O_2 + H_2 \rightarrow H_2O$
4. Electrode potential $\quad \Delta E_o = E_o$ acceptor $- E_o$ donor $= 0.814$ V $- (-0.421$ V$)$ $= 1.24$ V

The Electron Tower

A simple method of comparing reaction energetics is to imagine redox pairs as being part of a vertical tower (Brock and Madigan, 1991). The redox pairs are placed with the most negative reduction potential (most likely donors) at the top, moving down to the most positive reduction potential (most likely acceptors) at the bottom (Table 5.3). In forming coupled oxidation-reduction reactions it is easiest to remember that the reduced substance (the one on the right) of a redox pair, whose reduction potential is more negative, donates electrons to the oxidized substance (the one on the left) of a redox pair whose potential is more positive. In other words, imagine the electrons dropping from the donor, high up in the tower, to the acceptor at a lower level in the tower. The electrons cannot travel up the tower unless an energy input is exerted.

The difference in electrical potential between the two redox pairs is expressed as ΔE_o. The farther the electrons drop, the greater the amount of energy released; that is, ΔE_o is directly proportional to ΔG^o as explained by the Nernst equation.

$$\Delta G^o = -nF\,\Delta E_o \tag{5.8}$$

TABLE 5.3
The electron tower: standard reduction potentials at 25°C and pH 7 for selected environmentally important redox couples

Half reaction	E_o, V
$6CO_2 + 24H^+ + 24e^- = C_6H_{12}O_6 + 6H_2O$	–0.43
$2H^+ + 2e^- = H_2$	–0.41
$CO_2 + 6H^+ + 6e^- = CH_3OH + H_2O$	–0.38
$NAD^+ + 2H^+ + 2e^- = NADH + H^+$	–0.32
$CO_2 + HCO_3^- + 8H^+ + 8e^- = CH_3COO^- + 3H_2O$	–0.29
$CO_{2(g)} + 8H^+ + 8e^- = CH_{4(g)} + 2H_2O$	–0.25
$S_{(s)} + 2H^+ + 2e^- = H_2S_{(g)}$	–0.24
$SO_4^{-2} + 9H^+ + 8e^- = HS^- + 4H_2O$	–0.22
Pyruvate + $2H^+ + 2e^-$ = lactate	–0.19
$FeOOH_{(s)} + HCO_3^- + 2H^+ + e^- = FeCO_{3(s)} + 2H_2O$	–0.05*
$CH_3SOCH_3 + 2H^+ + 2e^- = CH_3SCH_3 + H_2O$	0.16
$NO_3^- + 10H^+ + 8e^- = NH_4^+ + 3H_2O$	0.36
$NO_3^- + 2H^+ + 2e^- = NO_2^- + H_2O$	0.42
$MnO_{2(s)} + HCO_3^- + 3H^+ + 2e^- = MnCO_{3(s)} + 2H_2O$	0.52*
$CHCl_3 + H^+ + 2e^- = CH_2Cl_2 + Cl^-$	0.56
$CCl_4 + H^+ + 2e^- = CHCl_3 + Cl^-$	0.67
$2NO_3^- + 12H^+ + 10e^- = N_2 + 6H_2O$	0.74
$Fe^{3+} + e^- = Fe^{2+}$	0.76
$O_{2(g)} + 4H^+ + 4e^- = 2H_2O$	0.82
$CCl_3CCl_3 + 2e^- = CCl_2CCl_2 + 2Cl^-$	1.13
$2HOCL + 2H^+ + 2e^- = Cl_2 + 2H_2O$	1.18

*Based on $[HCO_3^-] = 10^{-3}$ M.

where n = number of electrons transferred
F = Faraday constant = 96,630 J/V
$\Delta E_o = E_o$ (electron-accepting couple) – E_o (electron-donating couple)

Values of E_o for selected redox couples are given in Table 5.3.

The farther apart the two half reactions are on the electron tower, the more energy will be released. For example, for the two redox pairs H^+/H_2 and $1/2O_2/H_2O$ the difference in potential is $\Delta E_o = -1.24$ V, as calculated above. The free-energy change, with 2 electrons being transferred, is –240 kJ/mole. Compare this to the reaction involving the hydrogen redox pair and the NO_3^-/NO_2^- redox pair where $\Delta E_o = -0.84$ V. Again two electrons are transferred and the free energy is equal to –162 kJ/mole. Oxygen, at the bottom of the tower, is a powerful oxidizing agent. The electron donor in a biological oxidation-reduction reaction is the "energy source," which will be discussed in more detail in the following sections.

EXAMPLE 5.2. BALANCING OXIDATION-REDUCTION REACTIONS. Consider the metabolism of glucose by aerobic microorganisms. Write the balanced reaction that combines the redox pairs $CO_2/C_6H_{12}O_6$ and O_2/H_2O. What is the change in free energy for this reaction given the reduction potential values listed in Table 5.3?

Solution

1. Glucose is the energy source and the electron donor; it will be oxidized. Oxygen, on the other hand, is the electron acceptor; it will be reduced. The steps involved in balancing a redox reaction are:

The two half reactions

$$C_6H_{12}O_6 \rightarrow CO_2$$

$$O_2 \rightarrow H_2O$$

Balance the main elements other than oxygen and hydrogen

$$C_6H_{12}O_6 \rightarrow 6CO_2$$

$$O_2 \rightarrow H_2O \qquad \text{(no change)}$$

Balance oxygen by adding H_2O, then hydrogen by adding H^+

$$C_6H_{12}O_6 + 6H_2O \rightarrow 6CO_2 + 24H^+$$

$$O_2 + 4H^+ \rightarrow 2H_2O$$

Balance the charge by adding electrons

$$C_6H_{12}O_6 + 6H_2O \rightarrow 6CO_2 + 24H^+ + 24e^-$$

$$O_2 + 4H^+ + 4e^- \rightarrow 2H_2O$$

Multiply each half reaction by the appropriate integer that will result in the same number of electrons in each. Then add the two half reactions to come up with the balanced reaction.

$$C_6H_{12}O_6 + 6H_2O \rightarrow 6CO_2 + 24H^+ + 24e^-$$

$$6O_2 + 24H^+ + 24e^- \rightarrow 12H_2O$$

$$\overline{C_6H_{12}O_6 + 6O_2 \rightarrow 6CO_2 + 6H_2O}$$

2. Find the change in free energy for the reaction using Equation (5.8)

$$\Delta G^o = -nF\,\Delta E_o$$

where $n = 24$ electrons transferred per mole of glucose. Note that only 4 electrons were lost per mole of carbon (study the oxidation state for carbon as it goes from glucose to CO_2)

$$\Delta E_o = E_o \text{ (electron-accepting couple)} - E_o \text{ (electron-donating couple)}$$

$$\Delta E_o = 0.82 - (-0.43) = 1.25 \text{ V}$$

Therefore

$\Delta G^o = -(24 \text{ electrons/mole})\ (96.63 \text{ kJ/V})(1.25 \text{ V/electron}) = -2{,}899$ kJ/mole glucose oxidized.

A large amount of energy is released from the aerobic metabolism of glucose.

ELECTRON CARRIERS AND ENERGY STORAGE

The flow of electrons during a redox reaction generates energy. However, in microbially mediated reactions this electron flow does not occur in one step as implied by the overall reaction. In some cases the overall reaction may represent more than 20 individual reactions. The process of electron transfer from the primary donor to the terminal acceptor is mediated by electron carriers. There are two main types of electron carriers: those which are freely diffusible in the cell, and those which are attached to enzymes in the cell membrane. Among the most common electron carriers are the freely diffusible coenzymes NAD^+ (nicotinamide–adenine dinucleotide) and $NADP^+$ (NAD-phosphate), which are hydrogen-atom carriers. NAD^+ and $NADP^+$ always transfer two hydrogen atoms to the next carrier. For example:

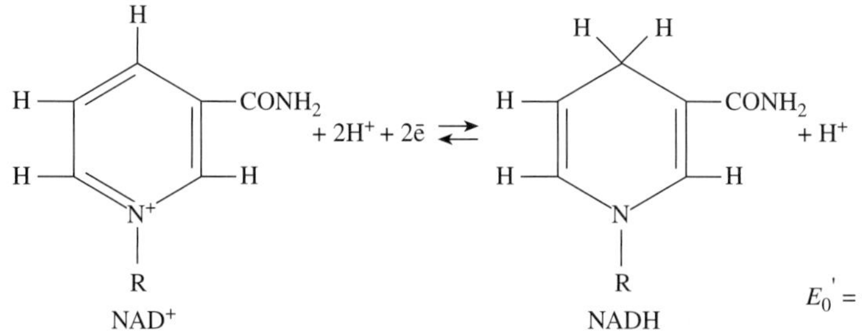

$$\underset{NAD^+}{\text{(pyridinium ring: H, H, H, } CONH_2\text{, H; } N^+\text{–R)}} + 2H^+ + 2\bar{e} \rightleftharpoons \underset{NADH}{\text{(dihydropyridine ring: H H, H, H, } CONH_2\text{, H; N–R)}} + H^+$$

$$E_0' = -0.32 \text{ V}$$

This transfer of hydrogen atoms is referred to as dehydrogenation. The reduction potential of the NAD^+/NADH redox pair is a relatively large negative value, placed relatively high on the electron tower, meaning that NADH is a good electron donor.

Energy released during an oxidation-reduction reaction is stored in the cell in the form of high-energy phosphate bonds in phosphate-containing compounds. When such compounds undergo hydrolysis, energy is released which can be used to drive endergonic reactions. Adenosine triphosphate (ATP) is the most important high-energy phosphate compound in living organisms. The free energy of hydrolysis for ATP is –30.6 kJ/mol (–10.2 kJ/phosphate bond).

DIVERSITY IN METABOLIC PROCESSES

As was noted at the beginning of the chapter, microorganisms require both an energy source and a carbon source to perform metabolic processes and to grow. Microorganisms can obtain energy from any of three sources: organic material, inorganic material, and light. When organic and inorganic materials are used as energy sources (electron donors), electron acceptors are needed to complete the oxidation-reduction reaction that results in the release of chemical energy. Oxygen is the most widely used electron acceptor. A number of other inorganic electron acceptors are also used (e.g., NO_3^-, NO_2^-, SO_4^{2-}, CO_2, Fe^{3+}). Internally generated organic compounds such as pyruvic acid may be used as electron acceptors when an external (exogenous) electron acceptor is unavailable. The reduced compounds produced are secreted into the surrounding environment of the cell.

Microorganisms have a limited capacity to use electron acceptors. Thus only a fraction of those organisms capable of using O_2 can also use NO_3^-. When light is used as an energy source, light energy must be converted to chemical energy before being put to use. Microorganisms also need a carbon source from which to construct new cellular material. The majority of organisms use organic compounds to build cell material, but some are able to use carbon dioxide as a sole carbon source.

Microbiologists have given metabolic processes names based on the type of energy the microorganisms utilize, the electron donor, and the electron acceptor required and the carbon sources used. For example, when organic material is used as an energy source and oxygen is used as an electron acceptor, aerobic respiration is said to occur. However, when an inorganic material is used as an energy source and oxygen is used as an electron acceptor, then lithotrophic metabolism is said to occur. In the following sections metabolic processes are discussed under three headings, based on the three energy sources possible: organic material, inorganic material, and light energy. Cometabolism is a special case of metabolism in which the organic compound does not directly serve as an electron donor or a carbon source. Instead, the compound is broken down incidentally by one group of organisms, but the degradation products may serve as carbon sources or electron donors for other species of organisms.

METABOLISM OF ORGANIC MATERIAL

Microorganisms metabolize organic material using two distinct pathways:

1. Fermentation, in which the organic material undergoes oxidation-reduction reactions in the absence of an external electron acceptor.
2. Respiration, in which the organic material serves as the electron donor, while an exogenous compound is utilized as the ultimate electron acceptor. In the case where molecular oxygen is used as the electron acceptor, aerobic respiration is said to occur. When some other electron acceptor is used (for example, NO_3^- or SO_4^{2-}), anoxic or anaerobic respiration is said to occur.

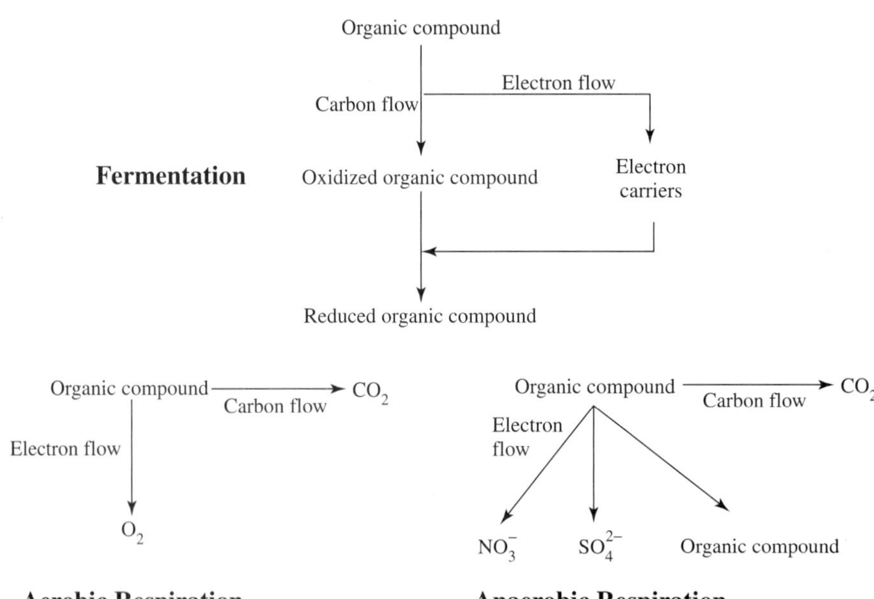

FIGURE 5.3
Electron and carbon flow in the metabolism of organic material. Adapted from Brock and Madigan (1991).

The major differences between the types of metabolism, in terms of the electron donor, the electron acceptor, and the carbon source are outlined in Figure 5.3.

Fermentation

Fermentations are a special kind of redox reaction in which organic compounds are used as both electron donors and electron acceptors. Fermentation may be described as an internally balanced oxidation-reduction reaction system. As noted above, the overall redox reactions are really summaries of a series of individual steps. Within the same molecule some atoms may become oxidized while others may become reduced. A good example is the fermentation of glucose to ethanol and carbon dioxide. In the most widely studied metabolic pathway, one 6-carbon glucose molecule is oxidized to two 3-carbon pyruvic acid molecules. Carbon dioxide is produced by decarboxylation of the pyruvic acid to form acetaldehyde. The acetaldehyde is then used as an electron acceptor for the oxidation of $NADH^+$ generated upstream in the sequence as shown in Figure 5.4.

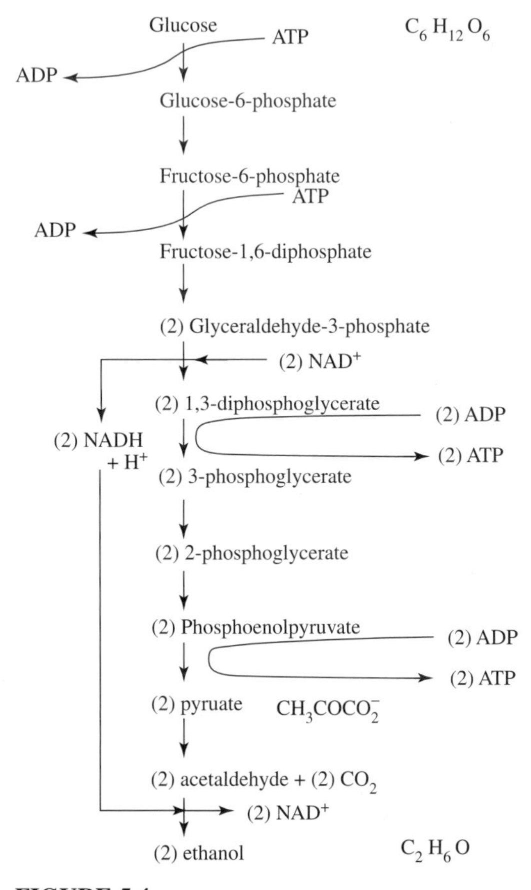

FIGURE 5.4
The Emden-Meyerhof-Parnas pathway of glucose fermentation.

In the overall reaction, given in Equation (5.9), two of the carbons in glucose are oxidized to carbon dioxide (oxidation state +4), while the remaining carbons are reduced to form ethanol (oxidation state –2).

$$C_6H_{12}O_6 \rightarrow 2C_2H_6O + 2CO_2 \quad (5.9)$$

As indicated in Figure 5.4, the oxidation-reduction reactions in a fermentation process are carried out in two subsequent groups. First, the parent compound is oxidized with the loss of electrons; then the product of this first step is reduced. For example, in the process of making cheese, lactic acid bacteria can ferment glucose ($C_6H_{12}O_6$) by oxidizing it to pyruvate first ($CH_3COCOOH$) and then reducing the pyruvate to lactate ($C_3H_6O_3$). The electron transfer between the different steps is carried out by coenzymes such as NAD^+/NADH, as in the case of ethanol fermentation.

Because only a fraction of the carbon atoms of a compound are oxidized, only a fraction of the potential energy available is released in a fermentation reaction. Most of the energy in the original compound remains in the reduced product. For example, if ATP is used as a measure of the energy released and conserved by microorganisms, then in a typical glucose fermentation pathway, a net of two ATPs are produced. When oxygen is the terminal electron acceptor and glucose is completely oxidized to CO_2, 38 ATPs are produced. Because the organic compound is not completely oxidized during fermentation, and because of the lack of efficiency in energy production, fermentation is not often utilized in bioremediation processes.

Respiration

Respiration involves the use of an exogenous (extracellular) electron acceptor and is divided into two groups, aerobic and anaerobic. The processes involved are essentially the same and the reaction sequences usually differ only in the final steps. Aerobic respiration supports a broader range of reactions, including oxidations important in bioremediation in which molecular oxygen is incorporated into molecules by oxygenase enzymes. Anaerobic respiration is carried out by obligate anaerobes (microorganisms that cannot function in the presence of oxygen) and by facultative anaerobes (microorganisms that can grow in both the presence and the absence of oxygen). With facultative anaerobes, oxygen, if present, will usually be preferentially used as an electron acceptor. Once oxygen has been depleted, however, the next available electron acceptor will be used.

The reactions in which electrons are transferred to the terminal electron acceptor are enzyme-mediated. Consequently a limited set of compounds and ions can be used. Oxygen is generally the best electron acceptor, i.e., the one that yields the most free energy in a completed reaction. To understand why, keep in mind that most electron donors have a reduction potential similar to that of the NAD^+/NADH couple (–0.32 V). Then, looking at the electron tower (Table 5.3), one can see that oxygen, near the bottom of the tower, is the most oxidizing element. Remember that the farther the drop from the electron donor to the electron acceptor on the tower, the higher the theoretical energy yield. Hence, if oxygen is not available or cannot be used, the next best electron acceptor would be Fe^{3+}, if the necessary enzymes are present, followed by NO_3^- and NO_2^-. Sulfate and carbon dioxide come next, and would be expected to yield less free energy per reaction for oxidizing the same substrate. Among the anaerobic electron acceptors, the use of NO_3^- and NO_2^- is quite common among soil bacteria. Few species use Fe^{3+} or SO_4^{2-}.

For the same organic substrate, microorganisms that use oxygen as an oxidizing agent can generate more energy than microorganisms that use nitrate, sulfate, or other electron acceptors. The larger amount of energy released in aerobic respiration allows aerobic microorganisms to grow at a faster rate and to dominate as long as oxygen is abundant. Where oxygen is absent, other electron acceptors are used in the order of the most energy-yielding first, to the least energy-yielding last. Nitrate and sulfate are the most commonly used electron acceptors because of their abundance in nature. Other electron acceptors, such as iron (Fe^{3+}) oxides and manganese (Mn^{3+}) oxides, are capable of producing more free energy per reaction but may not be accessible to microorganisms because of their insolubility. The use of carbon dioxide as an electron acceptor by methanogens (obligate anaerobic bacteria that generate methane) is of special importance because of the methanogen's enzyme system, which is capable of degrading some recalcitrant compounds. The following is a discussion of the use of nitrate and sulfate as electron acceptors by different groups of microorganisms in anaerobic respiration.

Nitrate respiration

Nitrate and ammonia, the most oxidized and the most reduced forms of inorganic nitrogen, respectively, are the most widespread inorganic nitrogen species in soil. As stated above, nitrate is the most common electron acceptor in anaerobic respiration. When reduced, NO_3^- is first reduced to NO_2^- and then to N_2O, NO, and ultimately N_2, all of which are gases and could escape out of the soil. This loss of nitrogen to the atmosphere, through nitrate reduction, is known as denitrification.

The use of nitrate as an electron acceptor to form gaseous nitrogen is also known as dissimilative nitrate reduction. This is to be distinguished from assimilative nitrate reduction, the process in which nitrate is reduced to ammonia and then used as a nitrogen source for production of amino acids and proteins, that is, for growth. The enzymes involved in dissimilative nitrate reduction are located near those used for oxygen respiration. Synthesis of nitrate reductase does not occur unless oxygen is limited; for this reason denitrification rarely occurs simultaneously with oxygen respiration. Assimilative nitrate reduction, on the other hand, can occur in the presence of oxygen. All denitrifying microorganisms are facultative anaerobes and preferentially use oxygen. Only bacteria are capable of dissimilative nitrate reduction, whereas bacteria, fungi, and plants can perform assimilative nitrate reduction.

Sulfate respiration

Sulfate, the most oxidized form of sulfur, can be used as an electron acceptor by a specialized group of microorganisms, the sulfate-reducing bacteria, to be converted to the most reduced form of sulfur, H_2S, and to elemental sulfur, S^o. Similar to nitrate, sulfate can undergo assimilatory and dissimilatory reduction. In assimilatory reduction, the sulfur is used to synthesize cell material. In dissimilatory reduction, it is used only as an electron acceptor to complete redox reactions for the generation of energy. The product, hydrogen sulfide, is lost as gas. Sulfate-reducing bacteria, which are obligate anaerobes, are the only group of microorganisms which are capable of dissimilatory sulfate reduction. Less than a dozen sulfate-reducing species have been identified.

Because sulfate is a weaker oxidizing agent than oxygen or nitrate, organisms growing on sulfate obtain less energy than those growing on nitrate or oxygen. As a result, growth yields are lower. For an electron donor, the sulfate-reducing bacteria can utilize organic material such as acetate, lactate, and pyruvate. Some can use inorganics, mainly hydrogen gas, in which case the metabolism would be lithotrophic.

Perhaps the most evident example of sulfate reduction is the production of hydrogen sulfide in sewage systems and the resulting corrosion of the sewage distribution system. One solution for this problem has been the addition of nitrate, which acts as a more favorable electron acceptor and eliminates the production of hydrogen sulfide (Morton et al., 1991).

METABOLISM OF INORGANIC MATERIAL: LITHOTROPHY

Lithotrophs are organisms that obtain energy from the oxidation of inorganic compounds. Lithotrophic metabolism is almost always coupled to the use of carbon dioxide as the sole carbon source, and hence the organisms are lithoautotrophs (Figure 5.5 is a schematic diagram of the electron and carbon flow in lithoautotrophic processes). These specialized groups of bacteria are known by the substrate that they oxidize. Examples are the hydrogen bacteria, sulfur bacteria, iron-oxidizing bacteria, and the nitrifying bacteria (ammonium and nitrite-oxidizing bacteria). In the following section we briefly discuss some of the reactions carried out by these groups of bacteria.

Hydrogen bacteria

Also known as the hydrogen-oxidizing bacteria, the hydrogen bacteria are a relatively diverse group of organisms that use hydrogen as an energy source. The majority of hydrogen bacteria are aerobic; that is, they use oxygen as the electron acceptor. However, hydrogen bacteria do not depend exclusively on hydrogen for their growth. These bacteria are facultative lithotrophs, which means that they also use organic compounds as energy sources in the absence of hydrogen.

As with most lithotrophs, hydrogen bacteria are mostly autotrophs, meaning that they can use carbon dioxide to form cell material. When growing on hydrogen and CO_2, the hydrogen bacteria perform the following general reaction:

$$6H_2 + 2O_2 + CO_2 \rightarrow CH_2O + 5H_2O$$

where CH_2O is an empirical formula representative of cell material and in which carbon has the same oxidation state (0) as carbohydrates. The more widely used empirical cell formulation $C_5H_7NO_2$ represents average cell composition more closely.

The ability to oxidize hydrogen is not exclusive to hydrogen bacteria. Many bacteria can use hydrogen as an energy source but depend on organic compounds

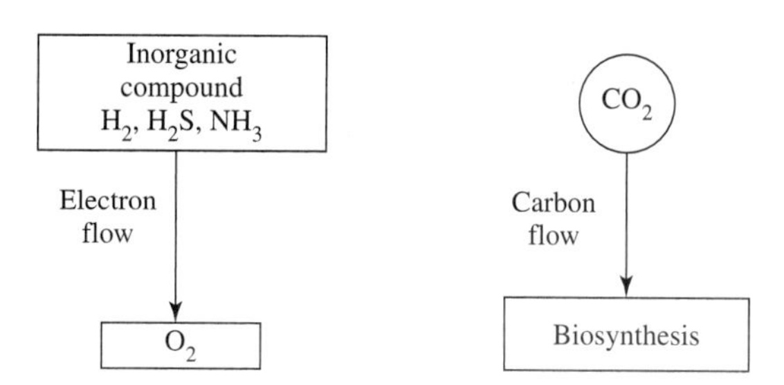

FIGURE 5.5
Electron and carbon flow in typical lithoautotrophic metabolism. Electrons stripped from inorganic electron donors are transferred to molecular oxygen. Products of the oxidation reactions would be H_2O, SO_4^{2-}, and NO_3^-, as indicated in Table 5.3.

for a carbon source; that is, they are not autotrophs. An example of such bacteria is *Escherichia coli,* the organism which is used as an indicator of contamination of water by human feces.

Sulfur bacteria

The sulfur bacteria use elemental sulfur (S^o) and hydrogen sulfide (H_2S), thiosulfate ($S_2O_3^{2-}$), and organic sulfides (R-SH) as energy sources. Like the hydrogen bacteria, the sulfur bacteria are facultative autotrophs and also metabolize organic compounds. They are generally known as the colorless sulfur bacteria, to distinguish them from the green and the purple sulfur bacteria, which are phototrophs and are briefly discussed below. *Beggiatoa* are a genus of sulfur bacteria that cannot grow autotrophically, requiring organic material as a carbon source. *Beggiatoa* are large-sized bacteria commonly found in marine and freshwater sediments rich with hydrogen sulfide.

The sulfur compounds most commonly oxidized by sulfur bacteria are hydrogen sulfide (H_2S), elemental sulfur (S^o), and thiosulfate ($S_2O_3^{2-}$). The overall oxidation reactions are as follows:

$$H_2S + 2O_2 \rightarrow SO_4^{2-} + 2H^+$$

$$S^o + H_2O + 3/2O_2 \rightarrow SO_4^{2-} + 2H^+$$

$$S_2O_3^{2-} + H_2O + 2O_2 \rightarrow 2SO_4^{2-} + 2H^+$$

The oxidation of hydrogen sulfide to sulfate occurs in stages, during which elemental sulfur is formed. In some bacteria a portion of the sulfur is deposited inside the cell to be oxidized and used as an energy source when the H_2S supply has been depleted.

Elemental sulfur is extremely insoluble. When present in the outside environment of bacteria, the microorganisms have to grow attached to the sulfur particles. By adhering to the surface of the particles, the bacteria can oxidize the sulfur atoms as they slowly go into solution. As some of the atoms become oxidized, more solubilize. Step by step, the sulfur particle is consumed.

Note that oxidation of sulfur compounds results in production of hydrogen ions as a final product, as seen in the reactions above. The addition of protons to the solution results in lowering the pH. Sulfur bacteria have been shown to reduce the pH in their environments to as low as 1 and continue to thrive. The low pH results in extremely corrosive conditions.

PHOTOTROPHIC METABOLISM

Photosynthesis is the process through which organisms can convert light energy into chemical energy. Most phototrophs depend on carbon dioxide as their sole carbon source, and hence they are autotrophs, or photoautotrophs (Figure 5.6 is a schematic diagram of the electron and carbon flow in photoautotrophic processes). Such organisms have a type of chlorophyll, similar to the light-sensitive pigment found in plants and algae, which enables them to absorb light and utilize it as an energy source. Examples of such organisms include the cyanobacteria and the purple and green bacteria.

Each group of phototrophic organisms has a particular type of chlorophyll which results in light adsorption at characteristic wavelengths. For example, chlorophyll a, the principal chlorophyll in plants, shows maximum absorption at wavelengths of 680 and 430 nm, for red and blue light, respectively. The chlorophyll of purple and green bacteria is of slightly different structure and is generally called bacteriochlorophyll. Several types of bacteriochlorophyll, each absorbing at a characteristic wavelength, exist. One type absorbs maximally at a range of 720 to 780 nm, while another type absorbs maximally at 850 nm and yet another type absorbs at 1,020 nm. The pigment diversity within microorganisms enables them to coexist in the same habitat.

The growth of photoautotrophic organisms depends on two sets of reactions: first, light energy is converted into chemical energy in the form of ATP, and then this chemical energy is used to reduce carbon dioxide to organic compounds (e.g., cell material). The process of converting CO_2 into organic material is known as CO_2 fixation.

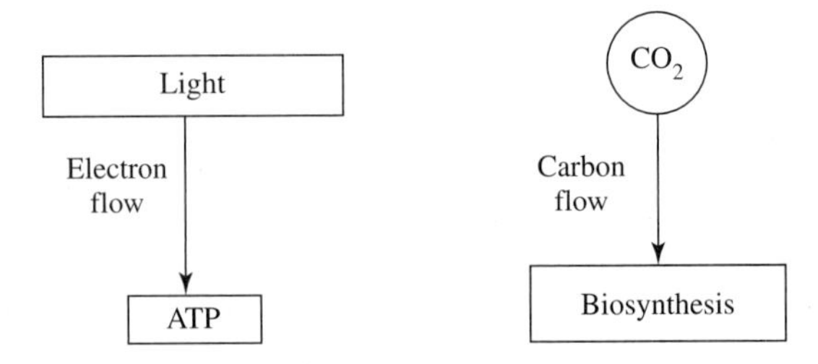

FIGURE 5.6
Electron and carbon flow in typical photoautotrophic metabolism. Adapted from Brock and Madigan (1991).

Carbon dioxide fixation in autotrophs is carried out through the Calvin cycle. The Calvin cycle is an energy-demanding process in which CO_2 is reduced to the oxidation state of other cell material. Reducing energy is provided by an external electron donor such as hydrogen sulfide, elemental sulfur, thiosulfate, hydrogen, or organic material. The oxidation of the external electron donor is coupled to the reduction of $NADP^+$ to NADPH, the coenzyme of major importance in the electron transport chain in the Calvin cycle.

An example of microorganisms growing on sulfur-containing compounds are the green and purple bacteria. When utilizing hydrogen sulfide, these bacteria produce elemental sulfur:

$$6CO_2 + 12H_2S \rightarrow C_6H_{12}O_6 + 6H_2O + 12S_o$$

In green bacteria, the sulfur formed is deposited outside the cell, whereas in purple bacteria, it is deposited inside the cell as sulfur granules.

COMETABOLISM

In some cases, microorganisms are capable of transforming an organic compound without being able to use that compound as a growth substrate or as an energy source. As such, those microorganisms would need another substrate as a carbon and energy source on which to grow. This special type of metabolism, the incidental metabolism of a nongrowth substrate, has been named cometabolism. A nongrowth substrate in this case will be defined as an organic compound that cannot serve as an energy source, or a significant source of nutrients for the degrading culture. Thus, in cometabolism, organisms use one substrate as a primary energy source and gratuitously metabolize another compound utilizing the enzymes which are synthesized to degrade the primary substrate.

The use of the term cometabolism, however, has been a source of dispute among scientists. Some have used the terms cooxidation, gratuitous biodegradation, or incidental metabolism. Some prefer restricting the use of the term cometabolism to cases in which the nongrowth substrate is being metabolized in the presence of the growth substrate. For cases where metabolism of the nongrowth substrate is occurring in the absence of the growth substrate, the term fortuitous metabolism may be more appropriate. Overall, however, the term cometabolism has been the most widely used, regardless of whether the growth substrate is present or absent (Alexander, 1994), and it will be the one used in this book too.

Cometabolism has been observed in the transformation of a number of important xenobiotic compounds, including dioxin, trichloroethene (TCE), and polychlorinated biphenyls (PCBs). Treatment systems for cometabolism of TCE have been successfully operated at laboratory scale but full-scale systems have not been stable or competitive with other forms of remediation. Use of cometabolism for degradation of dioxin and PCBs has not reached the process scale at this time.

PROBLEMS AND DISCUSSION QUESTIONS

5.1. Find the change in free energy for the reactions given below. Are the reactions exergonic or endergonic? In what direction will the reaction proceed spontaneously? Which reaction is likely to release the most free energy? Assume a pH of 7 for reactions *a* and *c,* and a pH of 2 for reaction *b.*

a. $CO_2 + 4H_2 \rightarrow CH_4 + 2H_2O$
b. $4Fe^{2+} + O_2 + 4H^+ \rightarrow 4Fe^{3+} + 2H_2O$
c. $4CO_2 + 8H_2O \rightarrow 4CH_4O + 6O_2$

5.2. Assuming the reactions listed in Problem 5.1 proceed spontaneously, name the electron donor and the electron acceptor for each reaction. How many electrons are lost or gained in each case?

5.3. Balance the following redox reactions. Show the half reactions and the number of electrons transferred in each. How many moles of nitrate and how many moles of oxygen are needed to oxidize one mole of benzene (C_6H_6)?

a. $C_6H_6 + NO_3^- \rightarrow CO_2 + N_2$
b. $C_6H_6 + O_2 \rightarrow CO_2 + H_2O$

5.4. Consider the metabolism of glucose by microorganisms using nitrate (NO_3^-) as an electron acceptor and by microorganisms using sulfate (SO_4^{2-}) as an electron acceptor.

Redox pair	E_o, **V**
$CO_2/C_6H_{12}O_6$	–0.43
NO_3^-/N_2	+0.74
SO_4^{2-}/H_2S	–0.22

a. Write out the balanced reactions in each case, starting with the redox pairs and following the steps for balancing a redox reaction as listed in Example 5.2.
b. Find the change in free-energy potential for each case.
c. Assuming all three electron acceptors (oxygen, nitrate, and sulfate) are present, which is the one most likely to be used last? why?

5.5. Based on their carbon source and energy source, and using Table 5.1, what metabolic group would you place the following organisms in: *Nitrosomonas,* Cyanobacteria, *Beggiatoa,* hydrogen bacteria, humans?
Name the energy source, the electron acceptor, and the carbon source in each case.

REFERENCES

Alexander, Martin (1994): *Biodegradation and Bioremediation,* Academic Press, CA.

Brock, T. D., and M. T. Madigan (1991): *Biology of Microorganisms,* 6th ed., Prentice-Hall, Englewood Cliffs, NJ.

Brock, T. D., M. T. Madigan, J. M. Martinko, and J. Parker (1994): *Biology of Microorganisms,* 7th ed., Prentice-Hall, Englewood Cliffs, NJ.

Cookson, J. T. (1995): *Bioremediation Engineering Design and Application,* McGraw-Hill, New York.

Morton, R. L., W. A. Yanko, D. W. Graham, R. G. Arnold (1991): "Relationships between Metal Concentrations and Crown Corrosion in Los Angeles County Sewers," *Research Journal of the Water Pollution Control Federation,* vol. 63, no. 5, pp. 789–798.

Stanier, R. Y., J. L. Ingraham, M. L. Wheelis, and P. R. Painter (1986): *The Microbial World,* 5th ed., Prentice-Hall, Englewood Cliffs, NJ.

Tate, R. L. (1995): *Soil Microbiology,* Wiley, New York.

CHAPTER 6

Biodegradation of Selected Compounds

A discussion of the biodegradation of selected organic compounds that are known to be significant in bioremediation is presented in this chapter. Further information as to compound degradability and pathways of degradation, microorganisms which have been shown to degrade them, and possible biotransformation products is available in other reviews (Gibson, 1984; Rochkind et al., 1986; Pitter and Chudoba, 1990; Vogel et al., 1987; U.S. EPA, 1983). Target compounds for bioremediation include petroleum hydrocarbons, solvents (methylethylketone, acetone, alcohols, methylene chloride), aromatics (benzene, toluene, xylene, polycyclic aromatic compounds, chlorobenzene), nitro and chlorophenols, phthalate esters, pesticides, and chlorinated aliphatic compounds (Skladany, 1992). These compounds can be roughly grouped as petroleum hydrocarbons and their oxidation products, halogenated aliphatic compounds, and halogenated aromatic compounds. Microbial degradation of only a fraction of these compounds has been studied thoroughly in laboratory culture and fewer still have been studied in natural ecosystems. Biodegradation rates and pathways for biotransformation determined in laboratory cultures do not necessarily reflect biodegradation in sewage, soil, or aquatic systems. In addition, the study of metabolic pathways generally identifies only those compounds which are excreted outside the cell and accumulate long enough for the intermediate to be above the detection limit of the analytical technique (Alexander, 1981).

BIODEGRADATION OF HYDROCARBONS

Over 2 billion metric tons of petroleum are produced per year worldwide (Bartha, 1986) and large amounts of petroleum products end up polluting both marine and terrestrial environments. Low-level routine discharges (urban runoff, effluents, oil treatment of roads, etc.) account for over 90 percent of the total petroleum hydrocarbon discharges. Accidents such as tanker disasters, pipeline breaks, and well blowouts account for less than 10 percent of these discharges. In general, petroleum hydrocarbons are intermediate between highly biodegradable and highly recalcitrant compounds. Petroleum compounds have entered the biosphere through seeps and erosion for millions of years and metabolic pathways for their degradation have evolved.

Hydrocarbons in crude petroleum are classified as alkanes (normal and iso), cycloalkanes, aromatics, polycyclic aromatics, asphaltines, and resins. Alkenes are generally not encountered in crude oil but may be present in small quantities in refined petroleum products due to the "cracking" process. Variations in chain length, in chain branching, in ring condensations, in interclass combinations, and the presence of oxygen, nitrogen, and sulfur-containing compounds account for the wide variety of petroleum hydrocarbons. The biodegradability of these compounds is greatly affected by their physical state and toxicity. Because petroleum is such a complex mixture, its degradation is favored by a mixed population of microorganisms with broad enzyme capabilities. In addition, the initial degradation of petroleum hydrocarbons often requires the action of oxygenase enzymes and so is dependent on the presence of molecular oxygen (Atlas, 1991). Aerobic conditions are therefore necessary for the initial breakdown of petroleum hydrocarbons. In subsequent steps nitrate or sulfate may serve as a terminal electron acceptor (Bartha, 1986), but oxygen is most commonly used.

Alkanes

The *n*-alkanes are the most biodegradable of the petroleum hydrocarbons. However, normal alkanes in the C_5 to C_{10} range are inhibitory to many hydrocarbon degraders at high concentrations because as solvents they disrupt lipid membranes. Alkanes in the C_{20} to C_{40} range (referred to as "waxes") are hydrophobic solids; their low solubility interferes with their biodegradation. In the degradation of alkanes, the monooxygenase enzyme attacks the terminal methyl group to form an alcohol as shown in Figure 6.1 (Pitter and Chudoba, 1990). The alcohol is oxidized further to an aldehyde and then to a fatty acid. The fatty acid is degraded further by β-oxidation of the aliphatic chain. Extensive methyl branching interferes with the β-oxidation process (Bartha, 1986) and may necessitate diterminal attack. In general, the degradation of alkanes produces oxidized products that are less volatile than the parent compounds. However, the parent alkanes are highly volatile and may be removed from soil primarily through stripping under aerobic conditions.

$$R-CH_2CH_3 + \tfrac{1}{2}O_2 \rightarrow R-CH_2CH_2OH \rightarrow R-CH_2CHO \rightarrow R-CH_2COOH \rightarrow$$
$$R'-CH_2COOH + CH_3COOH$$

FIGURE 6.1
Initial oxidation of alkanes.

Alkenes

Less is known about the biodegradation of alkenes than about alkanes (Pitter and Chudoba, 1990). Location of the unsaturated linkage is a factor. For example, 1-alkenes, where the unsaturated bond is on the first carbon, are more degradable than alkenes with an internal double bond. Two general pathways have been observed for the metabolism of 1-alkenes (Pitter and Chudoba, 1990; Britton, 1984). Either the double bond is oxidized, giving rise to a diol, or the saturated chain end is oxidized as shown in Figure 6.2.

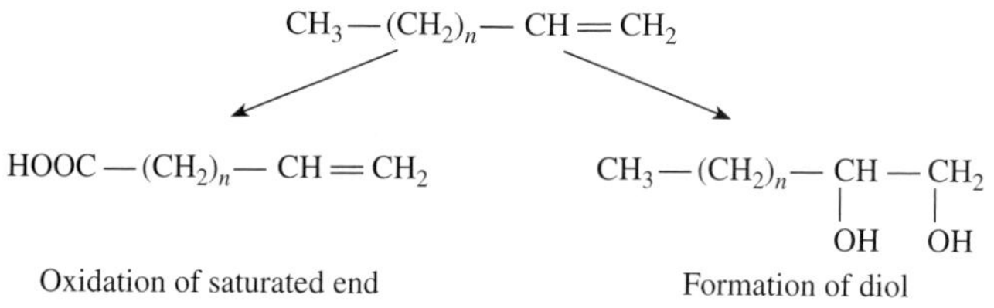

$CH_3—(CH_2)_n— CH=CH_2$

$HOOC—(CH_2)_n— CH=CH_2$ — Oxidation of saturated end

$CH_3—(CH_2)_n— CH(OH)—CH_2(OH)$ — Formation of diol

FIGURE 6.2
Metabolism of 1-alkenes.

Cycloalkanes

The cycloalkanes (alicyclic hydrocarbons) are less degradable than their straight-chain cousins the alkanes but more degradable than the polycyclic aromatics (PAHs) (Trudgill, 1984; Pitter and Chudoba, 1990). Biodegradability of cycloalkanes tends to decrease with increasing numbers of ring structures, as is the case with PAHs. At least part of the decrease in biodegradability is due to decreased solubility. Alkyl-substituted cycloalkanes are more readily degraded than nonsubstituted hydrocarbons, and cycloalkanes with long-chain side groups are more easily degraded than those with methyl or ethyl groups. Care must be taken to avoid confusing degradation with mineralization. Often a complex molecule will be partially degraded but a substantial portion may be recalcitrant. Conversely, alkyl substitution may increase solubility and hence make the cycloalkane structures more available for bacterial degradation.

Cycloalkanes are usually degraded by oxidase attack to produce a cyclic alcohol which is dehydrogenated to a ketone (Bartha, 1986) as shown in Figure 6.3. Alkylcycloalkanes undergo initial attack at the alkyl group, giving rise to a fatty

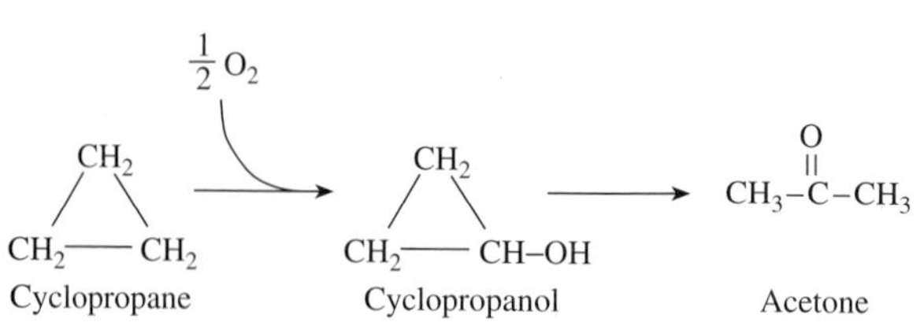

Cyclopropane (CH_2, CH_2—CH_2) $\xrightarrow{\frac{1}{2}O_2}$ Cyclopropanol (CH_2, CH_2—CH–OH) $\longrightarrow$ Acetone (CH_3–$C(=O)$–CH_3)

FIGURE 6.3
Degradation of cycloalkanes.

acid (Pitter and Chudoba, 1990). Cycloketones and cycloalkane-carboxylic acids are therefore the primary products of metabolism of cycloalkanes.

Aromatics

Aromatic compounds have structures based on the benzene molecule (Figure 6.4). Several aromatic compounds are present in petroleum, including 1-, 2-, 3-, 4-, and 5-ring compounds and alkyl-substituted aromatics. Aromatic compounds are more stable than other cyclic compounds owing to the sharing of delocalized electrons by the pi bonds. The simplest monoaromatic is benzene. Benzene, toluene, ethylbenzene, and the three xylenes (Figure 6.4), collectively known as BTEXs, are among the most water-soluble and the most mobile components of conventional gasoline. These volatile organic compounds are also some of the most potentially hazardous, especially benzene, being a carcinogen. Hence BTEXs are often used as indicators of soil and groundwater contamination, especially from leaking underground storage tanks. As discussed in a later section, oxygenates are currently being added to enhance gasoline combustion. These compounds comprise a significant volumetric fraction of gasoline, are quite soluble in water, and are very mobile in groundwater.

Biodegradation of an aromatic molecule involves two steps: activation of the ring, and ring cleavage. Activation involves the incorporation of molecular oxygen into the ring, that is, dihydroxylation of the aromatic nucleus. This step is achieved by enzymes known as oxygenases. Monooxygenases, characteristic of fungi and other eukaryotes, catalyze the incorporation of a single atom of oxygen to form an epoxide which can then undergo hydration to yield transdihydrodiols (Cerniglia, 1984; Rochkind et al., 1986). Dioxygenases, characteristic of bacteria, catalyze the incorporation of two atoms of molecular oxygen at one time to form a dihydrodiol. These dioxygenase reactions have been shown to occur for benzene, halogenated benzenes, toluene, *p*-chlorotoluene, xylenes, biphenyl, naphthalene, anthracene, phenanthrene, benzo[*a*]-pyrene, and 3-methylcholanthrene (Gibson, 1988).

Dihydrodiols (Figure 6.5) are further oxidized to dihydroxylated derivatives such as catechols, which are precursors to ring cleavage. Catechol can be oxidized either via the ortho pathway which involves cleavage of the bond between carbon atoms of the two hydroxyl groups to yield muconic acid, or via the meta pathway which involves cleavage of the bond between a carbon atom with a hydroxyl group and the adjacent carbon atom to yield 2-hydroxymuconic semialdehyde (Cerniglia, 1984). These compounds are then degraded to form acids which are readily utilized by microorganisms for cell synthesis and energy.

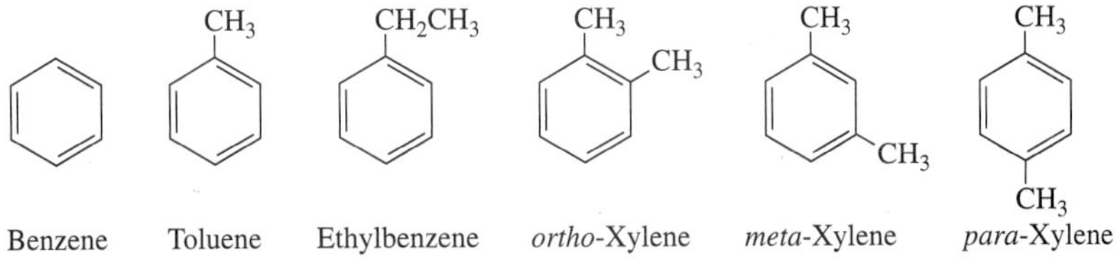

CH_3 CH_2CH_3 CH_3 CH_3 CH_3 CH_3 CH_3 CH_3 CH_3

Benzene Toluene Ethylbenzene *ortho*-Xylene *meta*-Xylene *para*-Xylene

FIGURE 6.4
Molecular structure of BTEXs.

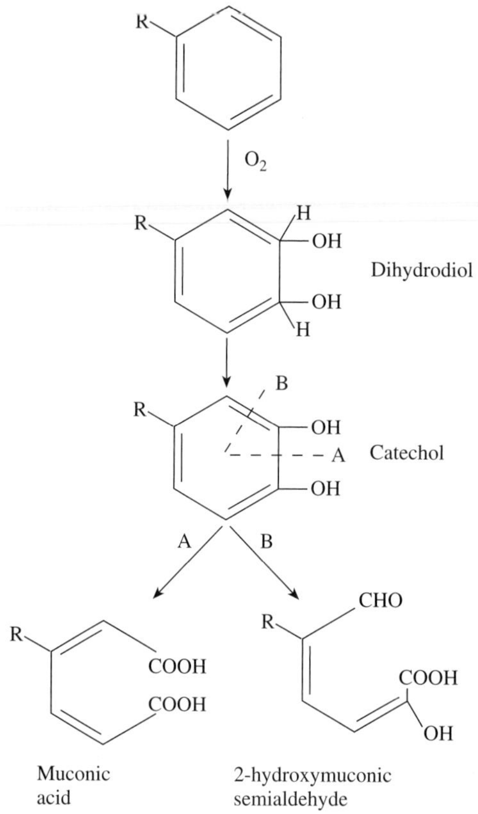

R
O_2
H
OH
OH
H
Dihydrodiol
B
OH
A
OH
Catechol
A B
CHO
COOH
COOH
COOH
OH
Muconic acid
2-hydroxymuconic semialdehyde

FIGURE 6.5
Degradation of aromatic compounds.

Polycyclic Aromatic Hydrocarbons

Also referred to as polynuclear aromatic hydrocarbons (PAHs, or PNAs), polycyclic aromatic hydrocarbons are produced during high-temperature industrial operations such as petroleum refining, coke production, and wood preservation (Park et al., 1990) and, as such, are common contaminants in industrial and uncontrolled hazardous-waste sites. Polynuclear aromatic hydrocarbons are major constituents of creosote, a product typically used in wood preservation. This group of compounds, which consists of 2 or more benzene rings, includes 16 priority pollutants (McEldowney et al., 1993; Dzomback and Luthy, 1984), some of which are suspected carcinogens. For PAHs in general, an increase in molecular weight and number of ring structures yields decreased solubility and volatility (Table 6.1) and increased adsorption capacity. A general empirical formula for PAHs is $C_{4n+2}H_{2n+4}$, where n = number of aromatic rings. For example, the 3-ring PAHs anthracene and phenanthrene, shown in Figure 6.6, have the same empirical formula, $C_{14}H_{10}$.

Polycyclic aromatic hydrocarbons are degraded, one ring at a time, by similar mechanisms as the ones used for aromatic compounds. Biodegradability of PAHs

TABLE 6.1
Properties of selected PAHs

Compound	Mol formula (mol weight)	Molecular structure	Solubility,* mg/L at 25°C	Vapor pressure,† mmHg (°C)
2-ring:				
Naphthalene	$C_{10}H_8$ (128)		31.7	0.082 (25)
3-ring:				
Anthracene	$C_{14}H_{10}$ (178)		0.045	1.9×10^{-4} (20)
Phenanthrene	$C_{14}H_{10}$ (178)		1.0	6.8×10^{-4} (20)
4-ring:				
Pyrene	$C_{16}H_{10}$ (202)		0.132	2.5×10^{-6} (25)
Benz(a)anthracene	$C_{18}H_{12}$ (228)		9.4×10^{-3}	1.0×10^{-8} (20)
Chrysene	$C_{18}H_{12}$ (228)		1.8×10^{-3}	6.3×10^{-9} (25)
5-ring:				
Benzo(*a*)pyrene	$C_{20}H_{12}$ (252)		1.2×10^{-3}	5.6×10^{-9} (25)
6-ring:				
Benzo(*g,h,i*)perylene	$C_{22}H_{12}$ (276)		0.7×10^{-3}	1.0×10^{-10} (25)

**Source:* U.S. EPA, 1984.
†*Source:* LaGrega et al., 1994.

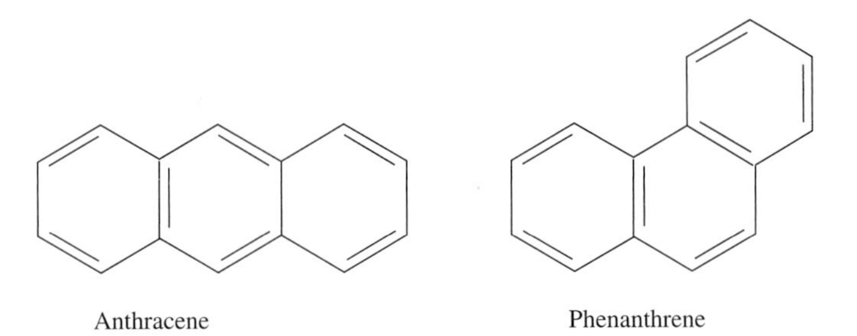

FIGURE 6.6
The two 3-ring PAHs, anthracene and phenanthrene.

tends to decrease with increased numbers of rings and with increasing numbers of alkyl substituents. The enzymes required for the procaryotic degradation of PAHs can be induced by the presence of lower-molecular-weight aromatics such as naphthalene (Atlas, 1991). Thus, the high-molecular-weight PAHs might be resistant to microbial degradation when lower-molecular-weight PAHs are not present. Fungal degradation of PAHs is environmentally significant because some of the products have been implicated as toxic forms in higher organisms (Cerniglia, 1984). Park et al. (1990) showed an increase in the volatility of certain PAHs (naphthalene, and 1-methylnaphthalene) as a result of biodegradation to lower-molecular-weight compounds.

Asphaltines and Resins

Asphaltines and resins are high-molecular-weight compounds containing nitrogen, sulfur, and oxygen. Asphaltines, and most resins, have complex structural arrangements composed of hydrocarbon chains and nitrogen, sulfur, and oxygen atoms linking polycyclic aromatic stacks which include nickel and vanadium. Compounds in these two groupings are recalcitrant to biodegradation because of their insolubility and the presence of functional groups that are shielded from microbial attack by extensive aromatic ring structures (Atlas, 1991). Relative and sometimes absolute amounts of asphaltines tend to increase during biodegradation of petroleum hydrocarbons because of their resistance to degradation and creation by condensation reactions. Some studies have reported removal of asphaltines by cometabolism in the presence of C_{12} to C_{18} *n*-alkanes (Leahy and Colwell, 1990).

Fuel Oxygenates

Fuel oxygenates are added to gasoline to enhance completeness of combustion. Oxygenates increase the oxygen content of gasoline, which results in improved combustion and reduction in the amounts of carbon monoxide (CO) and unburned hydrocarbons emitted. There are two principal types of oxygenated compounds that are added to gasoline: aliphatic alcohols (methanol and ethanol) and ethers. Addition of oxygenates to gasoline was mandated by provisions of the 1990 Clean Air Act Amendments (CAAA). The 1990 CAAA required the use of "oxygenated gasolines" on a seasonal basis, starting in the winter of 1992, in areas where the levels of carbon monoxide exceeded ambient air quality standards. Nine metropolitan areas that suffered from severe ozone pollution were required to use "reformulated gasoline," year round, beginning in 1995. Many other metropolitan areas in the United States have chosen to participate in the reformulated gasoline program.

Oxygenated and reformulated gasoline must contain at least 2.7 and 2.0 percent oxygen by weight, respectively (Squillace et al., 1996). To meet the oxygen requirement in oxygenated gasoline, compounds currently used (at percent volume in gasoline) include: methanol (CH_3OH) at 5.4 percent, ethanol (C_2H_5OH) at 7.8 percent, methyl *tert*-butyl ether [MTBE, $CH_3OC(CH_3)_3$] at 14.9 percent, ethyl *tert*-butyl ether [ETBE, $C_2H_5OC(CH_3)_3$] at 17.3 percent, and *tert*-amyl methyl ether [TAME, $(CH_3)_2CHCH_2CH_2OCH_3$] at 17.3 percent by volume (Feldman and Orchin, 1993). The most commonly used fuel oxygenate by far has been MTBE (Mormile et al., 1994; Squillace et al., 1996). It is relatively inexpensive, easy to produce, and blends well with gasoline without phase separation. Production of MTBE increased significantly in the last decade and a half. In 1995 MTBE was the third most produced organic chemical in the United States at approximately 8.0 billion kg (Kirschner, 1996).

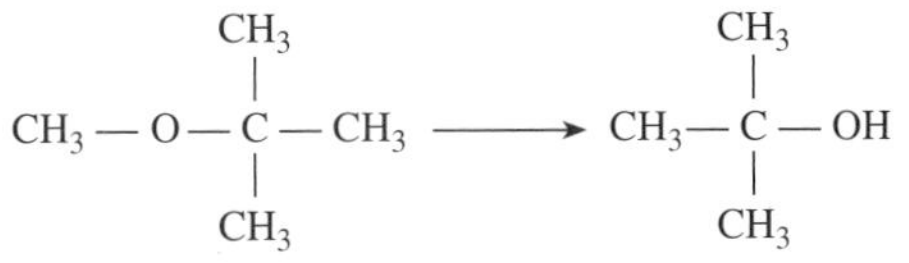

$$\begin{array}{c} CH_3 \\ | \\ CH_3 - O - C - CH_3 \\ | \\ CH_3 \end{array} \longrightarrow \begin{array}{c} CH_3 \\ | \\ CH_3 - C - OH \\ | \\ CH_3 \end{array}$$

Methyl *tert*-butyl ether — *tert*-Butyl alcohol

FIGURE 6.7
MTBE and the intermediate of biodegradation TBA.

Because of the solubility, mobility, and volatility of MTBE, the compound poses a potential risk to groundwaters. Methyl *tert*-butyl ether is more soluble than BTEX and has a considerably lower adsorption capacity (i.e., the K_{SD} is lower). Based on field studies, MTBE moves at about the same speed in groundwater as a conservative tracer (Barker et al., 1990). At the same time, MTBE seems to be persistent in groundwater, both under aerobic and anaerobic conditions (Barker et al., 1990; Suflita and Mormile, 1993; Mormile et al., 1994). The fact that MTBE is not easily degradable and may thus accumulate in groundwater poses an as yet undetermined health risk. Based on a limited number of studies that have shown MTBE to be a carcinogen in animals, the U.S. EPA has tentatively classified MTBE as a possible human carcinogen.

To date, few laboratory studies have been conducted on the biodegradation of MTBE. The ether bond in MTBE (Figure 6.7), as well as the methyl groups in the molecule, seem responsible for the biological stability of the compound. Long acclimation periods seem necessary before soil microorganisms begin MTBE biodegradation. A few anaerobic studies have been conducted in order to mimic groundwater conditions where oxygen is limited. Mormile et al. (1994), and Yeh and Novak (1994) observed no biodegradation under nitrate-reducing conditions, even after acclimation periods that exceeded 180 days. In both studies, however, MTBE biodegradation was observed under methanogenic conditions. Mormile et al. (1994) detected an intermediary product, *tert*-butyl alcohol (TBA), which seemed to persist and not biodegrade any further.

Salanitro et al. (1994) observed MTBE biodegradation under aerobic conditions after an acclimation period of two months. Using ^{14}C radiolabeled MTBE, they were able to show complete mineralization of the compound to carbon dioxide and cells. The intermediary product, TBA, was also detected during aerobic biodegradation. *Tertiary* butyl alcohol seemed to accumulate for a while, then biodegraded at a rate slower than that at which MTBE was degraded. *Tertiary* butyl alcohol can be of special importance since it also is an oxygenate that is sometimes used as a fuel additive.

Aerobic biodegradation of MTBE under controlled laboratory conditions also has been reported by Cowan and Park (1996), and Mo et al. (1997). Removal of MTBE in a field-operated, compost-based biofilter at the Joint Water Pollution Control Plant (JWPCP) of Los Angeles County Sanitation Districts in Carson, California, has been reported (Eweis et al., 1997a). Approximately 1 year acclimation period was needed to achieve greater than 95 percent removal (based on average inlet concentrations of 200 ppb_v) of MTBE. The acclimation period was cut down to three weeks when a culture isolated from the JWPCP biofilter and enriched for MTBE was inoculated into a pilot-scale biofilter with artificial support media. Bench and pilot-scale experiments showed that MTBE can serve as the sole carbon and energy source for the enriched culture, that MTBE concentrations as high as 150 ppm_v did not result in inhibition of growth, and that simultaneous biodegradation of MTBE and more easily metabolized substrates such as toluene was possible (Eweis et al., 1997b).

BIODEGRADATION OF HALOGENATED ALIPHATIC COMPOUNDS

Halogenated aliphatic compounds are common contaminants of groundwater and hazardous-waste sites. Industrially important halogenated aliphatics include chlorinated and brominated alkanes and alkenes in the C_1 to C_3 range. Chlorinated ethanes and ethenes are commonly used as cleaning solvents and in dry-cleaning operations and semiconductor manufacturing (Vogel et al., 1987). Brominated compounds are used as pesticides (e.g., ethylene dibromide or EDB and dibromochloropropane or DBCP) and halogenated methanes (e.g., $CHCl_3$, $CHCl_2Br$, $CHClBr_2$, and $CHBr_3$) are formed during the disinfection of water. Halogenated compounds are generally more resistant to microbial attack and tend to persist in the environment. The halogen atoms on the molecules increase the oxidation state of the carbon atom, and aerobic processes are energetically less favorable for highly halogenated compounds. Conversely, anaerobic degradation is then more favorable. Physicochemical processes such as stripping and adsorption are often more effective and reliable than bioremediation for halogenated compounds owing to their slow degradation rates. However, many halogenated compounds are rather easily degraded. Examples include methylene chloride (dichloromethane), chlorophenol, and ortho-, meta-, and para-chlorobenzoate.

Organic compounds generally act as electron donors; however, because of the electronegativity of the halogen substituents, polyhalogenated compounds can act as electron acceptors in reducing environments. Therefore, the greater the number of halogens in the molecule, the less biodegradable the compound will be in aerobic systems and the more degradable it will be in anaerobic systems. The biodegradation rate is also dependent on the type of halogen in the compound. Halogens can be ordered according to their decreasing electronegativities as follows: F, Cl, Br, I. Therefore, bromine, which is a less electronegative compound than chlorine, is more easily substituted. Cometabolism also plays an important role in the biotransformations of halogenated compounds. Trichloroethene (TCE), tetrachloroethene (PCE), and trichloromethane (TCM) (chloroform), for example, are compounds that are degraded by enzyme systems which are induced in response to a cometabolite (Strand and Shippert, 1986). In the case of TCE, degradation occurs as a result of cometabolism by either methanotrophs (Alvarez-Cohen and McCarty, 1991), aromatic degraders (Folsom et al., 1990), or ammonium oxidizers (Vannelli et al., 1989) through the action of monooxygenase or dioxygenase enzymes.

Microbially mediated reactions of chlorinated aliphatic compounds include substitutions, oxidations, and reductions (Pitter and Chudoba, 1990). Dehalogenation of the molecule is usually the first step with compounds containing a short alkyl chain. Where the alkyl chain is long, the halogen no longer influences the oxidation of the terminal carbon atom. In this case oxidation of the terminal methyl group is the first step resulting in a halogenated aliphatic alcohol. In substitution reactions, the halogen is substituted by a hydroxyl group:

$$R{-}X + H_2O \rightarrow R{-}OH + HX$$

An example of this is the dehalogenation of dichloromethane:

$$CH_2Cl_2 + H_2O \rightarrow [HOCH_2Cl] + HCl \rightarrow HCHO + 2H^+ + 2Cl^-$$

Intermediate products of the hydrolysis of dichloromethane and 1,2-dichloroethane are formaldehyde, 2-chloroethanol, and 1,2-ethanediol (Pitter and Chudoba, 1990). In mixed cultures these are further degraded to carbon dioxide. Oxidation by alpha-hydroxylation is also a possible mechanism but is less common:

$$R{-}CH_2{-}X + H_2O \rightarrow RCH(OH){-}X + 2H^+ + 2e^-$$

The aerobic degradation of chlorinated ethenes probably occurs by epoxidation. The epoxide is further hydrolyzed to carbon dioxide and HCl.

$$\mathrm{>C{=}C<}(X) + H_2O \rightarrow \mathrm{>C{-}C<}(\text{O bridge}, X) + 2H^+ + 2e^-$$

The third type of reaction, reductive dehalogenation, occurs in anaerobic environments. Either a halogen is substituted by a hydrogen atom or two halogen atoms are removed, giving rise to a double bond (dihalo-elimination):

$$R{-}X + H^+ + 2e^- \rightarrow RH + X^-$$

$$\mathrm{{-}C(X){-}C(X){-}} + 2e^- \rightarrow \mathrm{>C{=}C<} + 2X^-$$

Dihalo-elimination can occur in either anaerobic or aerobic environments. In the case of PCE and TCE, reductive dehalogenation results in the formation of vinylidene and vinyl chloride (Freedman and Gossett, 1989) which are also carcinogens and more volatile than the parent compounds.

BIODEGRADATION OF HALOGENATED AROMATIC COMPOUNDS

Halogenated aromatic compounds are also common contaminants of soil, groundwater, and hazardous-waste sites. Industrially important halogenated aromatics include solvents, lubricants, pesticides (e.g., DDT, 2,4-D, 2,4,5-T), plasticizers, polychlorinated biphenyls (PCBs), which were widely used as insulators in electrical transformers and capacitors; and pentachlorophenol, a wood preservative (Reineke

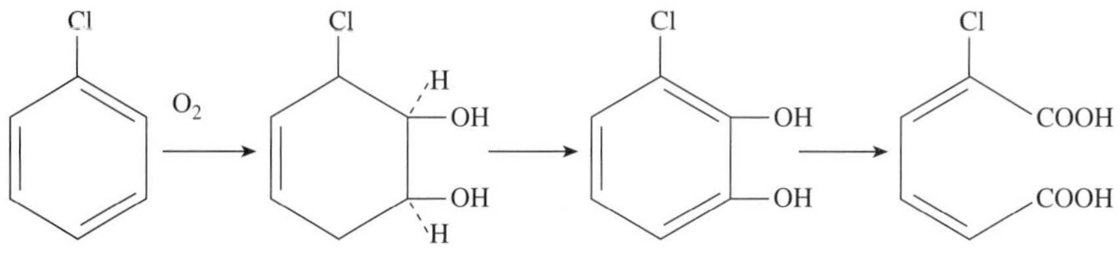

FIGURE 6.8
Ring cleavage of chlorobenzene.

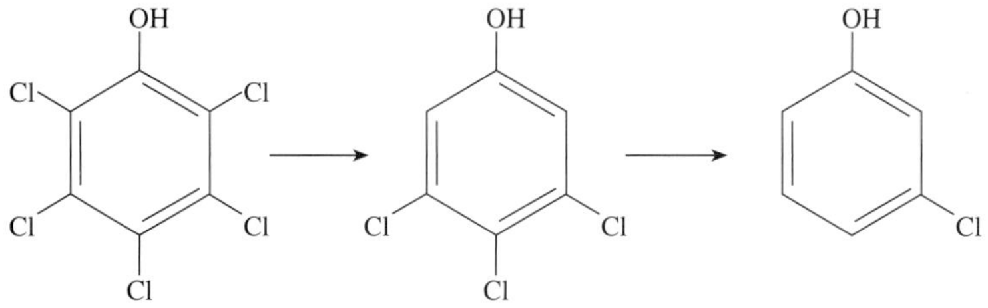

FIGURE 6.9
Reductive dehalogenation of pentachlorophenol.

and Knackmuss, 1988; Rochkind et al., 1986). As with alkylhalides, both the position and number of halogens are important in determining the biodegradability of halogenated aromatic compounds. Like alkylhalides, the more halogen substituents the compound has, the more likely it is to undergo reductive dehalogenation in reducing environments. Biodegradation of arylhalides may occur by dehalogenation of the ring structure by oxidation, reduction, or substitution; or ring cleavage can precede dehalogenation, generating halogenated aliphatic compounds.

Haloaromatics such as chlorophenoxy herbicides and chlorobenzenes are most often degraded by oxidation to halocatechols via chlorophenol with subsequent ring cleavage (Pitter and Chudoba, 1990). The ortho cleavage pathway which results in the formation of chloromuconic acid is shown in Figure 6.8. Cleavage of the meta bond also occurs and results in the formation of chlorohydroxymuconic semialdehyde. Dehalogenation may proceed spontaneously after cleavage of the aromatic ring (Reineke and Knackmuss, 1988).

Reductive dehalogenation has been shown to occur under methanogenic conditions for chlorinated benzoates (Suflita et al., 1982), PCBs (Thayer, 1991), PCP, the pesticide 2,4,5-T, chlorophenols, and 1,2,4-trichlorobenzene (Reineke and Knackmuss, 1988). The results of these transformations are products containing fewer chlorines than the parent compounds. These products are unlikely to be further degraded under anaerobic conditions but can be oxidized under aerobic conditions. For example, the monochlorinated benzoates are all biodegraded to CO_2 aerobically (Levitt et al., 1985). Reductive dechlorination of pentachlorophenol (PCP) is shown in Figure 6.9. The products are 3,4,5-trichlorophenol, 3,5-dichlorophenol, and 3-chlorophenol (Reineke and Knackmuss, 1988).

Substitution of the halogen by a hydroxyl group has been shown to occur for para-substituted monohalogenated benzoates, and PCP (Reineke and Knackmuss, 1988) as shown in Figure 6.10. The product of the PCP degradation is tetrachloro-*p*-hydroquinone, which can be degraded only under anaerobic conditions.

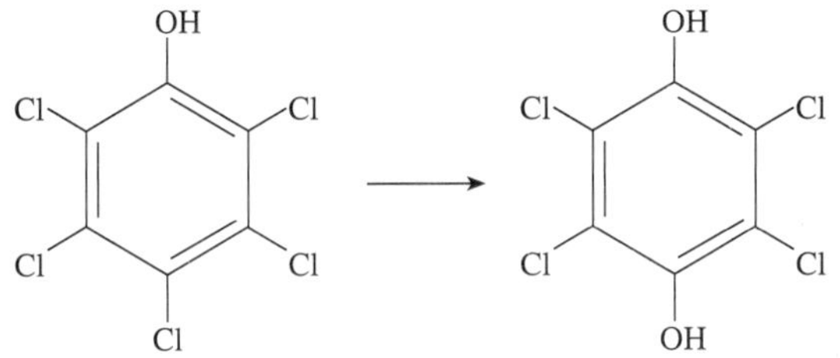

FIGURE 6.10
Hydrolysis of pentachlorophenol.

SUMMARY

Biotransformations of organic compounds may affect both their toxicity and their volatility. Biodegradation and volatilization are competing mechanisms, and the more degradable a compound is, the more likely it is to be degraded before volatilization occurs. Highly volatile parent compounds, however, such as gasoline hydrocarbons, may be preferentially stripped from the soil under certain conditions even though they are highly degradable. Some chemicals are transformed into products which are more degradable than the parent compounds and so have short lives in the environment. Others degrade to recalcitrant compounds that are more persistent than the parent compounds.

In general, aerobic transformations, such as the degradation of an alkane to a fatty acid or an aromatic compound to a catechol, add oxygen to the compound, making the product compounds less volatile, more soluble, and more degradable. Polyaromatic hydrocarbons, however, may degrade to more volatile products as a result of the cleaving of aromatic rings and formation of lower-molecular-weight compounds. This may be especially important in fungal metabolism of PAHs where extracellular enzymes are used to cleave aromatic rings. Transformation processes that make compounds more soluble can decrease their adsorption to surfaces and facilitate stripping from the liquid phase.

Some anaerobic transformations make compounds more volatile. Examples of this are the reductive dechlorination of TCE to dichloroethene and vinyl chloride or the dihalo-elimination reactions which convert 1,2-dichloroethane to ethene. If kept under anaerobic conditions in soil, these compounds might diffuse slowly into aerobic zones or possibly to the atmosphere. Some bioremediation schemes have been proposed, however, which would alternate anaerobic and aerobic conditions. In these treatment systems reductive dehalogenation followed by aeration and mixing might increase volatilization of compounds formed in the previous stage.

PROBLEMS AND DISCUSSION QUESTIONS

6.1. How do you expect a high soil organic matter (SOM) content to influence the biodegradation of hazardous contaminants?

6.2. Why do you suppose it is that biodegradation of certain contaminants in the soil slows down, if not stops completely, as the contaminant concentration starts to drop below a certain level? Use as an example high-molecular-weight PAHs in a creosote mixture.

6.3. It has been observed that some highly chlorinated compounds, such as tetrachloroethene, or DDT, are biodegraded faster in soils undergoing alternate wetting and drying cycles, as compared to soils that are constantly saturated or unsaturated. Why do you think that is? Are there any alternative physicochemical mechanisms for the removal of these compounds that are plausible?

6.4. The chemicals listed below are common hazardous-waste constituents. Comment on the probable biodegradability of each compound. What chemical characteristics may hinder or enhance biodegradation in each case? Given are some chemical properties of the compounds which may be of use.

Compound	Chemical structure	Solubility, mg/L	Vapor pressure, mmHg
Propane	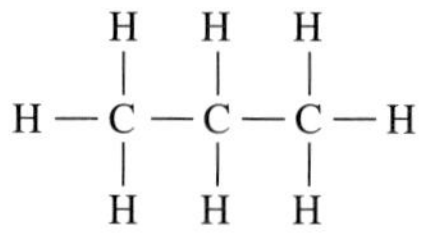 H H H H—C—C—C—H H H H	100	> 760
Picric acid	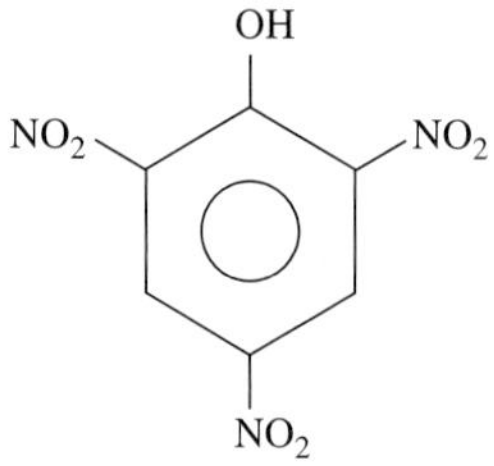 OH NO_2 NO_2 NO_2	10,000	1.0
Pentachlorophenol	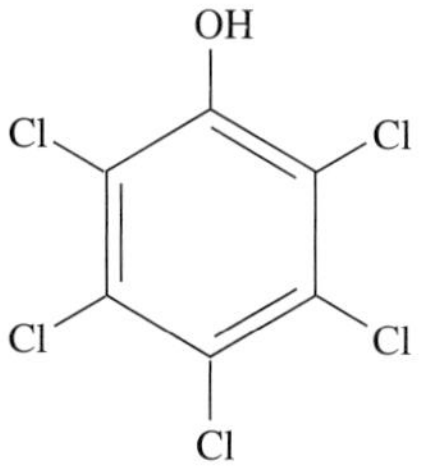 OH Cl Cl Cl Cl Cl	10	0.0001
Heptane	$CH_3[CH_2]_5CH_3$	50	40
Pyrene	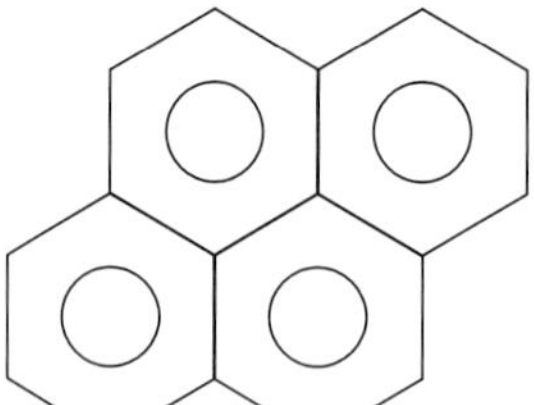	0.13	2.5×10^{-6}
Sodium Azide	N_3Na	soluble	NA
Aldicarb	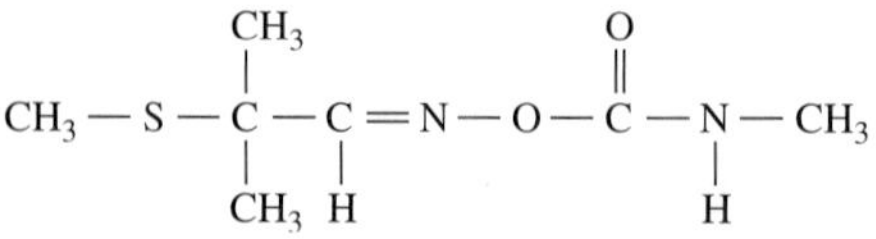 CH_3 O CH_3—S—C—C=N—O—C—N—CH_3 CH_3 H H	NA	NA

6.5. Petroleum hydrocarbon spills often occur in the ocean and on the ground surface. Briefly compare and discuss factors influencing the removal and natural degradation of the contaminants in each case.

REFERENCES

Alexander, M. (1981): "Biodegradation of Chemicals of Environmental Concern," *Science,* vol. 211, Jan. 9.

Alvarez-Cohen, L., and P. L. McCarty (1991): "Effects of Toxicity, Aeration, and Reductant Supply on Trichloroethene Transformation by a Mixed Methanotrophic Culture," *Applied and Environmental Microbiology,* vol. 57, no. 1, January, 1991, pp. 228–235.

Atlas, Ronald M. (1991): "Microbial Hydrocarbon Degradation-Bioremediation of Oil Spills," *Journal of Chemical Technical Biotechnol.,* vol. 52, pp. 149–156.

Barker, J. F., E. Hubbard, and L. A. Lemon (1990): "The Influence of Methanol and MTBE on the Fate of Persistence of Monoaromatic Hydrocarbons in Groundwater," *Proceedings of Petroleum Hydrocarbons and Organic Chemicals in Groundwater Prevention, Detection, and Restoration,* Oct. 31–Nov. 2, 1990, American Petroleum Institute and Association of Ground Water Scientists and Engineers, Houston, Texas, pp. 113–127.

Bartha, Richard (1986): "Biotechnology of Petroleum Pollutant Biodegradation," *Microbial Ecology,* vol. 12, pp. 155–172.

Britton, L. N. (1984): "Microbial Degradation of Aliphatic Hydrocarbons," in Gibson, D. T., (ed.), *Microbial Degradation of Organic Compounds,* Marcel Dekker, New York, pp. 89–130.

Cerniglia, C. E. (1984): "Microbial Metabolism of Polycyclic Aromatic Hydrocarbons," *Advances in Applied Microbiology,* vol. 30, pp. 31–69.

Cowan, R. M., and K. Park (1996): "Biodegradation of the Gasoline Oxygenates MTBE, ETBE, TAME, TBA, and TAA by Aerobic Mixed Cultures," *Proceedings of the 28th Mid-Atlantic Conference,* July 15–17, 1996, Buffalo, New York.

Dzomback, D. A., and R. G. Luthy (1984): "Estimating Adsorption of Polycyclic Aromatic Hydrocarbons on Soils," *Soil Science,* vol. 137(5) pp. 292–308.

Eweis, J. B., D. P. Y. Chang, E. D. Schroeder, K. M. Scow, R. L. Morton, and R. Caballero (1997): "Meeting the Challenge of MTBE Biodegradation," *Proceedings of the 90th Annual Meeting & Exhibition, Air and Waste Management Association,* Toronto, Canada, June 8–13.

Feldman, J., and M. Orchin (1993): "Determination of Methyl Tert Butyl Ether (MTBE) in Gasoline," *Analytical Letters,* vol. 26(2), pp. 357–365.

Folsom, B. R., P. J. Chapman, and P. H. Pritchard (1990): "Phenol and Trichloroethene Degradation by *Pseudomonas cepacia* G4: Kinetics and Interactions between Substrates," *Applied and Environmental Microbiology,* vol. 56, no. 1, May, pp. 1279–1285.

Freedman, D. L., and J. M. Gossett (1989): "Biological Reductive Dechlorination of Tetrachloroethylene and Trichloroethene to Ethylene under Methanogenic Conditions," *Applied and Environmental Microbiology,* vol. 55, no. 9, pp. 2144–2151.

Gibson, D. T. (1988): "Microbial Metabolism of Aromatic Hydrocarbons and the Carbon Cycle," in *Microbial Metabolism and the Carbon Cycle,* Hagedorn, S. R., R. S. Handson, and D. A. Kunz (eds.), Harwood Academic Publishers.

Gibson, D. T., (ed.) (1984): *Microbial Degradation of Organic Compounds,* Marcel Dekker, New York.

Gibson, D. T. (1984): "Microbial Degradation of Aromatic Hydrocarbons," in Gibson, D. T., (ed.), *Microbial Degradation of Organic Compounds,* Marcel Dekker, New York, pp. 181–252.

Kirschner, E. M. (1996): "Growth of Top 50 Chemicals Slowed in 1995 from Very High 1994 Rate," *Chemical & Engineering News,* vol. 74(15), April 8, pp. 16–22.

LaGrega, M. D., P. L. Buckingham, and J. C. Evans (1994): *Hazardous Waste Management,* McGraw-Hill, New York.

Leahy, J. G., and R. R. Colwell (1990): "Microbial Degradation of Hydrocarbons in the Environment," *Microbiology Review,* vol. 54, pp. 305–315.

Levitt, L. A., W. J. C. Pfeiffer, and E. D. Schroeder (1985): "Evaluation of Anaerobic Treatment Potential of Chemical Landfill Leachate Containing Organochlorine Compounds," *Proceedings,* International Conference: New Directions and Research in Waste Treatment and Residuals Management, University of British Columbia, Vancouver, B.C., pp. 215–231.

McEldowney, S., D. J. Hardman, and S. Waite (1993): *Pollution: Ecology and Biotreatment,* Longman Scientific and Technical, England.

Mo, K., C. O. Lora, A. E. Wanken, M. Javanmardian, X. Yang, and C. F. Kulpa (1997): "Biodegradation of Methyl t-Butyl Ether by Pure Bacterial Cultures," *Applied Microbiology and Biotechnology,* vol. 47, pp. 69–72.

Mormile, M. R., Shi, Liu, and J. M. Suflita (1994): "Anaerobic Biodegradation of Gasoline Oxygenates: Extrapolation of Information to Multiple Sites and Redox Conditions," *Environmental Science and Technology,* vol. 28, pp. 1727–1732.

Park, Kap S., Ronald C. Sims, and R. Ryan Dupont (1990): "Transformation of PAHs in Soil Systems," *Journal of Environmental Engineering,* vol. 116, no. 3, May/June pp. 632–640.

Pitter, P., and J. Chudoba (1990): *Biodegradability of Organic Substances in the Aquatic Environment,* CRC Press, Boca Raton, FL.

Reineke, W. (1984): "Microbial Degradation of Halogenated Aromatic Compounds," in Gibson, D. T. (ed.), *Microbial Degradation of Organic Compounds,* Marcel Dekker, New York, pp. 319–360.

Reineke, W., and H. J. Knackmuss (1988): "Microbial Degradation of Haloaromatics," *Annual Review of Microbiology,* Ornston, L. N., A. Balows, and P. Baumann (eds.), Annual Reviews, Palo Alto, CA.

Rochkind, Melissa L., James Blackburn, and Gary S. Sayler (1986): *Microbial Decomposition of Chlorinated Aromatic Compounds,* EPA/600/2-86/090.

Safe, S. H. (1984): "Microbial Degradation of Polychlorinated Biphenyls," in Gibson, D. T., (ed.), *Microbial Degradation of Organic Compounds,* Marcel Dekker, New York, pp. 361–398.

Skladany, G. J. (1992): "Overview of Bioremediation," in *Bioremediation: The State of Practice in Hazardous Waste Remediation Operations,* A seminar sponsored by Air and Waste Management Association and HAWC, January, 1992.

Strand, S. E., and L. Shippert (1986): "Oxidation of Chloroform in an Aerobic Soil Exposed to Natural Gas," *Applied and Environmental Microbiology,* vol. 52, no. 1, pp. 203–205.

Squillace, P. J., J. S. Zogorski, W. G. Wilber, and C. V. Price (1996): "Preliminary Assessment of the Occurrence and Possible Sources of MTBE in Groundwater in the United States," *Environmental Science and Technology,* vol. 30, pp. 1721–1730.

Suflita, J. M., A. Horowitz, D. R. Shelton, and J. M. Tiedje (1982): "Dehalogenation: A Novel Pathway for the Anaerobic Biodegradation of Haloaromatic Compounds," *Science,* vol. 218, pp. 1115–1117.

Suflita, J. M., and M. R. Mormile (1993): "Anaerobic Biodegradation of Known and Potential Gasoline Oxygenates in the Terrestrial Subsurface," *Environmental Science and Technology,* vol. 27, pp. 976–978.

Thayer, A. M. (1991): "Bioremediation: Innovative Technology for Cleaning Up Hazardous Waste," *Chemical and Engineering News,* vol. 69, no. 34, pp. 23–44.

Trudgill, P. W. (1984): "Microbial Degradation of the Alicyclic Ring: Structural Relationships and Metabolic Pathways," in Gibson, D. T., (ed.), *Microbial Degradation of Organic Compounds,* Marcel Dekker, New York, pp. 131–180.

U.S. Environmental Protection Agency (1963): *EPA Guide for Identifying Clean up Alternatives at Hazardous Waste Sites and Spills: Biological Treatment,* EPA 600/3-83/063.

U.S. Environmental Protection Agency (1984): *Health Effects Assessment for Polycyclic Aromatic Hydrocarbons (PAHs),* Cincinnati, Ohio, EPA/540/1-86/013.

Vannelli, T. M., M. Logan, D. M. Arciero, and A. B. Hooper (1989): "Degradation of Halogenated Aliphatic Compounds by the Ammonium Oxidizing Bacterium Nitrosomonas europaea," *Applied and Environmental Microbiology,* vol. 56, pp. 1169–1171.

Vogel, T. M., C. S. Criddle, and P. L. McCarty (1987): "Transformations of Halogenated Aliphatic Compounds," *Environmental Science and Technology,* vol. 21, no. 8, pp. 722–736.

Yeh, C. K., and J. T. Novak (1994): "Anaerobic Biodegradation of Gasoline Oxygenates in Soils," *Water Environment Research,* vol. 66(5), pp. 744–752.

CHAPTER 7

In Situ Treatment

In situ treatment is generally the most desirable approach for remediation of both contaminated groundwater and contaminated soil because excavation and disposal of the contaminated material is not required. In most cases the total cost of remediation is considerably less using in situ methods in comparison to excavation and treatment. In particular cases, the cost of in situ treatment may not be significantly different from alternative remediation methods, but ancillary issues such as interruption of business activities, inconvenience, disruption of the site, and impact on public image are important factors in the selection process. However, in situ treatment has a number of inherent disadvantages relative to excavation and on-site or off-site treatment and disposal, including the difficulty of providing oxygen and nutrients to the reaction sites, the difficulty of determining the amount of treatment taking place, the relatively low rates of treatment, and the potential to spread the contamination into new areas.

In situ treatment of contaminated aquifers is conducted differently from treatment of contaminated vadose zone soil because of the oxygen concentration and transport limitations associated with aquifers. Oxygen solubility in water is low and is usually controlled by Henry's law equilibrium with the ambient soil air. Typical equilibrium oxygen concentrations in groundwaters are below 8 mg/L and most shallow groundwaters are found to have dissolved oxygen concentrations between 4 and 7 mg/L. Aerobic biodegradation is the principal method of bioremediation, and the low oxygen concentrations found in aquifers will severely constrain the rate and extent of remediation unless oxygen, or an alternate electron acceptor, is added on a continual basis.

Nutrients are often present in less than the required stoichiometric concentrations in both aquifers and vadose zone soils. The nutrient most often limiting microbial growth, and hence microbial degradation of organic contaminants, is nitrogen. However, the solubility of nitrogen, as NH_4^+ and NO_3^-, in water is high and relatively high concentrations can be added without becoming inhibitory to microbial growth and respiration. Moreover, nitrate can be added as an electron acceptor as well as a growth nutrient. Because not all soil microorganisms are capable of respiring on nitrate, the range of contaminants degraded using nitrate respiration is more limited than that for oxygen. Additionally, some reactions (e.g., the breakdown of aromatics catalyzed by oxygenase enzymes) cannot occur without molecular oxygen.

IN SITU REMEDIATION OF AQUIFERS

In situ biodegradation in aquifers requires that oxygen, nutrients, and sometimes specific bacterial species be transported throughout the contaminated region. Three approaches are taken to add the necessary components: percolation of materials into the aquifer; pump, treat, and recirculation; and air sparging.

Percolation

In percolation, nutrients necessary for biodegradation are added as a solution at the soil surface and allowed to flow through the vadose zone to the top of the aquifer (see Figure 7.1). Adding nutrients by percolation is limited to shallow

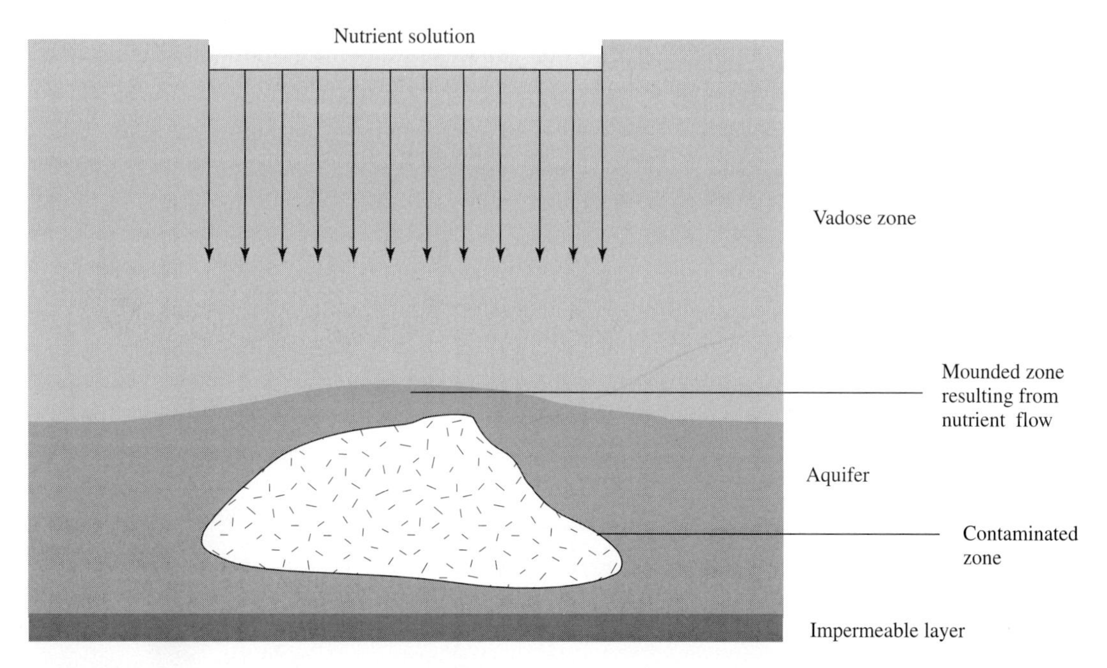

FIGURE 7.1
Percolation of nutrients into a contaminated aquifer. Mounding of the aquifer will normally occur and the contaminated zone may spread out as a result.

aquifers because of the time required to move materials into the contaminated zone. Oxygen cannot be added by percolation because of the limited concentrations that can be dissolved in water and the contact of the fluid with air in the vadose zone. Specialized bacteria could possibly be added by percolation but losses during transport would render such addition infeasible. Usually nutrients are added by filling an unlined pond or series of trenches to a shallow depth and letting the nutrient solution seep into the soil. Vertical transport of the nutrient solution is assumed to dominate the flow pattern. Some mounding of the aquifer can be expected, and spreading of the contaminated zone may result.

Availability of oxygen will limit the rate of aquifer remediation in percolation systems. If the only source of oxygen is diffusion from the phreatic surface, the bioxidation process will be very slow. Addition of nutrients may allow growth of facultative or obligate anaerobic bacteria. Fermentation products will usually be more mobile than the original contaminant compounds, and spreading of the contaminated zone may result from anaerobic conditions. Addition of NO_3^- as a nitrogen source and as an electron acceptor is attractive (see Chapter 5 for a discussion of electron acceptors) but, as noted above, the number of compounds degraded through nitrate respiration is limited. A number of important organic contaminants, most notably the alkanes and aromatics, are difficult to degrade in the absence of oxygen because the most common metabolic pathways require oxygen in the initial step. However, NO_3^- should be considered wherever possible.

Pump, Treat, and Reinjection

Pump, treat, and injection is a combination of ex situ pump and treat remediation and oxygen and nutrient injection for in situ bioremediation (see Figure 7.2). In pump and treat remediation, contaminated water is pumped from the contaminated

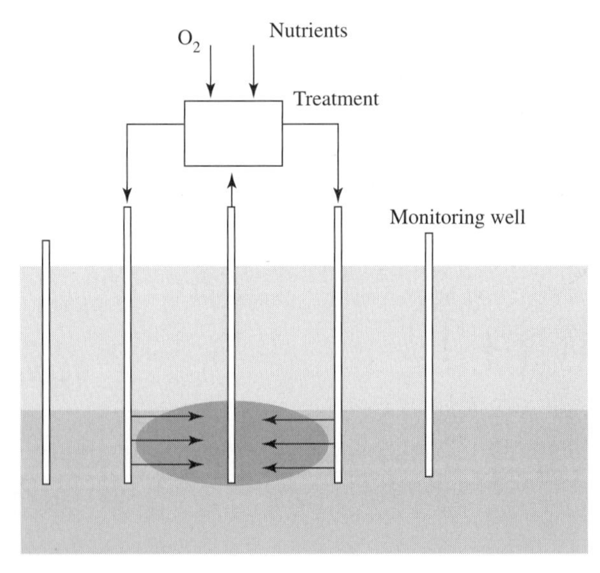

FIGURE 7.2
Schematic diagram of pump, treat, and reinjection operation.

zone, treated, and discharged to a community sewer, a receiving water, or a land disposal site. Ideally the treated water would be reinjected into the aquifer but because the reclaimed water is classified as a waste discharge, such a program may be prohibited. Injection of water to which oxygen, nutrients, and in some cases bacteria has been added is a method by which a continuous supply of these materials can be provided to the contaminated zone. Hydrophobic contaminants are often difficult to remove by standard pump and treat methods (see Chapter 3). Injection of nutrients and oxygen into the contaminated zone allows the development of in situ biodegradation and greatly increases the rate of aquifer remediation. Combining the pump and treat with oxygen and nutrient injection results in the development of a reaction zone within the aquifer.

Hydrogen peroxide (H_2O_2), or NO_3^-, can be added as a source of electron acceptor instead of atmospheric oxygen or pure oxygen. Considerably more experience has been reported for the use of H_2O_2 than for NO_3^- and use of H_2O_2 can be considered a standard procedure. Hydrogen peroxide is broken down into water and molecular oxygen in a reaction catalyzed by the enzyme catylase.

$$2\ H_2O_2 \rightarrow 2\ H_2O + O_2 \qquad (7.1)$$

The oxygen provided by addition of H_2O_2 is 0.47 g/g. Addition of H_2O_2 at concentrations greater than 100 mg/L may result in toxic conditions.

Design principles

Principal factors in designing a pump, treat, and injection system are determination of the conditions in the contaminated zone, selection of a pumping rate, estimation of the contaminant concentrations in the pumped water, estimation of the oxygen and nutrient requirements, and estimation of the time required for bioremediation. Inherent in the design of a bioremediation system is the need for containment of the contaminant, inclusion of monitoring wells, and a sampling program. Saturation of the injection water with oxygen using pressurized air or pure oxygen can provide oxygen concentrations greater than 20 mg/L. If higher concentrations are desirable, H_2O_2 can be added.

Remediation system consisting of physical, chemical, and biological treatment for solvent- and metal-contaminated groundwater at a Superfund site in New Jersey. *(Courtesy of OHM Remediation Services Corp.)*

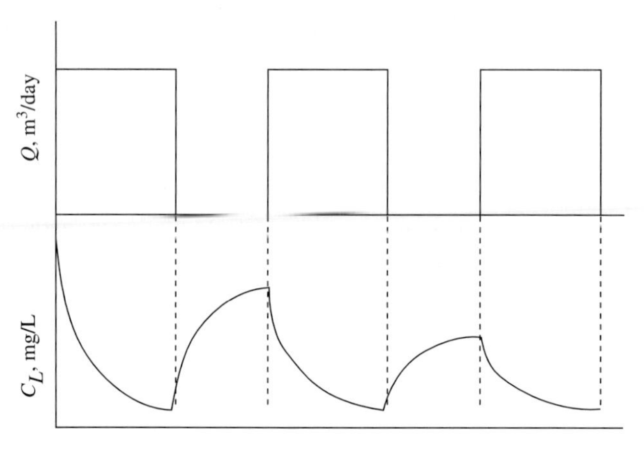

FIGURE 7.3
Liquid-phase concentration (C_L) as a function of intermittent pumping (Q) for hydrophobic contaminants. In cases where the bulk of the contaminant is sorbed onto the solid phase, disorption rates may be lower than the mass removal rate due to pumping. Also, where pure product exists, the mass transfer rate may be lower than the mass removal rate due to pumping. In such cases the liquid-phase concentration will decrease during pumping and rise during nonpumping periods.

Pumping rate The pumping rate selected will be a function of the available time for remediation, the hydraulic conductivity of the aquifer, the size of the contamination zone, and the distribution coefficient of the contaminant. Many pump and treat systems have been in operation for over a decade, and treatment times of several years are common. Long treatment times result in operating costs (e.g., electrical power) becoming significant design factors; consequently low pumping rates are desirable. In many cases the distribution of contaminants between the solid and solute phases is very unfavorable and the pumped water will have extremely low contaminant concentrations. Initial liquid-phase contaminant concentrations may be high but will drop to low values very quickly because the desorption rate and diffusion out of low permeability zones is less than the extraction rate. In such cases decreasing the pumping rate may be beneficial or intermittent pumping may be utilized, as suggested in Figure 7.3.

Contaminant concentration estimation Contaminant concentration in the pumped water will be a function of the mass of contaminants in the aquifer, the volume of the contaminated zone, and the soil distribution coefficient. Accurate estimates of the mass of contaminants present at a remediation site are difficult to make. In some cases the amount of material discharged can be calculated from records of tank levels or wastewater disposal. If operating procedures were routinely followed (e.g., tank cleaning and rinsate disposal procedures) estimates of the contaminant mass satisfactory for system design may be possible. Site investigation prior to system design should include analysis of core samples at selected points in the contaminated zone to determine liquid- and sorbed-phase contaminant concentrations.

EXAMPLE 7.1. DETERMINATION OF CONTAMINANT CONCENTRATION. Core samples from the aquifer shown in Figure 7.4 have been analyzed for total petroleum hydrocarbons (TPH) for various depths as shown in the table below. Estimate the total mass of contaminants in the aquifer and calculate the distribution coefficient. Bulk density of the drained aquifer material is 2,500 kg/m³ and the porosity ϕ is 0.4.

Soil and water TPH

	0 m		3 m		6 m	
Core	***S*, g/kg**	**C_L, mg/L**	***S*, g/kg**	**C_L, mg/L**	***S*, g/kg**	**C_L, mg/L**
d2	0.006	0.63	0.000	0.00	0.000	0.00
d4	0.000	0.00	0.000	0.00	0.000	0.00
e1	0.000	0.00	0.000	0.00	0.000	0.00
e3	0.010	1.10	0.009	0.95	0.000	0.00
e5	0.000	0.00	0.000	0.00	0.000	0.00
f1	0.011	1.22	0.013	1.40	0.005	0.04
f3	0.013	1.20	0.016	1.62	0.012	1.20
f4	0.008	0.07	0.010	1.05	0.003	0.03
g2	0.000	0.00	0.000	0.00	0.000	0.00
g4	0.000	0.00	0.000	0.00	0.000	0.00

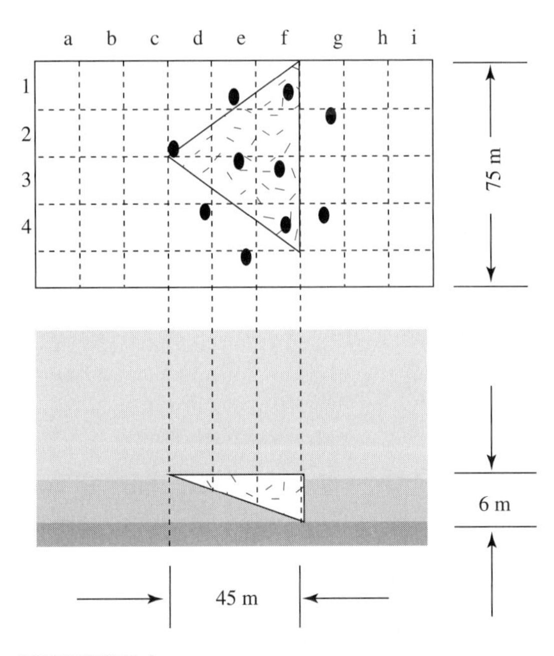

FIGURE 7.4
A plan view and a cross-sectional view of a petroleum hydrocarbon contamination zone in an underground aquifer.

Solution

1. Divide the contaminated zone into subsections and assign a mean concentration to each subsection based on the core samples.

	Volume, m^3			Concentration, g/kg Soil			Concentration, mg/L Water		
Section	**0–2**	**2–4**	**4–6**	**0–2**	**2–4**	**4–6**	**0–2**	**2–4**	**4–6**
d2	90	0	0	0.008	0.005	0.000	0.87	0.48	0.00
d3	90	0	0	0.008	0.005	0.000	0.87	0.48	0.00
e1	70	35	0	0.008	0.010	0.006	0.81	1.00	0.43
e2	420	210	0	0.010	0.009	0.000	1.10	0.95	0.00
e3	420	210	0	0.010	0.009	0.000	1.10	0.95	0.00
e4	70	35	0	0.03	0.006	0.001	0.37	0.67	0.01
f1	300	300	150	0.011	0.013	0.005	1.22	1.40	0.04
f2	450	450	225	0.012	0.014	0.008	1.21	1.51	0.62
f3	450	450	225	0.013	0.016	0.012	1.20	1.62	1.20
f4	300	300	150	0.008	0.010	0.003	0.07	1.05	0.03

2. Determine K_{SD} values for each section.

$$K_{SD} = \frac{s}{C_L}$$

Example calculation for section e3 at a depth of 2 to 4 m:

$$K_{SD} = \frac{(0.009 \text{ g/kg})(10^{-3} \text{ kg/g})}{0.95 \text{ mg/L}} = 9.5 \times 10^{-6} \text{ L/mg}$$

Tabulate values for the contaminated zone.

3. Determine total mass of contaminant for each section.
   a. Contaminant mass sorbed

$$M_S = s\,\rho_b V$$

Example calculation for section e3 at a depth of 2 to 4 m:

$$M_S = (0.009 \text{ g/kg})(2{,}500 \text{ kg/m}^3)(210 \text{ m}^3) = 4{,}725 \text{ g}$$

   b. Contaminant mass in solution

$$M_L = \Theta\, V C_L$$

Example calculation for section e3 at a depth of 2 to 4 m:

$$M_L = (0.4)(210 \text{ m}^3)(0.95 \text{ mg/L}) = 80 \text{ mg}$$

Note that despite the low value of K_{SD}, which is expected in an aquifer, over 98 percent of the contaminant mass is sorbed.
Tabulate values of contaminant mass for each section.

Section	K_{SD}, L/mg × 10^6			Mass sorbed, g			Mass in solution, mg		
	0–2	2–4	4–6	0–2	2–4	4–6	0–2	2–4	4–6
d2	9.2	10.4		1,800			31		
d3	9.2	10.4		1,800			31		
e1	9.88	10.0	13.95	1,400	875		23	14	
e2	9.09	9.5		10,500	4,725		185	80	
e3	9.09	9.5		10,500	4,725		185	80	
e4	2.7	9.0	10	175	525		10	9	
f1	9.02	4.3	12.5	8,250	4,500	1,875	146	168	24
f2	9.92	9.3	12.9	13,500	15,750	4,500	218	272	56
f3	10.83	9.9	10	14,625	18,000	6,750	216	292	108
f4	11.43	9.5	10	6,000	7,500	1,125	84	126	18
Avg	9.03	9.2	11.7	68,550	56,600	14,250	1,129	1,040	206

$$\bar{K}_{SD} = 9.7 \times 10^{-6} \text{ L/mg}$$

$$M_{S\text{ total}} = 139{,}400 \text{ g} = 139.4 \text{ kg}$$

$$M_{w\text{ total}} = 2{,}375 \text{ mg} = 2.4 \text{ g}$$

Field demonstration of in situ biological treatment of PCB-contaminated river sediments in the Upper Hudson River, New York. *(Courtesy of OHM Remediation Services Corp.)*

EXAMPLE 7.2. PUMP, TREAT, AND REINJECTION. A saturated soil zone has been contaminated with 5,000 L of diesel (4,000 kg) from a leaking underground storage tank. The soil is fairly high in organic matter (5 percent by weight), has a porosity ϕ = 0.4, wet bulk density ρ_{wb} = 1,700 kg/m^3, and a hydraulic conductivity K_C of 2 × 10^{-3} m/s. The K_{SD} of the soil and diesel is quite high (0.01 L/mg) and the diesel is not very

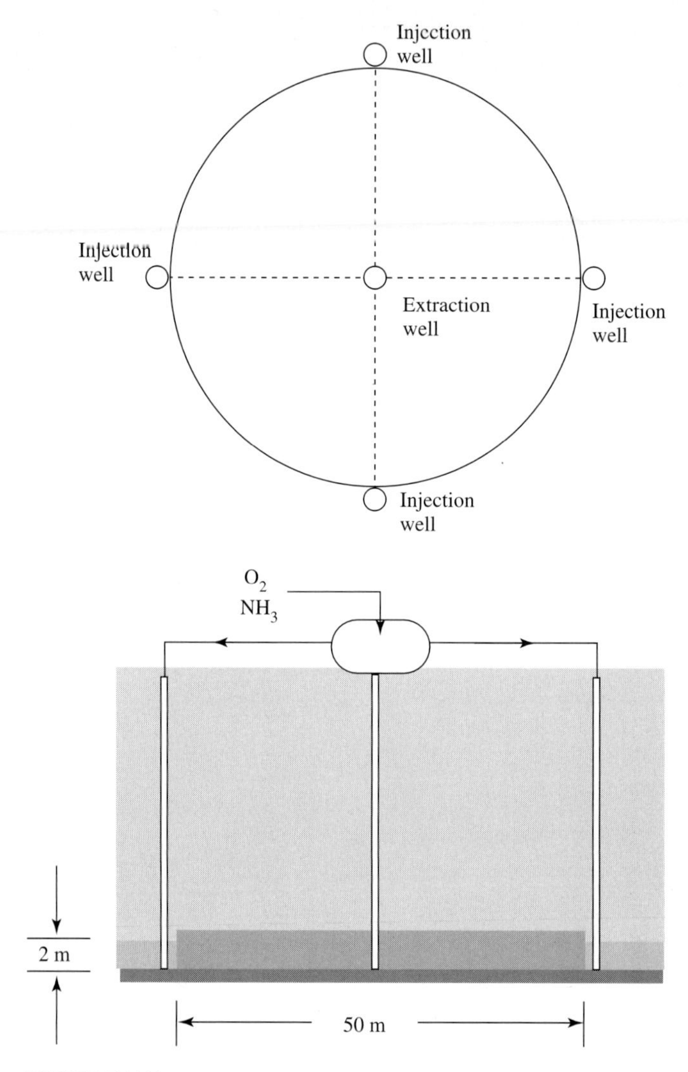

FIGURE 7.5
Schematic diagram of pump, treat, and reinjection system for Example 7.2.

mobile. As shown in Figure 7.5, the groundwater is not moving and contamination is spreading only by diffusion. The cross-sectional dimensions of the saturated zone are 2 m deep by 50 m in diameter.

a. Determine the liquid- and solid-phase concentrations of the diesel (assume that the concentrations are approximately constant at all points in each phase).

b. Determine an appropriate flow rate.

c. Assuming that the diesel is nearly immobile because of the large K_{SD} value, determine the rate of diesel degradation possible in kg/day, if the flow was pumped from the center of the contaminated zone using a 100-mm-diameter well, treated, enriched by the addition of 100 mg/L of H_2O_2 and 10 mg/L of NH_4Cl, and recirculated by injecting the enriched stream at four points around the periphery. Assume that degradation of diesel requires 2.5 g O_2 per g of diesel.

Solution

1. Determine liquid- and solid-phase diesel concentrations
   a. Determine volume of contaminated area and volume of water

$$V = (2\text{ m})\frac{\pi}{4}(50\text{ m})^2 = 3{,}925\text{ m}^3$$

$$V_L = \phi V = 0.4(3{,}925) = 1{,}570\text{ m}^3$$

   b. Determine mass of soil

$$M_T = V\rho_{wb} = V_L\rho_L + V_{\text{soil}}\,\rho_{\text{soil}} = V_L\rho_L + M_{\text{soil}}$$

$$M_T = (3{,}925\text{ m}^3)(1{,}700\text{ kg/m}^3) = (1{,}570\text{ m}^3)(1{,}000\text{ kg/m}^3) + M_{\text{soil}}$$

$$M_{\text{soil}} = 5.1 \times 10^6\text{ kg}$$

$$V_{\text{soil}} = (1 - \phi)V = 0.6(3{,}925\text{ m}^3) = 2{,}355\text{ m}^3$$

$$\rho_{\text{soil}} = \frac{M_{\text{soil}}}{V_{\text{soil}}} = \frac{5.1\times10^6\text{ kg}}{2{,}355\text{ m}^3} = 2{,}165\text{ kg/m}^3$$

This value is lower than might be expected. Typical soil densities are about 2,700 kg/m^3.

   c. Determine liquid-phase concentration

$$K_{SD} = \frac{s}{C_L} = \frac{\text{mass sorbed/mass of soil}}{C_L}$$

$$\text{Mass sorbed} = 4{,}000\text{ kg} - V_L C_L$$

$$C_L K_{SD} = \frac{4{,}000 - V_L C_L}{5.1\times10^6\text{ kg}}$$

$$K_{SD} = (0.01\text{ L/mg})(0.001\text{ m}^3\text{/L})(10^6\text{ mg/kg}) = 10\text{ m}^3\text{/kg}$$

$$C_L = (10\text{ m}^3\text{/kg})(5.1\times10^6\text{ kg}) = 4{,}000\text{ kg} - (1{,}570\text{ m}^3)C_L$$

$$C_L \approx \frac{4{,}000\text{ kg}}{5.1\times10^7\text{ m}^3} = 7.84\times10^{-5}\text{ kg/m}^3 = 0.078\text{ mg/L}$$

$$s \approx \frac{4{,}000\text{ kg}}{5.1\times10^6\text{ kg}} = 7.84\times10^{-4}\text{ kg/kg} = 784\text{ mg/kg}$$

2. The results above support the suggestion in the problem statement that the contaminants are essentially immobile. Assume that the biodegradation rate is equal to the rate of oxygen addition. This assumption implies a plug flow system with an oxygen-limited reaction rate. If the desorption rate is less than the rate of oxygen transport, the assumption will be invalid.

3. Determine the flow rate of water in the aquifer (QC_{O_2} = mass rate of O_2 addition). Assume ideal steady radial flow to the 0.1-m-diameter extraction well. Flow in the well can be described by Equation (7.2) (Bedient and Huber, 1988).

$$Q = \frac{\pi K_C(h_w^2 - h_o^2)}{\ln(r_w/r_o)} \tag{7.2}$$

$$= \frac{\pi(2\times 10^{-3}\text{ m/s})(86{,}400\text{ s/day})(h_w^2 - 2^2)}{\ln(0.05\text{m}/25\text{m})}$$

where K_C = hydraulic conductivity, m/s
h_w = head at well, m
h_o = head at radius r_o, m
r_w = radius of well, m
r_o = distance to observation well, m

h_w, m	Q, m³/day
1.9	39.2
1.8	76.3
1.7	111.5
1.6	144.7
1.5	175.8
1.4	204.9
1.3	232.1
1.2	257.2
1.1	280.3
1.0	301.4

4. Select a 0.3-m drawdown (h_w = 1.7 m) which gives a flow rate of 111 m³/day (20 gal/min). A drawdown of 15 percent of the depth is reasonable, but normally flow-rate selection will be based on cost information unavailable in this case.
5. Oxygen addition rate = (111 m³/day)(100 g H_2O_2/m³)(0.47 g O_2/g H_2O_2)
   = 5,217 g/day
6. Estimate bioremediation rate assuming that oxygen is rate-limiting.

$$\text{Rate of degradation} = \frac{O_2\text{ supply rate}}{2.5\text{ g }O_2\text{ per g diesel degraded}} = \frac{5{,}217\text{ g/day}}{2.5\text{ g/g TPH}}$$

$$= 2{,}087\text{ g/day} = 2.1\text{ kg diesel/day}$$

7. Estimate bioremediation time

$$t_b = \frac{4{,}000\text{ kg}}{2.1\text{ kg/day}} = 1{,}900\text{ days} = 5.2\text{ years}$$

The flow through time θ_H can be estimated as the contaminated zone volume divided by the flow rate:

$$\theta_H = \frac{\phi\pi r^2 h}{Q} = \frac{(0.4)\pi(25\text{ m})^2(2\text{ m})}{111\text{ m}^3/\text{day}} = 14.2\text{ days}$$

Note that the water being pumped will have very low diesel concentrations. External treatment using granular activated carbon (GAC) adsorption would reduce the concentration further. Water-quality regulations may require further treatment before reinjection of the groundwater into the aquifer. Adsorption of the remaining diesel onto GAC would be very effective. Discharge to a municipal sanitary sewer would be a more cost-effective solution, presuming that the material is shown to be biodegradable.

Advantages and disadvantages

The principal advantages of pump, treat, and reinjection are (1) relatively low overall project cost, (2) control of contaminants on the site, and (3) minimization of site disturbance. In most cases the project capital cost is relatively low because the equipment required is inexpensive and the depth of the contaminated zone is relatively shallow. Pump and treat require minimal maintenance and operational costs are typically low. Responsibility for contaminated material can be assumed to continue indefinitely. Material moved off-site remains, in part, the responsibility of the original owner. Therefore, an advantage exists in maintaining control of hazardous materials for which one is responsible. Most contaminated sites are located in areas where construction, removal of soil, and other remediation activities are a major inconvenience. Commercial and industrial activities may need to be shut down during on-site or off-site remediation. In situ remediation can usually be accomplished without shutting down commercial or industrial operations and with minimal disturbance during operation.

Disadvantages of pump, treat, and reinjection include (1) possible spreading of the contaminant plume during remediation, (2) difficulty in controlling the reinjection process, (3) difficulty in controlling the flow patterns in the aquifer due to inhomogeneity of the material, (4) biological fouling of the saturated zone, and (5) long treatment times.

Air Sparging

Bioremediation of aquifers is nearly always an oxygen-limited process. Air sparging of wells and aquifers is a method of adding oxygen to the contaminated zone. Two approaches have been used: forcing air into the aquifer (in situ aquifer sparging, or IAS), as shown in Figure 7.6*a;* and raising the water surface in the well by air-lift action, which causes mounding of the water table and circulation of aerated water near the well (in-well aeration, or IWA) as shown in Figure 7.6*b.* In IAS systems channelization is believed to be an important factor (Hinchee, 1994) and mass balances on oxygen and the contaminants have yet to be closed. Field tests of IAS systems have demonstrated that oxygen concentrations rise in the vicinity of the injection well (Brown et al., 1994). Movement of volatile contaminants through stripping can be expected and appropriate monitoring systems need to be installed to determine the fate of VOCs.

As in other in situ bioremediation systems, nutrient addition may be required with air sparging. Injection of ammonia with the air is a possible method of providing nitrogen. In cases where the contaminated aquifer is shallow and soils are sandy, addition of nutrients by percolation may be feasible.

Experience with air sparging is limited but reported results have been promising. A number of apparently successful remediations have been conducted (Brown et al., 1994; Billings et al., 1994). Demonstrating that bioremediation has occurred is difficult in soils and aquifers because of the heterogeneity of the subsurface. Decreases in contaminant concentration at monitoring and recovery wells are a good sign but may also result from nonequilibrium conditions or spreading of the contaminated zone. Increases in CO_2 concentrations and changes in the contaminant mixture consistent with biodegradation (e.g., BTEX compounds are usually biodegraded rapidly and a decrease in their concentration relative to TPH would support the assumption of biodegradation) are also a good sign. Core samples need to be analyzed to determine if sorbed organic concentrations are decreasing. Microbial counts would also be expected to rise above the initial condition during bioremediation.

Emission of VOCs in air sparging

Volatile compounds can be expected to move into the gas phase during air-sparging operations. In a number of cases IAS systems have been combined with soil vapor extraction to contain VOCs (Johnson et al., 1993; Billings et al.,

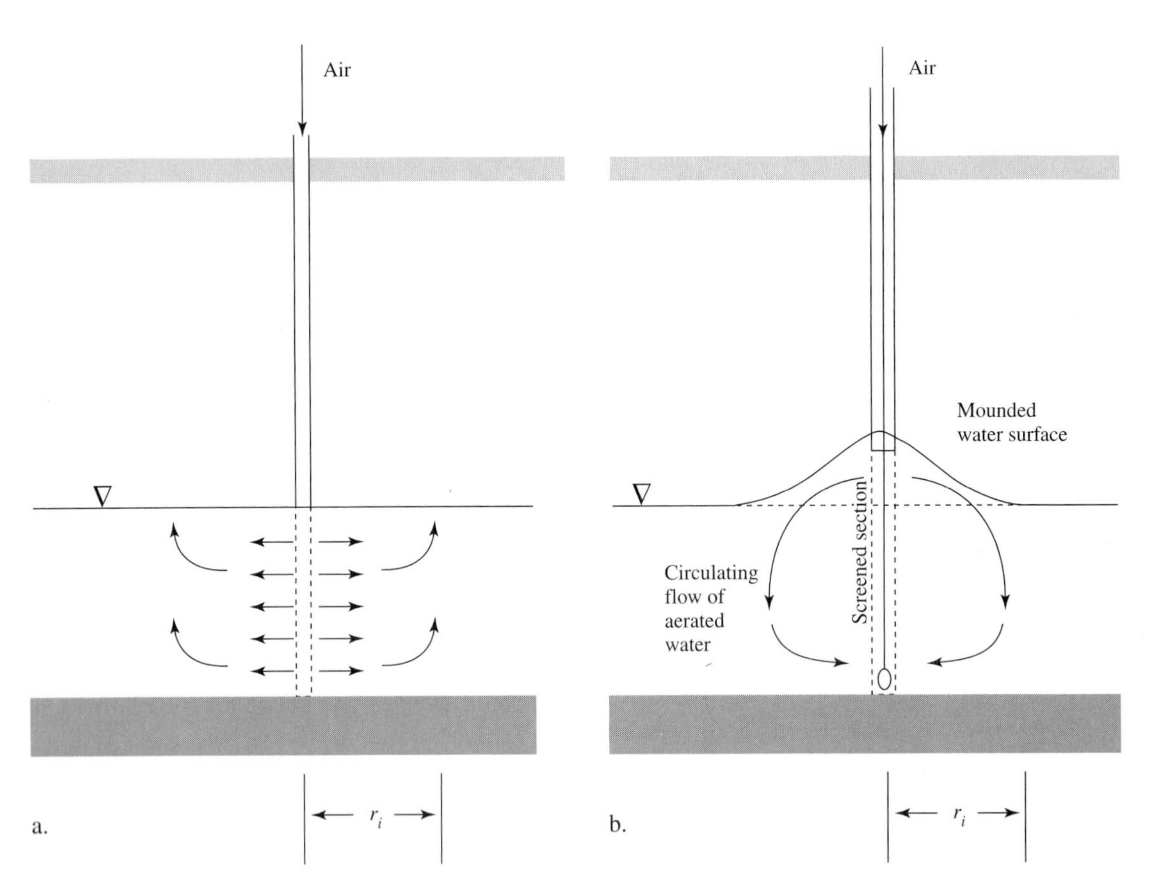

FIGURE 7.6
Schematic diagrams of (*a*) in situ aquifer sparging and (*b*) in-well aeration systems. Some mounding may occur with in situ aquifer sparging, but the general pattern is for injected air to follow large pores and crevices. Aeration occurs by interphase transfer and diffusion within the aquifer. In-well aeration results in transport of aerated water from the well into the mounded zone. The aerated water then circulates outward and downward with the pressure gradient. r_i is the radius of influence.

1994). In-well sparging will result in emission of VOCs, also, and off-gases may require treatment.

Design principles

Air-sparging systems are too new to have established design criteria. Wells may be vertical or horizontal. Screening should begin below the lowest point of contamination. The driving pressure required can be calculated from the hydrostatic pressure at the point of injection.

$$P_{\text{air}} = \rho_w g H \tag{7.3}$$

where P_{air} = air pressure required to empty well, Pa
ρ_w = density of water, kg/m^3
g = gravitational constant, 9.81 m/s^2
H = distance from the top of the screened section of the well to the water table, m

Wells are screened a distance below the water table to enhance contact between injected air and the water within the aquifer. Most applications of IAS have been with shallow aquifers and screen placement has begun at relatively shallow depths (1 to 15 m) below the water table. Well diameters are typically 10 to 50 mm and PVC casings are commonly utilized. Air-injection flow rates are typically 0.05 to 0.2 m^3/min. Wells used for IAS need not be vertical and examples exist where horizontal wells have been used as shown in Figure 7.7.

In general, the expected radius of influence for an IAS well is a few meters because of channeling of the air. Thus IAS will be most appropriate for small contamination sites. Maximum liquid-phase O_2 concentrations will be limited by the gas-phase concentration. Use of pure oxygen would be prohibitive in most situations owing to losses and air will be used in most cases. Saturation concentrations of oxygen in highly contaminated water would be lower than in clean water as a result of biological activity, decreasing mass transfer rates and the potential radius of influence. For these reasons, air sparging should be pilot tested before initiation of a large-scale operation.

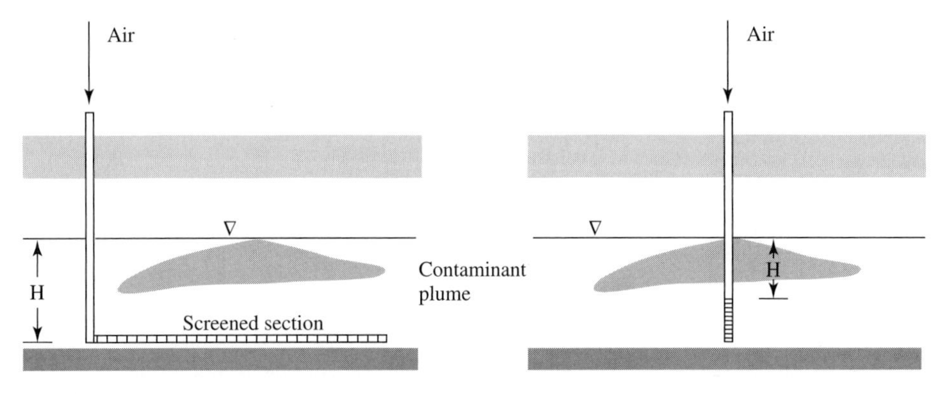

FIGURE 7.7
Horizontal and vertical well IAS systems.

Methods of in situ bioremediation of soils in the vadose zone can be divided into two types: treatment of VOCs and treatment of semi- and nonvolatile compounds. Remediation of soils contaminated with VOCs is simplest and the use of bioremediation is indirect. The most common approach to remediation of vadose zone soils contaminated with VOCs is soil vapor extraction (SVE), a process in which suction wells are used to draw air through the contaminated zone and the resulting contaminated gas is treated in some manner. Current methods of contaminated gas treatment include sorption on GAC, catalytic oxidation, and biodegradation in a biofilter or biostripper (see Chapter 10). Because SVE is such a widespread practice and because some bioremediation takes place in the process, conceptual development and process design are discussed here. Semi- and nonvolatile contaminants can be treated by bioventing, a process in which air and nutrients are added in a manner that biodegradation occurs at the site of contamination.

Soil Vapor Extraction

Soil vapor extraction is a process in which VOCs are removed from the vadose zone by installing wells and applying a negative pressure, as indicated in Figure 7.8. A single contaminated site (e.g., a gasoline service station) may require installation of a number of wells, as indicated in Figure 7.9. Most soil vapor extraction systems are relatively shallow (less than 10 m depth) and individual extraction

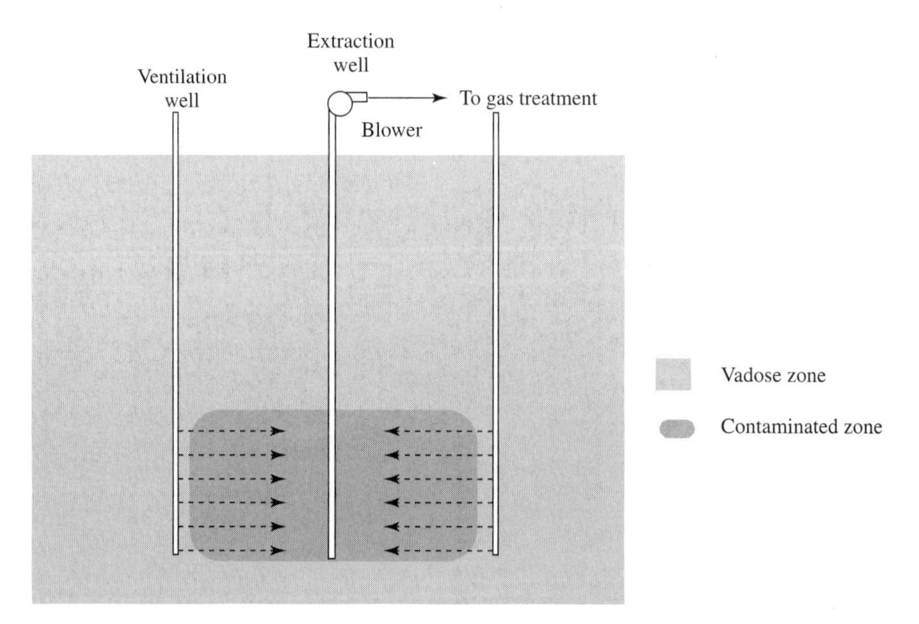

FIGURE 7.8
Schematic diagram of typical soil vapor extraction operation. Ventilation wells also serve as monitoring wells. The radius of influence of the extraction well depends on the soil characteristics but is normally less than 10 m. Gas flow rates are usually less than 5 m^3/min.

wells typically have radii of influence of 5 to 10 m. Vacuums of 0.1 to 0.2 atm (75 to 150 mmHg or 1 to 2 m H_2O) are typically applied and extraction flow rates generally are between 1 and 6 m^3/min. Vacuum as a function of distance from the extraction well is a function of the flow rate and the nature of the soil make up.

Volatile contaminants in the vadose zone are partitioned among all three phases and extraction rates should be developed that account for the limiting transfer rate. In most cases desorption from the solid phase will be rate-limiting. Systems may be operated continuously at a low rate or intermittently at higher rates. In either case maintaining high extract contaminant concentrations is desirable to (1) minimize the total extraction cost and (2) maximize gas-treatment efficiency. Application of a high vacuum and extraction of air at a high rate will result in a rapid decrease in contaminant concentration.

Distribution of VOCs between phases can be estimated if the exact mixture of contaminants is known, and the partition coefficients between the soil and water and the soil and air are available, as well as the Henry's law coefficients for each of the materials. However, extraction operations will result in nonequilibrium conditions, and an empirical approach utilizing the steps below is appropriate:

- Determine nature of contaminants and target compounds from site records.
- Estimate quantities of sorbed VOCs from core samples.
- Estimate quantities of dissolved VOCs from soil moisture measurements, liquid extract analysis, and solubility data.
- Estimate quantities of vapor-phase VOCs using vapor samples.
- Conduct pilot SVE tests to determine well flow characteristics and suitable extraction rates.
- Design system based on pilot studies.

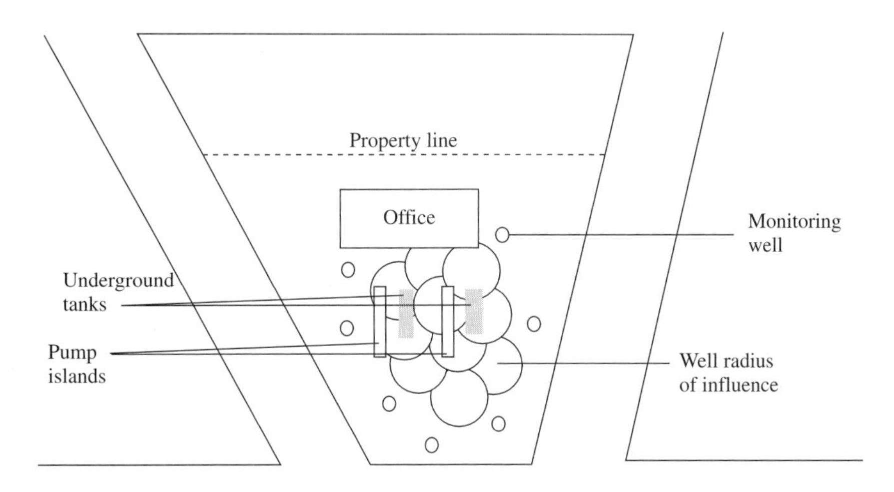

FIGURE 7.9
Sketch of soil vapor extraction and monitoring well system at gasoline service station site. Monitoring wells can serve as ventilation wells during operation. The radii of influence of the extraction wells overlap to ensure coverage of the entire contaminated zone. Wells should be connected such that extraction can be individually controlled.

EXAMPLE 7.3. SOIL VAPOR EXTRACTION. A leaking underground storage tank has discharged 10 m^3 of 1,1,1-TCA to a sandy clay soil. The contaminated soil is found to extend over a diameter of approximately 30 m and a depth of 10 m, beginning 2 m below the ground surface as shown in Figure 7.10. Data on the site and from pilot testing of a single extraction well located at the site is given below (Figure 7.11). A short-term SVE program was run to select the design extraction rate and the data are shown in Figure 7.12. A 3.0 m^3/min extraction rate was selected. Estimate the time necessary to remove 95 percent of the TCA.

Characteristics of 1,1,1-TCA

Vapor pressure = P_v = 100 mmHg at 20°C

$H = 0.56$ at 20°C

MW = 133.4 Solubility = 4,400 mg/L

$\rho_{TCA} = 1{,}325$ kg/m^3

$K_{OW} = 10^{-2.5}$

Soil characteristics

Wet bulk density = $\rho_{wb} = 2{,}090$ kg/m^3

$\Theta = 0.10$ m^3/m^3 $\phi = 0.25$

$K_{SD} = 10^{-5}$ m^3/g (for 1,1,1-TCA)

$T_{soil} = 20$°C

Solution

1. The contaminated zone is small and an assumption of uniform distribution of the TCA will be made.
2. Determine mass of TCA in contaminated zone.

$$M_{TCA} = \rho_{TCA} V_{TCA} = (1{,}325 \text{ kg/m}^3)(10 \text{ m}^3) = 13{,}250 \text{ kg}$$

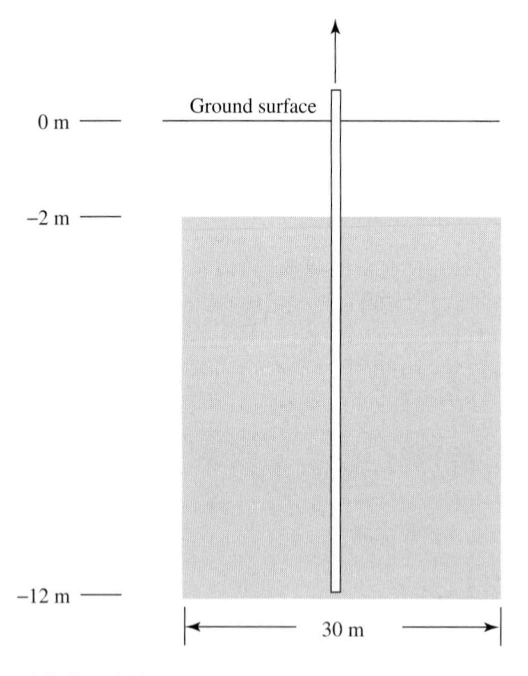

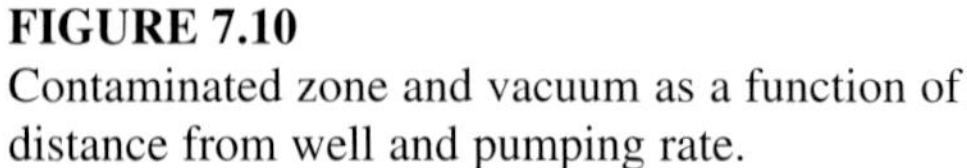

FIGURE 7.10
Contaminated zone and vacuum as a function of distance from well and pumping rate.

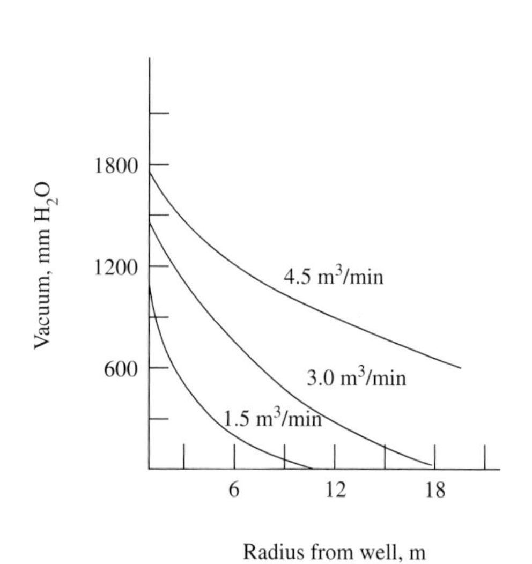

FIGURE 7.11
Vacuum applied and radii of influence at different airflow rates.

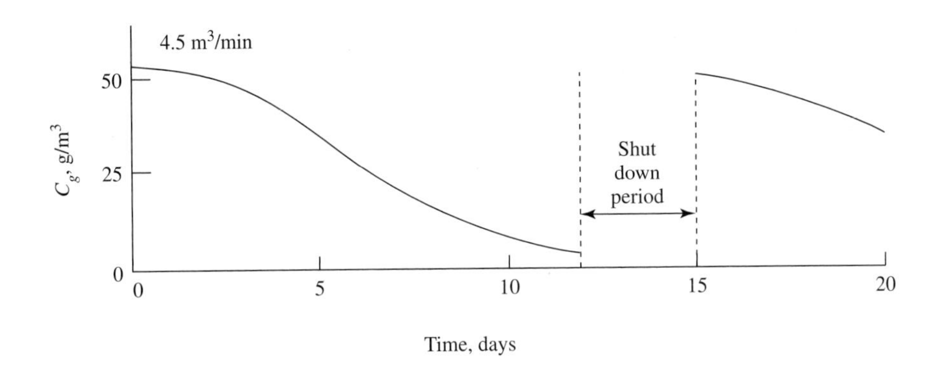

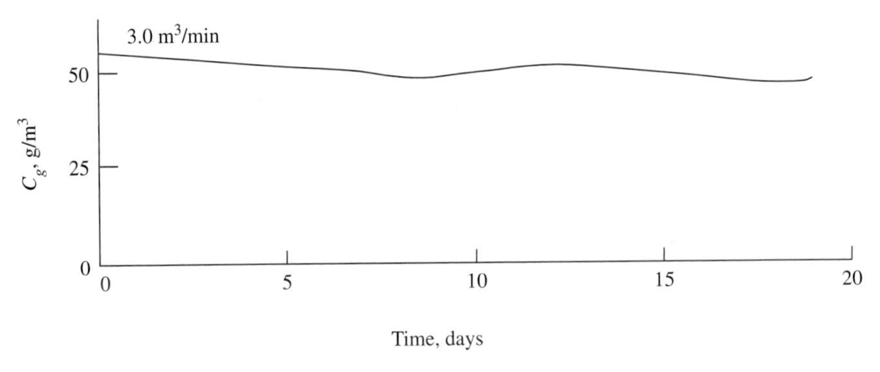

FIGURE 7.12
Results of short-term SVE tests: Change in gas-phase contaminant concentration over time.

3. Determine the dry bulk density of the soil ρ_b.

$$\rho_{wb}V = \rho_b V + \Theta\, \rho_w V + (\phi - \Theta)\rho_{\text{air}} V$$

$$2{,}090 \text{ kg/m}^3 = \rho_b + 0.1\ (1{,}000 \text{ kg/m}^3) + 0.15\ (1.28 \text{ kg/m}^3)$$

$$\rho_b = 1{,}990 \text{ kg/m}^3$$

4. Estimate initial distribution of TCA between phases

$$V = \frac{\pi}{4}(30 \text{ m})^2(10 \text{ m}) = 7{,}069 \text{ m}^3$$

$$M_{\text{TCA}} = M_{\text{soil}} + M_{\text{water}} + M_{\text{gas}}$$

$$= s\rho_{b\text{ soil}} V + \Theta V C_e + (\phi - \Theta) V H C_e$$

$$C_e = \frac{s}{K_{SD}}$$

$$s = \frac{M_{\text{TCA}}}{\rho_{b\,\text{soil}} V + \Theta V/K_{SD} + (\phi - \Theta)VH/K_{SD}}$$

$$= \frac{13{,}250 \text{ kg}}{(7{,}069 \text{ m}^3)[(1{,}990 \text{ kg/m}^3) + 0.1/(0.01 \text{ m}^3/\text{kg}) + (\phi - \Theta)H/(0.01 \text{ m}^3/\text{kg})]}$$

$$= \frac{13{,}250 \text{ kg}}{7{,}069 \text{ m}^3[(1{,}990 \text{ kg/m}^3) + 10 \text{ kg/m}^3 + 8.4 \text{ kg/m}^3]}$$

$$= 9.33 \times 10^{-4} \text{ kg/kg}$$

$$M_{\text{soil}} = (9.33 \times 10^{-4} \text{ kg/kg})(1{,}990 \text{ kg/m}^3)(7{,}069 \text{ m}^3) = 13{,}129 \text{ kg}$$

$$M_{\text{water}} = \frac{\Theta V s}{K_{SD}} = \frac{0.1(7{,}069 \text{ m}^3)(9.33 \times 10^{-4} \text{ kg/kg})}{0.01 \text{ m}^3/\text{kg}} = 66 \text{ kg}$$

$$M_{\text{gas}} = (\phi - \Theta)VHC_e = (\phi - \Theta)VH\frac{s}{K_{SD}}$$

$$= 0.15(7{,}069 \text{ m}^3)(0.56)\frac{(9.33 \times 10^{-4} \text{ kg/kg})}{0.01 \text{ m}^3/\text{kg}} = 55 \text{ kg}$$

5. Estimate gas-phase concentration of TCA

$$C_g = \frac{55 \text{ kg}}{0.15(7{,}069 \text{ m}^3)} = 0.053 \text{ kg/m}^3 = 53 \text{ g/m}^3$$

Note that the saturation concentration in the gas phase, based on the vapor pressure, is

$$C_{gS} = \frac{MW\ P_v}{RT}$$

$$= \frac{(133.4 \text{ g/mol})(100 \text{ mmHg})/(760 \text{ mmHg/atm})}{(0.08205 \text{ L}\cdot\text{atm/mol}\cdot\text{K})(293 \text{ K})} = 0.73 \text{ g/L} = 730 \text{ g/m}^3$$

The mass transfer rate from the free product to the vapor phase will be high owing to the large difference between saturation and actual vapor-phase TCA concentrations.

6. Estimate the time necessary to remove 95 percent of the TCA.

$$\frac{dM_{\text{soil}}}{dt} = -C_g Q_g = -H\frac{s}{K_{SD}} Q_g = -H\frac{M_{\text{soil}}}{\rho_b V K_{SD}} Q_g$$

$$\frac{dM_{\text{soil}}}{M_{\text{soil}}} = -\frac{HQ_g}{\rho_b V K_{SD}} dt$$

Integrating between 0 and t and M_o and M_{soil} gives

$$\ln\left(\frac{M_{\text{soil}}}{M_o}\right) = -\frac{HQ_g}{\rho_b V K_{SD}} t$$

$$t_{95} = \frac{\rho_b V K_{SD} \ln(20)}{HQ_g}$$

$$= \frac{(1{,}990 \text{ kg/m}^3)(7{,}069 \text{ m}^3)(10^{-2} \text{ m}^3/\text{kg})\ln(20)}{0.56(3 \text{ m}^3/\text{min})} = 2.51 \times 10^5 \text{ min}$$

$$= 174 \text{ days}$$

Note that the estimate is based on the assumption that phase equilibrium will continue to exist throughout the SVE process if a constant gas pumping rate of 3 m^3/min is applied. A second significant assumption is homogeneity throughout the contaminated zone. Short circuiting, changes in the K_{SD} value due to drying, temperature change, or presence of liquid TCA, and failure of the short-term test to model actual equilibrium conditions would result in a longer time period to achieve the desired removal. Short circuiting is a particular source of error in tight formations, such as clays, where mass transfer limitations develop. Finally, doubt always exists about the actual amount of contaminant in the soil.

An assumption was made in Example 7.3 that liquid TCA did not exist in the contaminated zone. The nonhomogeneous distribution of contaminants that nearly always occurs often takes the form of fingers of contaminant flowing from the source (the leak in this case). Pores in the finger may be filled with free product, i.e., liquid TCA. In such a case application of the equilibrium relationships defined by the soil-water distribution coefficient K_{SD} and Henry's law give poor estimates of the actual phase distribution of contaminants.

Advantages and disadvantages

Soil vapor extraction is a excellent process selection for cases where volatile compounds are the principal contaminants. Mixtures of contaminants, such as gasoline, are very effectively treated as well as single constituents. Movement of air through soil is much easier than through water because of the large difference in viscosity and diffusivity, and for this reason application of SVE to relatively tight soils is possible. As stated above, SVE is a remediation process, not a treatment process. In most cases the extracted contaminants cannot be discharged to the atmosphere and some method of gas treatment must be used. In cases where the contaminants are biodegradable (e.g., H_2S, NH_3, petroleum VOCs, methylene chloride, styrene) treatment by biofiltration is possible (see Chapter 10). For less easily degraded materials (e.g., TCE, PCE, TCA) catalytic oxidation or storage on GAC may be necessary. Because the extraction wells are open systems, the effective radius of influence is small. In cases where contamination extends to considerable depths the cost of drilling a large number of closely spaced wells may become prohibitive.

Bioventing

Bioventing is used for the treatment of less volatile biodegradable contaminants (e.g., contaminants having a dimensionless Henry's law coefficient < 0.1) in the vadose zone and is an alternative to excavation and ex situ soil treatment. In bioventing, transformation and degradation of the contaminants in the vadose zone are carried out at the point of contamination. Process configurations are similar to those used for soil vapor extraction but the airflow rate is limited to that necessary for biodegradation to occur (i.e., oxygen supply). In keeping with the point of contaminant treatment concept, one objective in bioventing is to minimize contaminant migration.

Common limitations on effective application of bioventing include lack of nutrients in the subsurface, low moisture content of the soil, and difficulty in achieving airflow through the contaminated zone. Addition of nutrients and moisture is

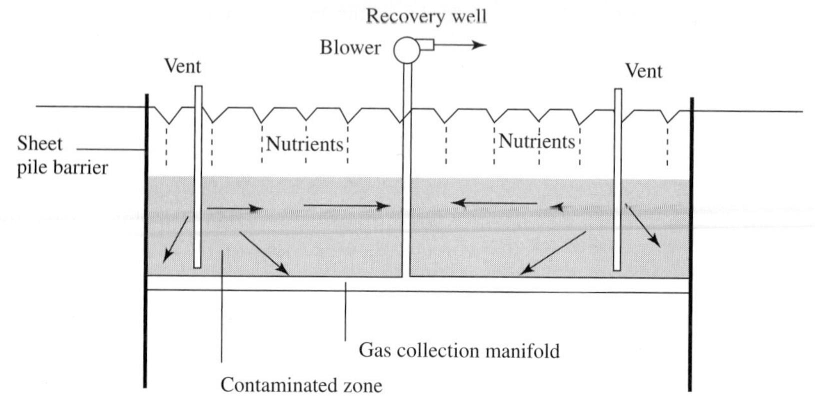

FIGURE 7.13
Schematic of a bioventing system for a shallow contaminated zone. Sheet pile is used to isolate the contaminants. A vapor-collection manifold may be installed, in addition to the recovery well, to collect the gas drawn through the contaminated zone.

normally accomplished by infiltration from trenches, as suggested in Figure 7.13. Maintaining moisture content near the field capacity and nutrient concentrations sufficient to support an active microbial population is a significant problem. Monitoring off-gas O_2 and CO_2 concentrations provides an indication of the level of biological activity on a day-to-day basis. When biodegradation is progressing, off-gas O_2 concentrations should be depressed and CO_2 concentrations should be elevated. However, regular soil sampling is required, also. Forced ventilation, in which air is blown into the contaminated zone and air recovery is passive, is often used in bioventing systems, but preferred aeration strategy has not evolved in the limited experience with bioventing systems. However, physical characteristics of the soil, depth of the contaminated zone, and potential to transport contaminants out of the zone will be factors in selection of the method of aeration.

Site characterization

As has been stated above, characterization of the contaminated site is the single most important factor in remediation. Knowledge of the type, concentration, and location of the contaminants is essential in setting up a remediation program. Site characterization includes development of data on soil characteristics, moisture contents, nutrient availability, and hydraulic conductivity of the soil. Nutrient availability can be estimated by standard tests used for agricultural soils (e.g., available nitrogen) and by setting up laboratory bioremediation studies using soil from the contaminated site.

Bioventing sites are usually relatively small in area and shallow in depth. Installation of barriers to guide flow, such as sheet pile, and extensive monitoring, sampling, and venting systems are often economically feasible.

System design

The principal design factors in bioventing are to ensure that oxygen, moisture, and nutrients are provided throughout the contaminated zone. Oxygen is supplied by blowing or drawing air through the contaminated zone. Because of cracks, fissures, and nonhomogeneous soil characteristics, ideal flow should not be expected. Airflow monitoring with tracer gases may be helpful in setting up a system of vent and recovery wells to ensure oxygen availability throughout the contaminated zone.

Addition of moisture and nutrients to the vadose zone is difficult because soil permeabilities often differ over small distances and simply adding nutrient solutions at the soil surface does not ensure homogeneous distribution in the contaminated zone. Saturating the soil will result in loss of ability to oxygenate the contaminated zone. A periodic application of nutrient solutions can be used. Moisture sensors should be installed to determine if the application rate, duration, and interval are appropriate. Ridge and furrow irrigation, trenches, or spray irrigation can be used to supply moisture and nutrients. Airflow may need to be stopped during this period because of increased head losses.

Emission of toxic VOCs

Bioventing is applied where contaminants have low volatility. However, some emissions can be expected and vented gases should be either recycled or treated at the surface. Biofiltration of off-gases is an economical approach that includes biodegradation (see Chapter 10). Sorption onto GAC and combustion are common alternatives to biofiltration. Production of toxic VOCs is not likely in aerobic systems because most biodegradation products are less volatile than the parent compounds. However, anoxic conditions may exist in bioventing systems (see Example 7.4) and anaerobic degradation products may be quite toxic. An example is the production of vinyl chloride as a result of anaerobic transformations of TCE.

EXAMPLE 7.4. BIOVENTING. A 4,000-m^3 diesel-contaminated zone has been discovered at a former bus depot site. The contaminated zone is approximately 20 m × 20 m square and 10 m deep beginning at a depth of 4 m. Based on a sampling program, the estimated mass of diesel oil present is approximately 80,000 kg ($\approx$ 100 m^3). The soil is a silty sand. Site layout and results of air extraction versus vacuum for a test well are given in Figure 7.14. Data from laboratory evaluation of diesel biodegradation in core samples is given in Table 7.1. Assuming that nutrient limitations and moisture content will not be factors and that the air extraction rate will be 0.4 m^3/min, estimate the (1) time required to degrade and/or remove 90 percent of the TPH and (2) the extraction well head O_2 and CO_2 concentrations at 10-day intervals. From laboratory measurements of oxygen consumption and of CO_2 production resulting from diesel biodegradation, values of 3.3 g O_2/g TPH and 3 g CO_2/g TPH may be used. Soil bulk density is 2,100 kg/m^3.

Solution

1. For this analysis, ignore the overlap between the radii of influence and assume that flow is uniform and approximately horizontal. Each of the four wells will be assumed to be independent.

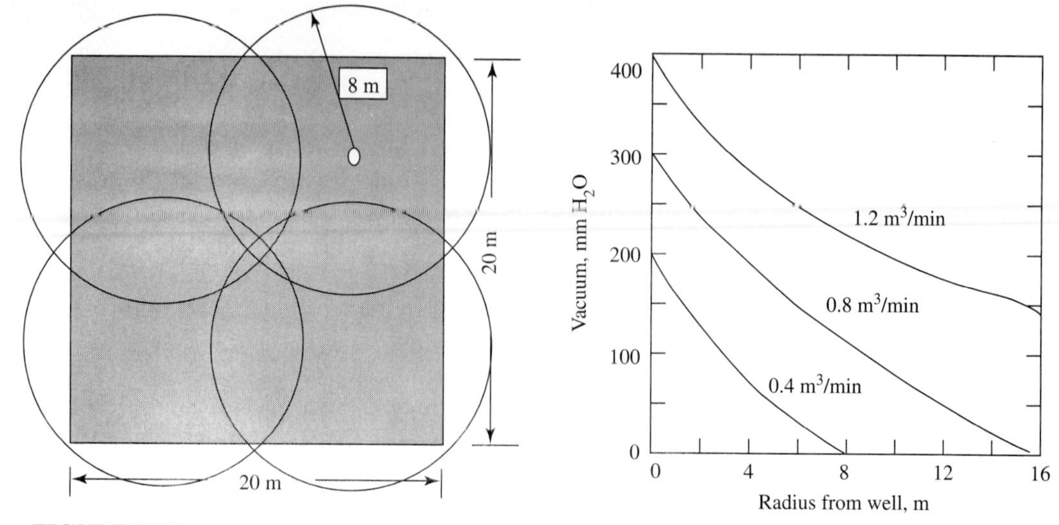

FIGURE 7.14
Site layout of diesel-contaminated zone with radius of four 0.4 m³/min wells inscribed and airflow versus head loss data for Example 7.4.

TABLE 7.1
Biodegradation of diesel in core samples

Time, days	TPH, mg/g
0	10.0
5	9.4
10	9.2
15	8.9
20	8.2
30	7.9
40	7.6
50	6.3
75	5.6
100	4.4

2. From Figure 7.15, the rate of degradation can be described as a first-order decay.

$$\frac{dC_{\mathrm{TPH}}}{dt} = r_{\mathrm{TPH}} = -kC_{\mathrm{TPH}}$$

$$C_{\mathrm{TPH}} = C_{\mathrm{TPH}_o} e^{-kt}$$

where r_{TPH} = rate of TPH biodegradation, g/kg · day
k = rate coefficient per day = 0.008 per day
C_{TPH} = diesel concentration, g TPH/kg soil

3. Estimate the initial TPH value C_{TPH_o} assuming a homogeneous diesel distribution.

$$C_{\mathrm{TPH}_o} = \frac{80{,}000 \text{ kg}}{(4{,}000 \text{ m}^3)(2{,}100 \text{ kg/m}^3)} = 0.0095 \text{ kg/kg} = 9.5 \text{ g/kg}$$

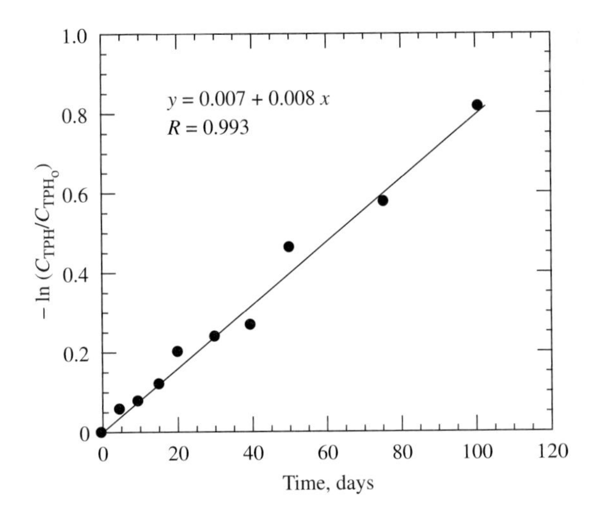

FIGURE 7.15
Plot of ln TPH fraction remaining as a function of time for Example 7.4.

4. Estimate initial oxygen uptake rate for single- and four-well systems.

$$r_{O_2} = (3.3 \text{ g } O_2/\text{g TPH})(r_{TPH})(\text{mass of soil})$$

Single well:

$$r_{O_2} = (3.3)(4{,}000 \text{ m}^3)(2{,}100 \text{ kg/m}^3)(-0.008 \text{ per day})(9.5 \text{ g/kg})$$
$$= -2.1 \times 10^6 \text{ g/day}$$

Four-well system (shown above)

$$r_{O_2} = -5.3 \times 10^5 \text{ g/day}$$

5. Determine oxygen supply rate M_{O_2}, g/day

$$M_{O_2} = \frac{Q_g f_{O_2} MW_{O_2}}{V_m}$$

where Q_g = volumetric gas flow rate, m^3/day
f_{O_2} = mole fraction of O_2 in air, 0.21
MW_{O_2} = molecular weight of O_2, 32 g/mole
V_m = molar volume of a gas, 0.0224 m^3/mole

Q_g	M_{O_2}
m^3/min	g/day
0.4	1.7×10^5
0.8	3.4×10^5
1.2	5.2×10^5

Oxygen will be rate-limiting unless (*a*) a substantially higher airflow rate is used or (*b*) four wells are used at a pumping rate of 1.2 m^3/min. An economic analysis would be required to determine the optimal alternative. Two scenarios will be considered in determining the time required for treatment, oxygen-limiting and TPH-limiting.

6. Scenario I: oxygen-limiting, $Q_g = 0.4\ m^3/min$.
   a. Determine the TPH concentration at which oxygen is no longer rate-limiting. From the stoichiometry given, 3.3 g O_2 required per g TPH removed, the maximum TPH removal rate is

$$r_{TPH_{max}} = -\frac{M_{O_2}}{3.3V\rho_b} = -\frac{1.7\times10^5\ \text{g/day}}{3.3(1{,}000\ \text{m}^3)(2{,}100\ \text{kg/m}^3)}$$

$$= -2.5\times10^{-2}\ \text{g TPH/kg}\cdot\text{day} = -(0.008\ \text{per day})C_{TPH_{max}}$$

$$C_{TPH_{max}} = \frac{0.025}{0.008} = 3.1\ \text{g TPH/kg soil}$$

   b. Estimate time required to lower C_{TPH} to 3.1 g/kg.

$$t = \frac{\Delta C_{TPH}(3.3\ \text{g O}_2/\text{g TPH})(V_{soil})\rho_b}{M_{O_2}}$$

$$= \frac{(9.5-3.1)(3.3)(1{,}000)(2{,}100)}{1.7\times10^5} = 261\ \text{days}$$

   c. Estimate additional time required to achieve 90 percent removal. At 90 percent removal, $C_{TPH} = 0.95$ g/kg

$$\Delta t = -\frac{1}{k}\ln\left(\frac{0.95}{3.1}\right) = 148\ \text{days}$$

   d. Total time = 409 days
7. Scenario II: TPH-limiting, $Q_g = 1.2\ m^3/min$ with four wells.

$$t = -\frac{1}{k}\ln\left(\frac{0.95}{9.5}\right) = 288\ \text{days}$$

   The time difference is relatively small, and a flow rate of 0.4 m^3/min will be selected.
8. Oxygen and CO_2 concentrations in off-gas.
   a. Oxygen concentration will be very low until TPH becomes limiting at 261 days.
   b. CO_2 production rate will be constant until TPH from $t = 0$ to $t = 261$ days. Determine the concentration of oxygen in inlet air.

$$C_{O_2i} = \frac{0.21\ \text{L O}_2}{\text{L air}}\ \frac{1\ \text{mole O}_2}{0.0224\ \text{m}^3}\ 32\ \text{g O}_2/\text{mole} = 300\ \text{g/m}^3$$

   If oxygen is completely used, then CO_2 concentration in outlet gas is:

$$C_{CO_2} = \frac{3\ \text{g CO}_2/\text{g TPH}}{3.3\ \text{g O}_2/\text{g TPH}}(300\ \text{g/m}^3) = 273\ \text{g/m}^3$$

c. Determine C_{O_2} in outlet gas as a function of time for $t > 261$ days. Assume pseudo-steady-state conditions—(reactions are slow relative to time scale).

$$V\frac{dC_{O_2}}{dt} = Q_g(C_{O_2i} - C_{O_2}) + \rho_b V_{soil} r_{O_2} = 0$$

$$r_{O_2} = (3.3 \text{ g } O_2/\text{g TPH}) r_{TPH}$$

$$= -3.3\, k\, C_{TPH} = -0.026\, C_{TPH}$$

$$C_{TPH} = C_{TPH_{t=261}} \exp[-0.008(t-261)]$$

$$C_{O_2} = C_{O_2i} - \frac{\rho_b V_{soil}}{Q_g}(0.026)(3.1 \text{ g/kg})\exp[-0.008(t-261)]$$

$$\cong 300\{1 - \exp[-0.008(t-261)]\}$$

d. Determine C_{CO_2} in outlet gas as a function of time for $t > 261$ days. Neglect inlet CO_2.

$$0 = -Q_g C_{CO_2} + \rho_b V_{soil} r_{CO_2}$$

$$r_{CO_2} = -(3 \text{ g } CO_2/\text{g TPH}) r_{TPH} \qquad \text{(Note: } CO_2 \text{ is being produced)}$$

$$= 0.024\, C_{TPH}$$

$$C_{CO_2} = \frac{\rho_b V_{soil}}{Q_g}(0.024)(3.1 \text{ g/kg}) \exp[-0.008(t-261)]$$

$$= 273 \exp[-0.008(t-261)]$$

The solutions are presented in Table 7.2 and Figure 7.16.

TABLE 7.2

Predicted soil TPH concentrations and off-gas O_2 and CO_2 concentrations for Example 7.4

Time, days	TPH, g/kg	O_2, g/m^3	CO_2, g/m^3
261	3.10	0	273
270	2.88	21	254
280	2.66	42	235
290	2.46	62	216
300	2.27	80	200
310	2.09	97	184
320	1.93	113	170
330	1.78	127	157
340	1.65	141	145
350	1.52	153	134
360	1.40	164	124
370	1.30	175	114
380	1.20	184	105
390	1.10	193	97
400	1.02	201	90
410	0.94	209	83

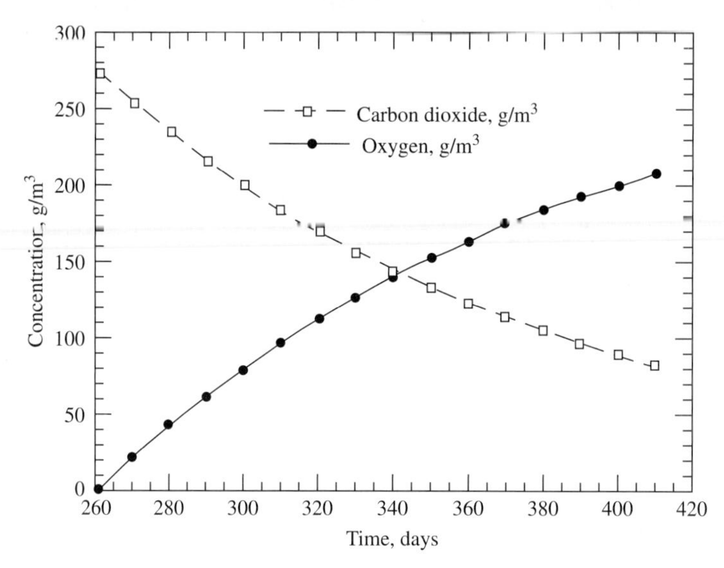

FIGURE 7.16
Off-gas O_2 and CO_2 concentrations as a function of time for Example 7.4.

Temperature control

Biodegradation rates in bioventing systems can be expected to be quite temperature-sensitive. One of the attractions of forced-air ventilation is that the compressed air will be warm and increase reaction rates in the soil. Soil warming by covering the soil with black plastic has been used with nominal success, as has active soil warming with buried heating tapes. Note that heating increases solubility and volatility of contaminants, and determination of an optimum temperature may include factors other than biological transformation rates.

PROBLEMS AND DISCUSSION QUESTIONS

7.1. Estimate the average contaminant concentration in a shallow aquifer contaminated with TCE based on the sampling results shown in Figure 7.17 and Table 7.3.

7.2. The aquifer of Problem 7.1 has an organic fraction f_{oc} of 0.006, a dry bulk density, ρ_b of 2,100 kg/m^3, and a porosity ϕ of 0.35. Estimate the total amount of TCE present.

7.3. A saturated soil zone has been contaminated with 50 L of gasoline (40 kg) from a leaking underground storage tank. The soil is fairly high in organic matter (5 percent by weight), has a porosity $\phi = 0.4$, bulk density $\rho_b = 1{,}500$ kg/m^3, and a hydraulic conductivity of 2×10^{-3} m/s. The K_{SD} of the soil is quite high (0.01 L/mg) and the gasoline is not very mobile. As shown in Figure 7.18 the saturated soil drains to a small pond 250 m from the point of contamination, with a loss in head of 10 m. The cross-sectional dimensions of the saturated zone are 2 m deep by 50 m wide.

a. If the contamination is confined to the first 50 m below the point of entry, determine the liquid- and solid-phase concentrations of the gasoline (assume that the concentrations are approximately constant at all points in each phase).

b. Determine the flow rate of the groundwater into the pond in m^3/day.

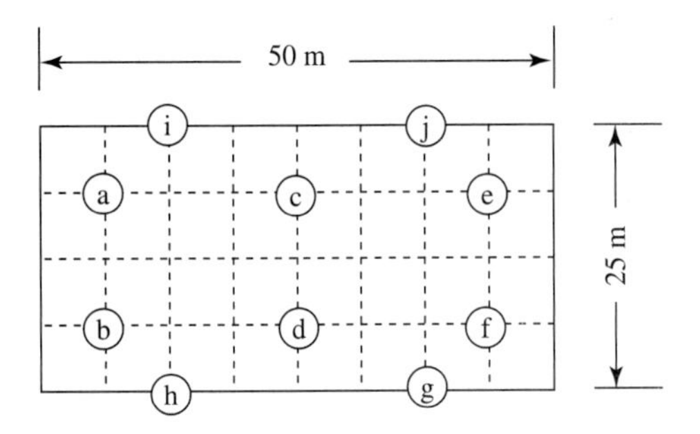

FIGURE 7.17
Sampling-site locations for Problem 7.1.

TABLE 7.3
Liquid-phase TCE concentrations in aquifer of Problem 7.1, μg/L

	Depth, m		
Location	**5**	**10**	**15**
a	20	50	75
b	10	30	65
c	180	300	350
d	340	650	700
e	60	90	110
f	30	55	70
g	3	5	7
h	15	25	30
i	1	3	6
j	0	2	3

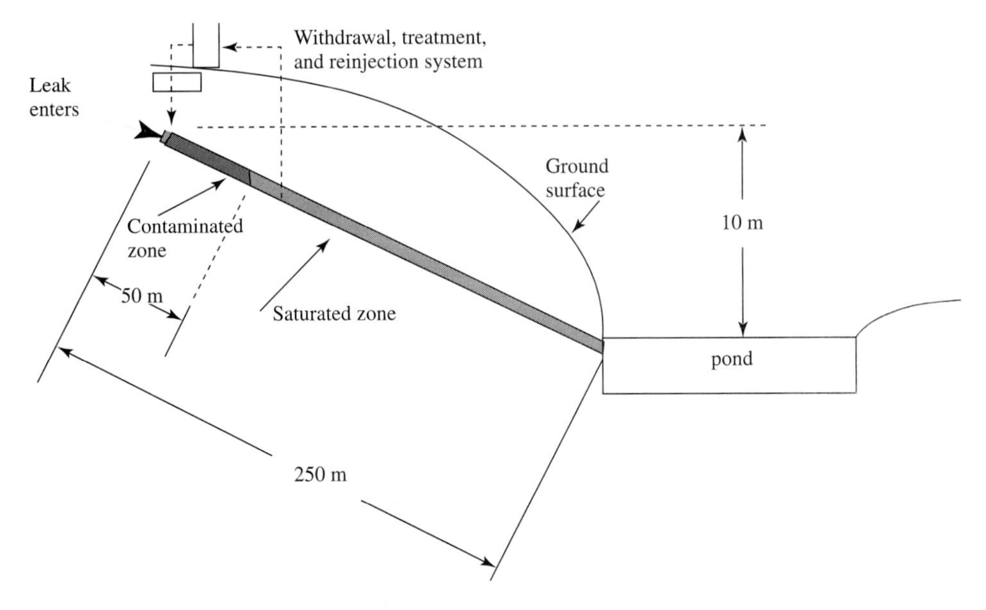

FIGURE 7.18
Definition sketch for Problem 7.3.

c. Assuming that the gasoline is nearly immobile because of the large K_{SD} value, determine the rate of gasoline degradation possible in kg/day if the flow was pumped from the soil downstream of the contaminated zone, treated, aerated by the addition of 100 mg/L of H_2O_2, and recirculated. Assume that degradation of gasoline requires 2.5 g O_2 per gram of gasoline and that the pumping rate will be equal to the flow rate calculated in (*b*).

7.4. The crosshatched portion of the aquifer shown in Figure 7.19 has been contaminated with 25,000 L of diesel (20,000 kg as TPH) from a leaking underground storage tank. The aquifer has a porosity $\phi = 0.3$, bulk density $\rho_b = 2{,}200$ kg/m^3, and a hydraulic conductivity K_C of 9×10^{-4} m/s. The K_{SD} of the soil and diesel is 10^{-5} L/mg and the diesel can be considered nonvolatile. Assume ideal radial flow and oxygen distribution and estimate the time required to reduce the total diesel mass to 1,000 kg for two cases:

a. No reaction, removal by pump and treat only, and a pumping rate of 0.15 m^3/min.

b. A first-order degradation of the diesel with a rate coefficient of 0.002 per day based on liquid-phase concentration. Total injection flow rate is equal to the extraction rate of 0.15 m^3/min, and the injection water will contain 25 mg O_2/L through addition of H_2O_2. The BOD_U of diesel is approximately 3 g/g.

7.5. An aquifer contaminated with isopropyl benzene (IPB) is to be treated in situ using a pump, treat, and reinject system. Experiments in laboratory-scale systems have resulted in the conclusion that biodegradation follows first-order kinetics based on the liquid concentration with a rate coefficient k of 0.05 per day. The aquifer volume contaminated is 50,000 m^3 and the shape is approximately cylindrical (80 m diameter and 10 m deep). Aquifer porosity ϕ is 0.25, the organic fraction f_{oc} is 0.005, and the bulk dry density ρ_b of the soil is 2,000 kg/m^3. Because the site is shallow, a decision has been made to use 8 extraction wells and 16 injection wells, as shown in Figure 7.20, in the belief that the system will behave approximately as an ideal stirred-tank reactor. The extraction rate will equal the reinjection rate at 500 m^3/day. The initial liquid-phase concentration of isopropyl benzene is 25 mg/L.

a. Write a mass balance equation using the homogeneous spatial concentration ($C_{\text{out}} = C_{\text{reactor}}$) stirred-tank assumption.

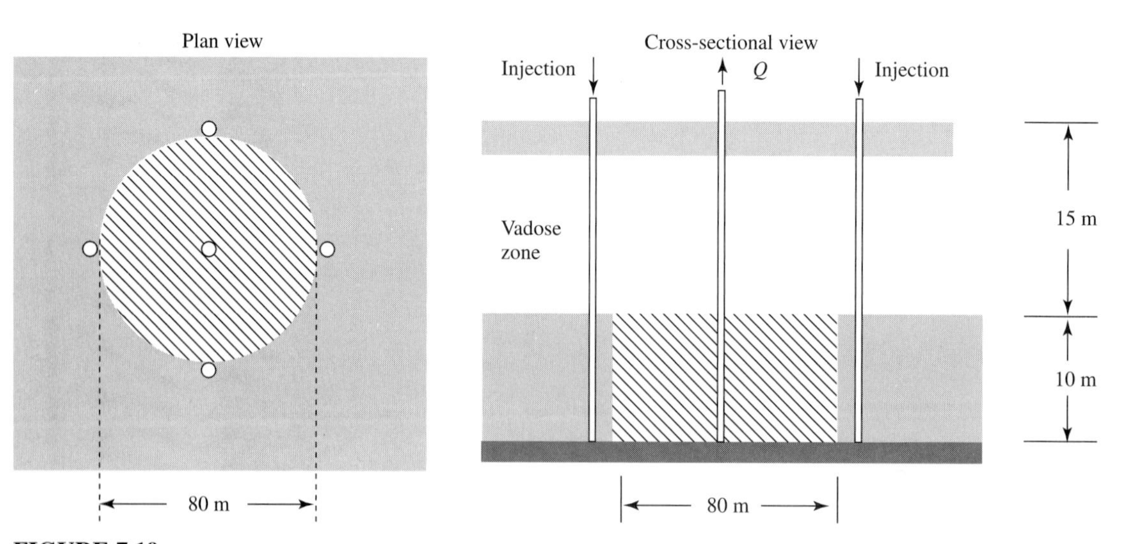

FIGURE 7.19
Definition sketch for Problem 7.4.

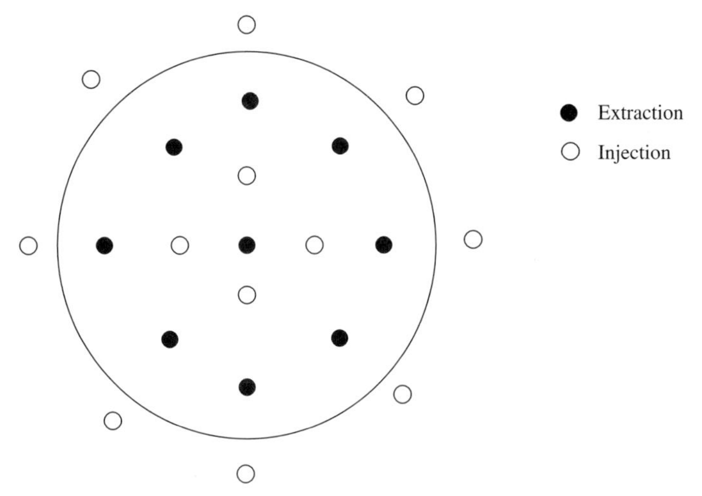

FIGURE 7.20
Extraction and injection well layout for Problem 7.5.

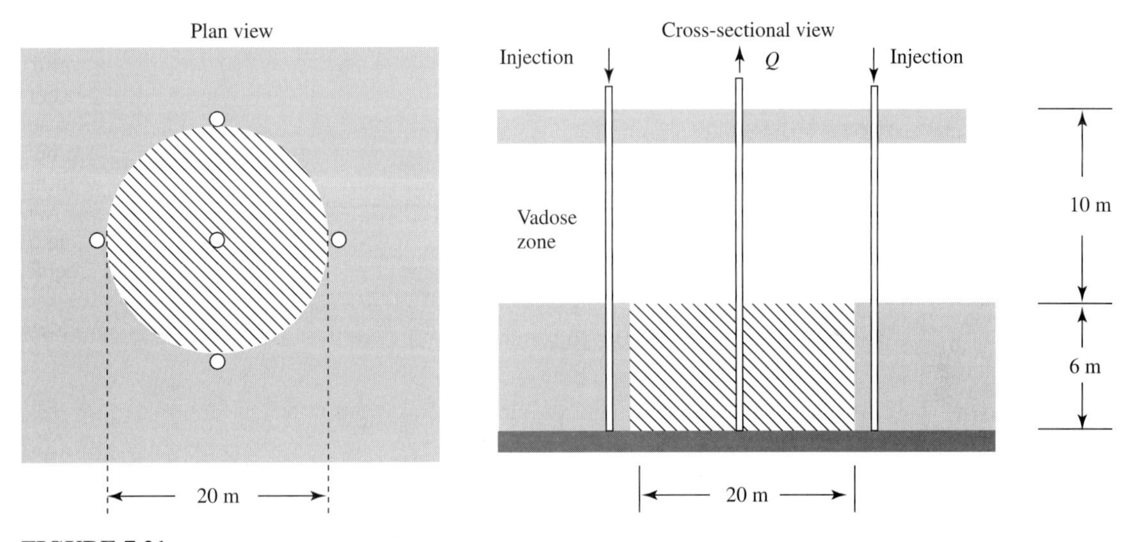

FIGURE 7.21
Definition sketch for Problem 7.6.

b. Determine the time required for the concentration to drop to 1 mg/L.

$MW = 120.19$ $\quad H = 0.42$ at 20°C $\quad \rho_{IPB} = 862$ kg/m^3

Solubility = 50 mg/L $\quad K_{OW} = 4{,}570$

7.6. The crosshatched portion of the aquifer shown in Figure 7.21 has been contaminated with gasoline. Measurements on core samples indicate a total concentration of 25 g/kg as TPH. The aquifer has a porosity $\phi = 0.35$, bulk density $\rho_b = 2{,}100$ kg/m^3, and a hydraulic conductivity K_C of 1.5×10^{-3} m/s. The K_{SD} of the soil and diesel is 0.0002 L/mg and the diesel can be considered nonvolatile. Assume ideal radial flow and oxygen distribution and estimate the time required to reduce the total diesel mass by 90 percent. Consider two cases:

a. No reaction, removal by pump and treat only, and a pumping rate of 0.15 m^3/min.
b. A first-order degradation of the diesel with a rate coefficient of 0.002 per day based on liquid-phase concentration. Total injection flow rate is equal to the extraction rate of 0.15 m^3/min, and the injection water will contain 25 mg O_2/L through addition of H_2O_2. The BOD_U of diesel is approximately 3 g/g.

7.7. For the contaminated aquifer of Problem 7.4 consider a situation where desorption rate can be described by

$$r_{\text{desorp}} = K_{ds}(C_{\text{equ}} - C_{\text{TPH}})$$

where r_{desorp} = rate of TPH desorption, kg/$m^3 \cdot$ day
K_{ds} = desorption rate coefficient, per day
C_{equ} = concentration defined by the soil distribution coefficient

Determine the effect of mass transfer rate limitations on the time for remediation if $K_{ds} = 10^{-5}$ per day. Note that two coupled ordinary differential equations will result (one for the soil and one for the water). Numerical integration will be necessary.

7.8. A groundwater aquifer has been contaminated with gasoline (the composition of gasoline can be found in Appendix F) and a proposal has been made to bioremediate the site using air sparging with gas recovery by vapor extraction. The vapor injection well has a horizontal screened section below the contaminated zone. As shown, the vapor extraction well will have twice the flow rate of the injection well. Assume that (1) the

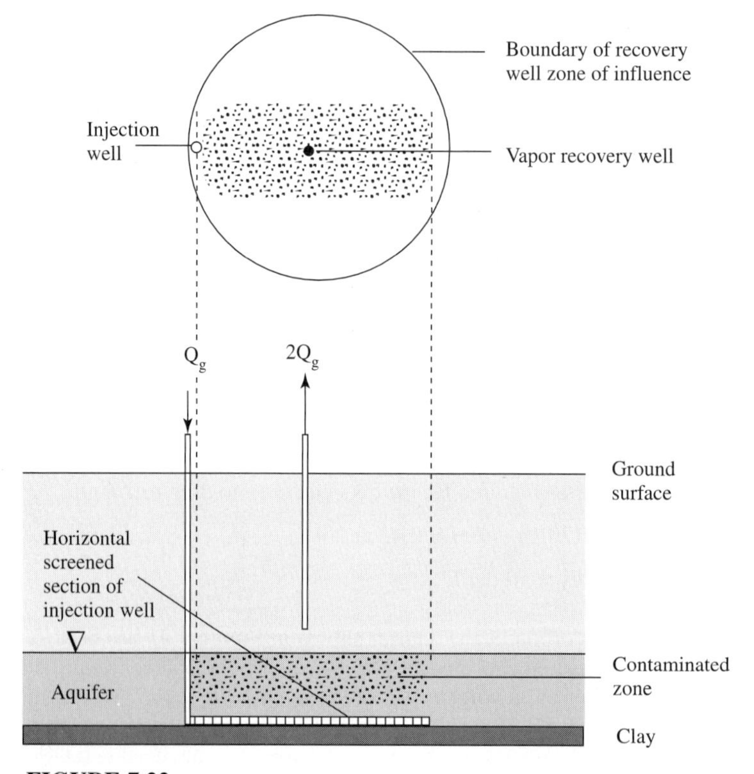

FIGURE 7.22
Definition sketch of contaminated aquifer and proposed air-sparging system for Problem 7.8.

groundwater is not moving and (2) nutrients are not limiting. A diagram of the proposed system is shown in Figure 7.22.

Using a relative scale, sketch

a. The ideal vapor extraction gas concentrations of carbon dioxide and oxygen as functions of time.

b. The probable vapor extraction gas concentrations of carbon dioxide and oxygen as a function of time.

c. The ideal vapor-phase total petroleum hydrocarbon (TPH) concentration as a function of time.

d. The probable vapor-phase total petroleum hydrocarbon (TPH) concentration as a function of time.

Note that air is approximately 21 percent O_2 and 0.03 percent CO_2.

7.9. A 10,000-m^3 soil zone contaminated with 200 mg/kg phenanthrene has been found near a chemical plant. The soil has a dry bulk density ρ_b of 2,000 kg/m^3, a porosity of 0.35, an organic fraction f_{oc} of 0.04, and a volumetric moisture content Θ of 0.15. Information on phenanthrene is given below.

$$\text{Empirical formula: } C_{14}H_{10}$$
$$\text{MW} = 178.2$$
$$\text{Solubility} = 1.0 \text{ mg/L}$$
$$\log K_{ow} = 4.46$$
$$H = 0.001$$

Laboratory experiments have been conducted on phenanthrene degradation in which moist air containing NH_3 was blown through small samples taken from the contaminated zone. Results of the experiments are given on the graph in Figure 7.23.

a. Is the site appropriate for application of bioventing? Explain!

b. Develop a hypothesis for the results of the biodegradation experiment.

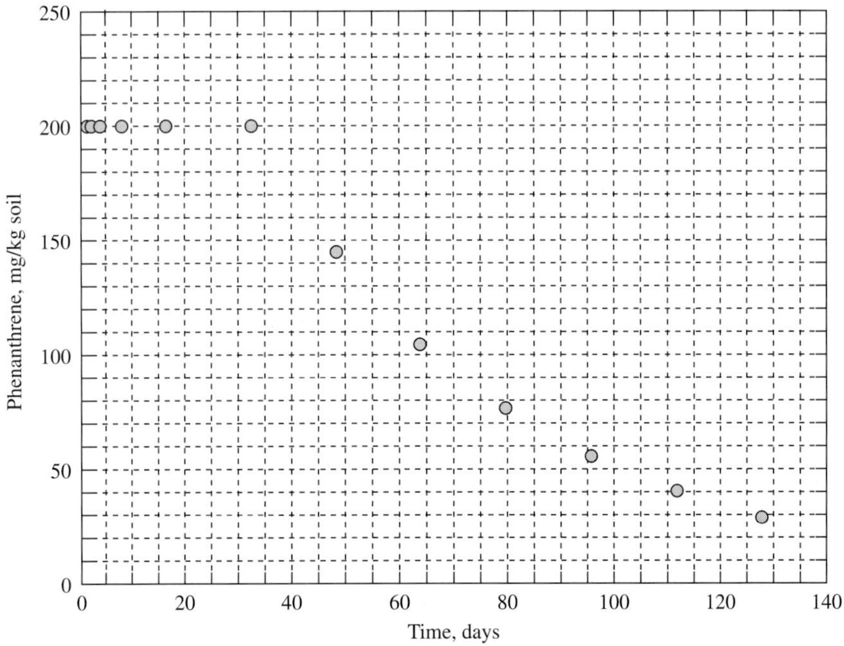

FIGURE 7.23
Results of laboratory experiments in phenanthrene degradation.

c. Determine a usable rate of biodegradation model.
d. Determine the initial mass fraction of phenanthrene in the liquid phase.
e. Estimate the rate at which air will need to be supplied to prevent anaerobic conditions from developing.

7.10. A contaminated soil zone has been found behind a pesticide manufacturing plant. The site contains a mixture of chemicals, and analyses have been run using total organic carbon (TOC) and chemical oxygen demand (COD) as measures of concentration. The zone is approximately cylindrical in shape, beginning at a depth of 1 m and extending downward to a depth of 11 m with an average diameter of 50 m. Concentrations of organic material in the zone vary somewhat, but average values are believed to be satisfactory measures of the spatial distribution; that is, a uniform concentration distribution can be assumed. Vapor-phase concentrations of the contaminants are negligible. Porosity of the soil ϕ is 0.30 and the field capacity is 90 g H_2O per kg dry soil. Dry bulk density of the soil ρ_b is 2.1 kg/L. Select a gas flow rate and estimate the time necessary for treatment.

Soil water distribution analyses

A 1-kg sample of dry soil was mixed with selected volumes of water and allowed to come to equilibrium. The TOC and COD of the water at equilibrium were determined. Prior to adding the water the contaminant was extracted from the soil using a solvent and the sorbed contaminant TOC was determined to be 2.40 g. The equilibrium liquid concentrations are given below.

V_L, L	C_L, mg/L	COD, mg/L
1	39.3	125
2	38.7	124
4	37.5	120
8	35.3	113
16	31.5	100
32	26.1	84

Biodegradation rate analysis

Core samples were placed in cylindrical bench-scale reactors, saturated with a nutrient solution, and allowed to drain and air-dry to a moisture content of 18 percent by volume (approximately the same as the field capacity). Air was then drawn through the cores on a continuous basis. Samples were sacrificed at selected time periods, and the total TOC values were measured on a dry-mass basis. Because the contaminants are only slightly volatile, contaminant originally in solution can be assumed to be measured.

Time, days	TOC, mg/kg
0	2,400
5	2,330
10	2,260
20	2,120
40	1,840
80	1,280

Soil vapor extraction test results

Soil vapor extraction tests were conducted at vacuums of 100, 200, and 300 mm H_2O. Radii of influence and volumetric flow rates were:

Vacuum, mm H_2O	Radii, m	Q_g, m^3/min
100	6	0.35
200	12	0.56
400	16	0.87

7.11. Leaking underground gasoline storage tanks in California have contaminated a large number of sites with fuel containing approximately 15 percent MTBE by volume. Assume that gasoline can be characterized by three compounds, MTBE, benzene, and octane. Discuss the applicability of the two approaches listed below for bioremediation of the soils.

a. Soil vapor extraction followed by biological treatment of the extracted gases. (Processes for treatment of vapor-phase contaminants are discussed in Chapter 10. However, the actual process is not important as long as one exists.)

b. Bioventing.

Compound	H	log K_{ow}	Solubility, mg/L
MTBE	0.03	0.08	40,000
Benzene	0.20	2.12	1,780
Octane	0.14	5.18	72

7.12. A laboratory soil column has been set up as shown below. Air is added to the column after being moisturized by passing it through a fine bubble diffuser. Temperature in the column is maintained just below the saturation temperature for the air, and the moisture content Θ of the column is stable at 0.15. The porosity of the soil ϕ is 0.3 and the bulk density ρ_b is 2,600 kg/m^3. A contaminant has been added to the dry soil prior to construction of the system to give a uniform concentration of 1,000 mg/kg. Properties of the contaminant are at the ambient temperature of 20°C:

$$H = 0.010$$

$$K_{SD} = 10^{-2} \text{ L/mg}$$

Clean air ($C_g = 0$) is forced through the column at a rate of 0.0942 m^3/min (superficial velocity of 3 m/min) and the outlet gas contaminant concentration is monitored. Results of the monitoring are shown in Figure 7.24.

a. Estimate the gas-liquid transfer rate coefficient value K_La.

b. Determine if the system is biodegradation rate or desorption rate limiting.

c. Depending on the answer to (*b*), determine either the biodegradation rate (g/$m^3 \cdot$ min) or the desorption rate coefficient (min^{-1}).

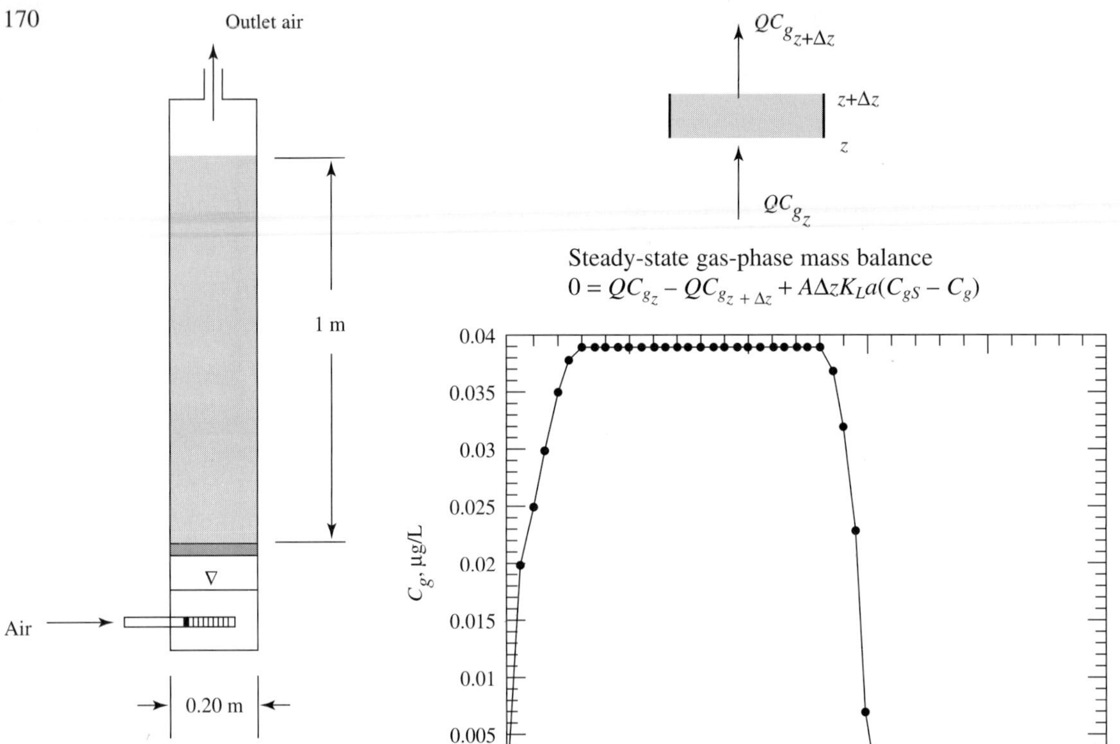

FIGURE 7.24
Laboratory setup and experimental results for Problem 7.12.

REFERENCES

Bedient, P. B., and W. C. Huber (1988): *Hydrology and Floodplain Analysis,* Addison-Wesley, Reading, MA.

Billings, J. F., A. I. Cooley, and G. K. Billings (1994): "Microbial and Carbon Dioxide Aspects of Operating Air-sparging Sites," *Air Sparging for Site Remediation,* ed. by R. E. Hinchee, Lewis Publishers, Boca Raton, FL.

Brown, R. A., R. J. Hicks, and P. M. Hicks (1994): "Use of Air Sparging for In Situ Bioremediation," *Air Sparging for Site Remediation,* ed. by R. E. Hinchee, Lewis Publishers, Boca Raton, FL.

Hinchee, R. E. (1994): "Air Sparging: State of the Art," *Air Sparging for Site Remediation,* ed. by R. E. Hinchee, Lewis Publishers, Boca Raton, FL.

Johnson, P. C., C. C. Stanley, M. W. Kemblowski, D. L. Byers, and J. D. Colthart (1990): "A Practical Approach to the Design, Operation, and Monitoring of In Situ Soil-Venting Systems," *Ground Water Management Review,* spring, pp. 159–178.

Johnson, R. L., P. C. Johnson, D. B. McWhorter, R. E. Hinchee, and I. Goodman (1993): "An Overview of In Situ Air Sparging," *Ground Water Monitoring Review,* vol. 13, no. 4, pp. 127–135.

Rittman, B. E., E. Seagren, B. A. Wrenn, A. J. Valocchi, C. Ray, and L. Raskin (1994): *In Situ Bioremediation,* 2d ed., Noyes Publications, Park Ridge, NJ.

CHAPTER 8

Solid-Phase Bioremediation

Solid-phase treatment will be the term used in this book to describe processes applied to the ex situ treatment of soil under unsaturated conditions. Such treatment is distinguished from in situ bioremediation (Chapter 7) and from slurry-phase treatment where the soil is mixed with water and mechanically stirred in a bioreactor (Chapter 9). Solid-phase bioremediation can be divided into two general classifications: land treatment and composting. The critical difference between the two processes is the mode of aeration. Land treatment follows principles similar to agricultural landfarming. Tilling equipment, similar to that used in farming, is employed to turn and aerate the soil. As such, only relatively shallow layers of soil, as deep as the turning blades, can be treated. One result is a large area requirement. Soil composting, like refuse composting, involves the formation of piles of degradable material. Pile depth is limited by the equipment used but can be up to 5 m without great difficulty. Consequently land requirements are considerably less than those associated with land treatment. Compost piles are aerated either actively, by mechanically turning the pile, or passively, through embedded perforated pipes. Additives such as nutrients to stimulate microbial growth and organic material to support more intense microbial activity are commonly used in solid-phase treatment systems. Addition of organic material, usually in the form of manure, is usually a requirement to generate the heat desired in compost systems.

LAND TREATMENT

Also known as solid-phase treatment, land treatment, or *landfarming,* is a term commonly used in agricultural practices to describe the method used by farmers for centuries to decompose organic nonhazardous waste. The method simply consists of spreading the waste and tilling the soil to incorporate the waste into the soil matrix and to provide aeration necessary to optimize microbial activity. The same technique, which often uses similar farming equipment, is used to biodegrade hazardous waste.

The use of land treatment in hazardous-waste management dates back to the early 1900s (King et al., 1992), at the height of the industrial revolution, when it was used more as a disposal technique than a remediation process. At the time, it was common for petroleum refineries to dump the sludge generated in processing onto nearby land. Often that land happened to be at the bank of a river or stream which was used as a water source for the refinery. The volume of sludge applied would decrease with time mainly because of volatilization of the lighter compounds and leaching and runoff of the heavier ones. Adsorption, photodecomposition, and biodegradation also contributed to sludge removal, but to a lesser extent. With increased understanding of the reasons why volume reduction occurred came the need to optimize the process so that the reduction could be achieved more quickly and effectively. With time, soil tilling, fertilizer amendments, and pH and moisture control were practiced to augment the biological removal of the waste.

By about 1950, land treatment became a popular method for the treatment of hazardous and industrial wastes applied in the form of liquid, solid, or sludge. In the 1970s, however, with the introduction of more stringent environmental regulations, the fate and transport of contaminants in land treatment operations had to be reconsidered. The old practice of land treatment which encouraged volatilization and leaching as a means of removing contaminants from the soil was no longer allowed. Instead, the current practice is to eliminate leaching, minimize volatilization, and allow primarily for biological removal of organics.

Process Description

Land treatment is one of the simplest bioremediation processes available. A treatment site is constructed, as will be described below. The contaminated medium, soil, sludge, sediment, or liquid waste, is spread out in relatively thin layers and tilled. New contaminated material is added on a regular basis. Tilling achieves aeration, as well as mixing, which can improve mass transport and bioavailability. To enhance microbial activity, nutrients, moisture content, and pH are monitored regularly and adjusted as necessary.

The focus of this chapter is primarily on the treatment of contaminated soil, which can be carried out in situ or ex situ depending on the site, type of contamination, and concentration level of contaminants in the soil. In situ treatment, which is somewhat unusual, is applicable if an impermeable layer exists under the contaminated soil which would eliminate the threat of groundwater contamination. Such conditions may exist where spills have occurred, such as from an aboveground storage tank, where the site is already set up with engineered controls to prevent groundwater contamination. This was the case at a site in western Pennsylvania, for example (Leavitt, 1992), where a tank implosion resulted in the release of approximately 3,785 m^3 (1 million gal) of No. 2 fuel oil. After emergency response, an estimated 549 m^3 (145,000 gal) of fuel remained in the soil. Contamination levels in the soil averaged between 10,000 and 20,000 mg total petroleum

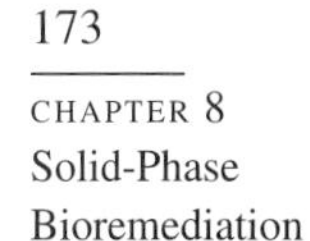

Biological land treatment of over 100,000 yd^3 of PHC-contaminated soils and sludges, in southern California. *(Courtesy of OHM Remediation Services Corp.)*

hydrocarbons (TPH) per kg of soil, but were as high as 100,000 mg TPH/kg in some places. The fact that concentration levels were not inhibitory to microbial activity allowed for successful in situ land treatment. For other examples of in situ land treatment, refer to Flathman et al. (1995), and Fogel (1994).

In most cases the soil must be excavated to contain the spread of contamination or because the site is to be used for different purposes and cannot be tied up for the duration of treatment. A common example is contamination resulting from leaking underground storage tanks. Ex situ treatment is carried out in a prepared bed system that can be constructed at an appropriate location remote from the site of contamination. Excavating and treating the soil ex situ may also be necessary if the contamination level is inhibitory to biodegradation. That is often the case with sediments in lagoons or ponds that had previously been used as disposal pits for liquid waste. In such a case, the contaminated medium can be mixed with clean soil to dilute the waste and lower the concentration levels. A prepared-bed system, or engineered landfarming, as it is sometimes called, employs the same principles as in situ land treatment but is constructed with engineered controls to minimize transport of contaminants and to maximize treatment efficiency.

Waste Application

When applying contaminated soil to a land treatment unit (LTU), the volume increase due to fluffing and soil disturbance during excavation must be considered. The volume of soil after spreading in a LTU is approximately 1.25 and 1.4 times the volume of soil to be excavated (King et al., 1992).

Soil application

The depth to which contaminated soil is spread in a LTU is critical to facilitating oxygen diffusion and hence maintaining aerobic conditions. That depth is dependent on soil properties, tilling equipment, and amount of soil to be treated. Soils that have high permeability and are relatively easy to drain can be spread to a deeper depth compared to soils high in clay and which may have lower permeability. In some cases, soils are amended with bulking material or organic matter to increase porosity and permeability, and thus increase treatment efficiency.

The depth of the contaminated soil layer should generally be smaller than the depth reached by the tilling equipment (15 to 50 cm), if effective treatment is to be achieved. Soil that is not tilled and turned regularly will be oxygen-limited since oxygen diffusion through the surface is relatively slow. This does not mean that biodegradation will not occur in soils that are not tilled. However, biodegradation rates are likely to be slower and the time to complete bioremediation will be longer. Treatment to depths of 150 cm below the surface has been observed (LaGrega et al., 1994). For example, with in situ land treatment where the contaminated zone may be relatively deep, the soil can be treated in layers. Once the top layer is effectively remediated, it can be excavated and removed, thus allowing for tilling at deeper depths. The excavated soil can be stored temporarily and then used as fill once the site is cleaned.

The depth to which contaminated soil can be spread in a LTU is dependent to some extent on the surface area available and the amount of soil to be treated. If all of the soil cannot be treated in one lift, then multiple applications may be necessary. The first lift can have a relatively large depth, but the successive ones should be shallower. Once the first lift is treated, a new lift can be applied with a depth that would allow some of the soil from the previous lift to be incorporated, through tilling, with the freshly applied soil. This allows for introduction of some of the already acclimated bacteria from the first lift into the second lift, thus cutting down on the lag period. An added advantage is dilution of the waste in the second lift as it gets mixed in with the cleaner soil from the first lift. The result is generally a shorter treatment period for the second lift compared to the first.

As a general rule, it is advised that soil depth in a single lift be greater than 15 cm (6 in) but not exceed 61 cm (24 in). It is preferable if it is less than 30 cm (12 in), though, if effective treatment is to be achieved. For clayey soils in particular, it is recommended that the depth not exceed 23 cm (9 in) due to limited permeability and soil workability (U.S. EPA, 1993).

EXAMPLE 8.1. LAND TREATMENT OF PAHS. Land treatment has been chosen as the bioremediation process to treat soil from an abandoned wood-treating facility contaminated with PAHs. The volume of soil to be excavated for treatment is estimated at 10,500 yd^3. A 6.8-acre land treatment unit has been constructed for remediation purposes. If the soil is a mixture of silty clay and sandy clay, estimate the number of lifts that should be applied, and the appropriate soil depth for each lift in centimeters.

Solution

$$\text{Volume of soil to be excavated} = (10{,}500\ \text{yd}^3)\ (0.7646\ \text{m}^3/\text{yd}^3) = 8{,}028\ \text{m}^3$$

$$\text{Increased volume due to “fluffing”} = (1.4)\ (8{,}028\ \text{m}^3) = 11{,}240\ \text{m}^3$$

$$\text{LTU surface area} = (6.8\ \text{acres})\ (4.047 \times 10^3\ \text{m}^2/\text{acre}) = 27{,}520\ \text{m}^2$$

If soil is to be applied in one lift, the resulting soil depth

$$h = \frac{\text{soil volume}}{\text{surface area}} = \frac{11{,}240\ \text{m}^3}{27{,}520\ \text{m}^2} = 0.41\ \text{m} = 41\ \text{cm}$$

This is probably too high a depth considering that the soil is high in clay. Suggest using 2 lifts. Let the first lift have a depth of 23 cm (maximum recommended for clay soils), so that the second lift would have a depth of 18 cm.

Liquid waste application

Although the focus of this chapter is on soil treatment, a few words on liquid-waste application are necessary. Liquid waste is often generated in industrial operations such as in metal processing where oil-in-water emulsions are used as coolants. When liquid waste is applied to a land treatment unit, the soil in that unit acts as the treatment medium or the bioreactor. Multiple applications are possible, and the same bed may be used for years without requiring soil replacement.

Liquid waste may be applied using sprinkle or spray irrigation, or by overland flow using drainage canals. If the potential for volatilization is high, subsurface injection may be employed. A typical injection depth is 130 mm (5 in) (U.S. EPA, 1983; U.S. EPA, 1990). The application rate of a liquid waste depends on the viscosity of the fluid and the type of contaminants. The application rate should be lower than the infiltration rate to avoid runoff, and it should be lower than the biodegradation rate to ensure treatment and minimize leaching. Biodegradation rates are temperature-dependent, and that should be taken into account when deciding on the application rate.

A problem commonly associated with liquid-waste application to soil is the accumulation of heavy metals. Examples of such metals are lead, cadmium, mercury, zinc, and copper. As the metal concentration increases with each successive application, a level toxic to microorganisms may be attained. At that point the contaminated soil may have to be disposed of in a landfill and replaced with fresh soil if the LTU is to be used again.

Land Treatment Unit Construction

There are six principal components of a land treatment bed: an impermeable layer, a drainage system, a soil treatment zone, berms and swales, a water-storage pond, and a monitoring system. A schematic diagram showing these elements is given in Figure 8.1.

Impermeable layer

An impermeable layer is necessary to prevent infiltration of contaminated liquid into previously uncontaminated soil and nearby groundwater. If the LTU is to be constructed at the site of the excavated soil, compaction may be necessary to

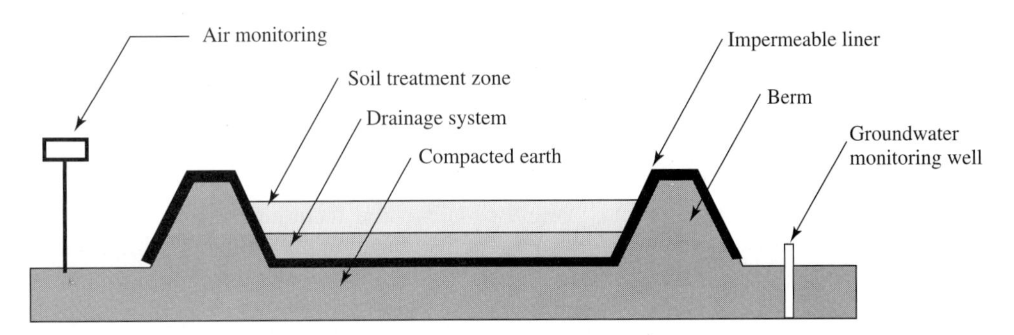

FIGURE 8.1
Schematic diagram of the principal elements in a land treatment unit.

decrease permeability of the native soil under the treatment bed. An impermeable layer can then be placed on top of the compacted earth. Synthetic liners, such as 60- to 80-mil-thick high-density polyethylene (HDPE), are commonly used. Equally common are compacted clay layers. A hydraulic conductivity of less than 1×10^{-7} cm/s is generally recommended for a clay liner. Several feet of clay (typically two to three) may be necessary to ensure adequate protection against infiltration. Depending on the site, an already existing impermeable layer may be used to construct a prepared bed. For example, a paved or asphalted parking lot may be a good choice, especially if at an abandoned contamination site.

Drainage system

A drainage system is necessary to collect any leachate due to water irrigation or waste application, but more importantly, to collect leachate due to rainwater infiltration. One method is to embed perforated pipes in a bed of sand or gravel on top of the impermeable liner. The drainage pipes usually divert water into a collection sump before the water goes into a storage pond. A layer of sand is typically placed over the drainage pipes to facilitate drainage into the pipes and to protect the drainage system from the impact of heavy machinery, such as tilling equipment. Sand layers 23 to 25 cm (9 to 10 in) thick are generally satisfactory [Genes and Cosentini, 1993; U.S. EPA, 1995 (3)]. The sizing of the drainage pipes depends on the amount of infiltration expected during heavy rainfall.

Soil treatment zone

In some cases, pretreatment of the soil may be necessary before application. Rocks and other large, nominally contaminated, debris that will interfere with tilling can be removed by screening. At the same time, large materials such as wood, which may adsorb the contaminants, can be removed and treated separately. Uncontaminated soil may be placed on top of the drainage system before the contaminated soil is applied to provide a deeper treatment medium. Leachate treatment may occur in the uncontaminated soil zone before discharge to the drainage system, or the uncontaminated soil may dilute the waste if tilling allows for mixing between the contaminated and uncontaminated layers.

The soil in the LTU should be sloped to prevent ponding and to divert runoff water into the drainage sump. A slope of 0.5 to 1.0 percent is recommended. A larger slope is not advised since it could increase the chance of surface erosion.

Berms and swales

Berms and swales are required to protect against cross contamination between treatment units or loss of contaminants from the system. Because land treatment systems are exposed to precipitation, contaminants may be mobilized during rainy periods and berms and swales prevent uncontrolled discharges. Rainwater falling within the LTU should be contained and collected in the drainage system, while water flowing outside the LTU should be prevented from entering the site. Berm walls can be constructed using clean soil, clay, concrete, or some other impervious material. If soil is used, a liner is required to prevent seepage through the walls. A minimum freeboard above the treatment zone of 30 cm (1 ft) is recommended. If

TABLE 8.1
Water storage capacity and disposal method of some field-scale applications of LTUs

LTU surface, m^2	Storage capacity, m^3	Site location	Contaminant	Water-disposal method*	Reference
27,520	3,790	Midwest	PAHs	Irrigation/discharge to WWTP	Genes and Cosentini, 1993
27,560	3,790	Missouri	PAHs	Irrigation/discharge to WWTP	U.S. EPA, 1995
16,190	2,840	Florida	PAHs	Irrigation	U.S. EPA, 1995
2,540	280	New Jersey	Diesel	Irrigation/discharge to WWTP	Troy et al., 1994

*WWTP: Wastewater-treatment plant. No pretreatment was needed in any of the references listed.

multiple soil lifts are used, the berm wall should be about 30 cm higher than the top of the last lift. The slope of the berm walls should be less than 45° to ensure stability. A smaller slope is preferred, but the resulting increase in land usage may be a problem.

Storage pond

A retention pond or a collection sump is usually required to collect and retain excess leachate resulting from rainfall. Water can be stored temporarily and reapplied to the LTU through the irrigation system during dry periods, or it may be discharged to a nearby wastewater-treatment plant during wet periods. Treatment prior to discharge may be necessary if the concentration levels of the contaminants are too high, but that is seldom the case. Most organic contaminants are hydrophobic and tend to adsorb to soil. Discharge may be costly if the water has to be trucked to a wastewater-treatment plant, in which case on-site biological treatment may be a cheaper option.

Storage capacity required depends on the site location and the expected amount of rainfall for that area. A backup storage system may be necessary for extreme weather conditions. King et al. (1992) recommend designing the pond to hold all the water from a 3-in rainfall across the entire area of the LTU. Extra allowance should be made for geographic areas that experience greater amounts of rainfall. One inch of rainfall may be expected to produce between 9 and 25 L/m^2 area (10,000 and 27,000 gal/acre) (U.S. EPA, 1993). Table 8.1 lists some examples from field applications of land treatment processes.

Monitoring system

In most cases, regulatory agencies require monitoring for air emissions and groundwater contamination. Depending on contaminant volatility, air emissions monitoring may not be required or may be terminated when air emission rates have been established to be very low. The threat of migration of contaminants from the LTU to groundwater, especially through leaks in the impermeable layer, necessitates groundwater monitoring. Monitoring wells may be installed at each side of the LTU, and contamination may be monitored at more than one depth. At the same time, lysimeters may be installed under the impermeable layer, during construction, to check for leaks and migration of contaminants.

Process Control

Before a land treatment process is started, laboratory tests are usually conducted to determine what conditions would be optimum for biodegradation and how such conditions can be corrected and maintained in the field. For example, soil tests are conducted to determine the field capacity of the soil and the optimal moisture content range for microbial activity. Soil structure is also determined to find the appropriate tilling depth and to decide whether soil amendments, such as organic matter or bulking agents, are needed to improve soil workability. Tests are conducted to find the C:N:P ratio and to determine if nutrient amendments are necessary. At the same time, the pH is measured to determine if adjustment will be necessary to maintain microbial activity. Microbial activity is measured through biodegradability studies, to determine if the indigenous population is capable of degrading the contaminants, especially if the contaminants are relatively persistent. With the more easily degradable petroleum hydrocarbons, a microbial count may be enough simply to ensure the existence of a viable population.

Monitoring is continued throughout the bioremediation process to ensure that the treatment system is performing correctly and that material is not being lost from the site. The minimum period between sampling times should correspond to the period between contaminant application.

Soil tilling

Tilling is necessary to aerate the soil and to incorporate the waste into the soil matrix. Through tilling, contact between microorganisms, nutrients, and contaminants is increased, thus improving the chances of biodegradability. Tilling and turning of the soil also redistributes the contaminants, resulting in better homogeneity in the concentration levels (Fogel, 1994).

The depth to which a soil is tilled has a large impact on treatment efficiency. As noted above, the primary reason for tilling is to increase the rate of oxygen diffusion, which is necessary for aerobic biodegradation, and which diminishes at depths greater than 30 cm (12 in), depending on the soil structure and texture. On the average, fine soils with low permeability are harder to drain and aerate than coarser soils. Hence the zone of incorporation in clayey soils, for example, should be lower than that for the more sandy soils.

The depth to which oxygen can penetrate the soil, due to diffusion alone and under steady-state conditions, can be estimated using simple mass-balance analysis. To simplify, assume that the soil is dry with all the pores being filled with air, and that porosity is uniform throughout the section considered. Also assume that the rate of oxygen consumption is zero-order. Then applying the mass-balance equation on the control volume in Figure 8.2, using the boundary conditions (BC) shown:

$$\text{Rate of mass accumulation} = \text{Rate of mass entering} - \text{Rate of mass leaving} + \text{Rate of mass generation}$$

$$0 = -\phi AD \frac{dC}{dz}\bigg|_z - \left(-\phi AD \frac{dC}{dz}\right)\bigg|_{z+\Delta z} + A\Delta z r_o \qquad (8.1)$$

$$\frac{d^2C}{dz^2} = -\frac{r_o}{\phi D} \qquad (8.2)$$

$BC\ 1$ $\qquad \frac{dC}{dz} = 0 \text{ at } z = L$

$BC\ 2$ $\qquad C = 0 \text{ at } z = L$

where D = diffusion coefficient, m^2/s
C = oxygen concentration, g/m^3
z = distance from the surface, cm
A = surface area, m^2
ϕ = porosity, m^3/m^3
r_o = rate of oxygen consumption, $g/m^3 \cdot s$
L = depth at which the oxygen concentration reaches zero, m

Integrating Equation (8.2) between $C = C_o$ to $C = 0$ and $z = 0$ to $z = L$ gives

$$\frac{dC}{dz} = -\frac{r_o}{\phi D} z + \alpha \qquad (8.3)$$

where α is an integration constant.

Applying BC 1: $\qquad \alpha = \frac{r_o}{\phi D} L$

Substituting into Equation (8.3) results in:

$$\frac{dC}{dz} = -\frac{r_o}{\phi D}(z - L) \qquad (8.4)$$

Integrating a second time:

$$C = C_o - \frac{r_o}{\phi D}\left(\frac{z^2}{2} - Lz\right) \qquad (8.5)$$

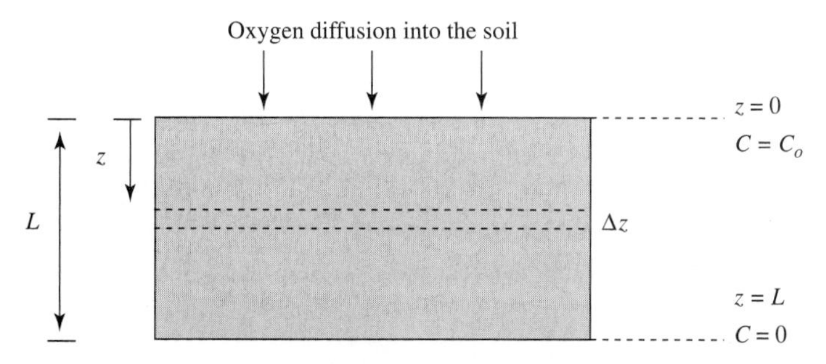

FIGURE 8.2
Oxygen diffusion within a control volume in a land treatment unit.

Applying BC 2:

$$L = \sqrt{\frac{-2\phi D C_o}{r_o}} \tag{8.6}$$

EXAMPLE 8.2. DEPTH OF OXYGEN PENETRATION INTO SOIL. Using Equations (8.1) through (8.6), estimate the depth to which oxygen can diffuse in a dry soil with a porosity of 0.35. Assume a rate of oxygen consumption of 150 $g/m^3 \cdot h$ and an oxygen diffusion coefficient of 1.89×10^{-5} m^2/s. Assume an ambient temperature of 25°C.

Solution

1. Convert the reaction rate r_o to units of $g/m^3 \cdot s$.

$$r_o = -(150 \text{ g/m}^3 \cdot \text{h})\frac{1 \text{ h}}{3{,}600 \text{ s}} = -4.17 \times 10^{-2} \text{ g/m}^3 \cdot \text{s}$$

2. Apply Equation (8.6).

$$L = \sqrt{\frac{-2\phi D C_o}{r_o}}$$

$$\phi = 0.35 \text{ m}^3/\text{m}^3$$

$$D = 1.89 \times 10^{-5} \text{ m}^2/\text{s}$$

$$C_o = \text{atmospheric oxygen concentration}$$

$$= \frac{0.21 \text{ L O}_2}{\text{L}_{\text{air}}}\left(\frac{1 \text{ mole O}_2}{22.4 \text{ L O}_2}\right)(32 \text{ g O}_2/\text{mole}) = 0.30 \text{ g/L}$$

$$= 300 \text{ g/m}^3 \text{ at } 0°\text{C}$$

$$= \frac{273 \text{ K}}{298 \text{ K}} 300 \text{ g/m}^3 = 275 \text{ g/m}^3 \text{ at } 25°\text{C}$$

$$L = \sqrt{\frac{-2(0.35)(1.89 \times 10^{-5} \text{ m}^2/\text{s})(275 \text{ g/m}^3)}{-4.17 \times 10^{-2} \text{ g/m}^3 \cdot \text{s}}} = 0.295 \text{ m}$$

Soil tilling should be carried out regularly to facilitate oxygen infiltration into the soil pores. Typically, tilling is conducted every week or once every 2 weeks. Too much tilling may be damaging since it tends to destroy the soil aggregates (structure) and results in compacting the layers underneath the tilling zone. Tilling is best carried out when the soil is relatively dry. In general, tilling should be carried out at least 24 h after irrigation or significant rainfall (U.S. EPA, 1993). It is also recommended that tilling be carried out in different directions each time. For example, the tiller can move lengthwise, crosswise, or diagonally across the land treatment unit, at alternate times. This type of movement is important to achieve optimum overall mixing and contaminant redistribution since most of the mixing occurs along the tractor's line of travel.

Nutrient addition

The addition of inorganic nutrients is often necessary to compensate for the lack of nitrogen or phosphorus in the natural soil. Unless laboratory tests are conducted

TABLE 8.2
Examples of nutrient additions from various filed applications

Desired C:N:P	Nutrient added (empirical formula)	Reference
100:2:0.4	Ammonium phosphate ($NH_4H_2PO_4$)	Genes and Cosentini, 1993
200:10:1	Ammonium nitrate (NH_4NO_3)	U.S. EPA (2), 1995
10:3:3		Fogel, 1994
100:10:1		Troy et al., 1994
	Diammonium phosphate [$(NH_4)_2HPO_4$]	Flathman et al., 1995
	Ammonium nitrate (NH_4NO_3)	Calabrese et al., 1993
	Potassium phosphate (KH_2PO_4)	

to find the optimum C:N:P ratio necessary to degrade the waste at hand, a ratio of 100 to 300:10:1 is often used. Laboratory tests may show that a higher concentration of nitrogen and phosphorus works better to degrade the waste. Examples of ratios that have been used in the field are listed in Table 8.2. McGinnis et al. (1991) and Fogel (1994) showed that the addition of more nutrients can result in higher biodegradation rates. The addition of high concentrations of nutrients, however, can result in two problems. One is the increase in cost, the other the resulting effect on salinity and osmotic pressure. If high concentrations are necessary, then the nutrients can be added in steps, instead of all at once, at the start of treatment.

Most often, inorganic nutrients such as the ones listed in Table 8.2 are used, although other organic complex nutrients, such as animal manure, can also be used. Synthetic fertilizers that are highly soluble can be introduced with the irrigation water. Nitrates, however, if added in high concentrations, can easily leach out with the drainage water. As such, slow-release nutrient pellets can be used to ensure a constant supply over time.

Moisture content

Maintaining optimum moisture contents and scheduling irrigation accordingly is very important in landfarming operations. Optimum moisture content is usually measured relative to the field capacity of the soil in question. Field capacity is defined as the moisture content of a soil that had been saturated with water and left to drain freely for 24 h. Coarser soils that are easy to drain generally have a higher moisture content at field capacity compared to the finer soils. For example, at field capacity, the moisture content of a sandy soil may be as low as 5 percent (on a dry-weight basis), compared to 30 percent for clay soil on the same basis (U.S. EPA, 1993). Unless otherwise tested under laboratory conditions, the moisture content optimum for aerobic biodegradation can generally be set at a range of 60 to 80 percent of field capacity.

To maintain the necessary moisture content level, water is added through different forms of irrigation, including sprinkler and trickling systems. The system used, however, should be easy to remove or to work around whenever a new batch of waste is to be applied to the land treatment unit or whenever the soil is to be tilled. Hand-moved sprinklers as well as permanently installed ones can be used. In general, the rate of application of water should not exceed the infiltration rate into the soil if surface runoff and erosion are to be avoided. Low application rates are also necessary to avoid water ponding, which can lead to anaerobic conditions.

Temperature control

Although irrigation is used principally to obtain a desired moisture-content level, water may be added to regulate the soil temperature. Thermal conductivity of the soil matrix is increased by adding water, thus reducing the daily variations in soil temperature. Sprinkle irrigation protects against frost formation in the winter and cools the soil in the summer. Another method used to modify the soil temperature is the addition of mulches. Certain types of mulches, typically organic material such as manure and sawdust, are also used sometimes to control the moisture content since they improve the water-holding capacity of the soil. Other examples of mulch materials used to control temperature variability are compost, wood chips and bark, asphalt emulsion, and gravel or crushed stones (Dupont et al., 1988). In some cases, a cover is used over the site to control the emissions of volatile compounds, thus causing the soil temperature to increase.

In geographic areas of extreme weather conditions, mulch addition or covering may not be enough to elevate the soil temperature to the point where aerobic activity can proceed. Frequent rainfall can also result in saturated soil, thus minimizing the chance of aerobic biodegradation. To be conservative in design, the LTU should be considered inactive during harsh winter conditions (Portier and Christiansen, 1994; Flathman et al., 1995; U.S. EPA, 1995). Tilling and sampling can be discontinued until the weather improves. In many areas the inactive period may extend as long as 5 months.

Control of pH

Because most soils are acidic, pH adjustment is often needed to enhance biodegradation. To increase and stabilize the soil pH, calcium or calcium/magnesium-containing compounds can be added to the soil. This process is known as liming, and examples of the compounds used are calcium oxide (lime), calcium hydroxide, calcium carbonate, magnesium carbonate, and calcium silicate slags. Should the soil pH be high because of a high carbonate concentration or because of the presence of hazardous wastes that are high in pH, acidification may be necessary. Acidification, or the reduction of the soil pH, can be achieved by adding elemental sulfur or sulfur-containing compounds such as sulfuric acid, liquid ammonium polysulfide, and aluminum and iron sulfates (Dupont et al., 1988).

Addition of oxygen in chemical form

Some vendors of commercial soil treatments have utilized chemical oxidants, e.g., CaO_2, as a source of oxygen to reduce tilling requirements. However, sound data regarding efficacy and economics of such a process are lacking.

Sampling protocol

A land treatment unit is usually divided into cells, and several subsamples are collected from each cell at one time. The U.S. EPA (1993) recommends that an area no larger than 4,000 m^2 (1 acre) be assigned per cell. Table 8.3 lists examples of cell sizes and number of samples taken per cell from different field applications. The subsamples are usually combined into one composite sample, representative of that cell.

Sampling is carried out to monitor contaminant biodegradation, as well as moisture content, nutrient levels, and if necessary, microbial activity. At each sam-

TABLE 8.3
Examples of cell sizes and numbers of samples collected per cell from various field applications

Total area, m^2	Cell area, m^2	Soil depth, mm	No. samples per cell	Reference
60	2.2	300	5–8	Flathman et al., 1995
360	4	600	NA	Ellis, 1994
2,540	318	350–450	2	Troy et al., 1994
9,300	1,825	200	20–30	Bleckman et al., 1995
16,190	2,020	100–300*	4	U.S. EPA(a), 1995
27,560	1,920–2,495	180–230*	NA	U.S. EPA(c), 1995

*Multiple lifts employed.
NA = not available.

pling event, only a few cells are sampled to cut down on the cost of analysis. Alternate cells are sampled at the next sampling event. However, for site closure, every cell must be sampled to demonstrate that contaminant concentrations have dropped to a level that would meet regulatory requirements.

Contaminants Treated by Landfarming

Land treatment has probably been the process most applied in the bioremediation of hazardous waste, yet research on this process has been relatively limited (LaGrega et al., 1994). In the field, land treatment has been applied to a variety of wastes, particularly at petroleum refinery sites and with creosote-contaminated sludges and soils (Ryan et al., 1991; Nyer, 1992; U.S. EPA, 1993). Examples cited in the literature of the successful use of landfarming in the treatment of different types and concentrations of contaminants are listed below. Note that the treatment efficiency is usually reported in terms of the reduction in concentration of the parent contaminants. Proof that biodegradation did actually occur is generally not provided. The extent of degradation and the toxicity of the products of degradation are not investigated. At the same time, these studies generally do not account for abiotic losses including volatilization.

Bogart and League (1988) report on the success of enhanced landfarming in reducing creosote concentrations in soil from 6,200 and 3,000 ppm to 800 and 100 ppm, respectively, within 30 days. The enhanced treatment involved bioaugmentation with bacteria grown in high-nutrient broth as well as tilling for aeration, irrigation, and supply of dissolved nutrients. Land treatment has also been successfully used to treat pesticides. A reduction in the concentration of 2,4-dichlorophenoxyacetic acid (2,4-D) from about 42 to 4 ppm within 77 days has been reported (Fiorenza et al., 1991). Hanstveit (1988) reported reductions of 73 percent in benzene, toluene, and xylene (BTX) concentrations; 36 percent in oil and grease; and 86 percent in total PAHs were achieved over a 4-month period. Nyer (1992) reported that pentachlorophenol (PCP) concentrations were reduced by 95 percent while PAH concentrations were reduced by about 50 to 75 percent over a 5-month period in a land treatment system.

Advantages and Disadvantages

Advantages of land treatment systems include low capital and operating costs and effectiveness in treating wastes with relatively high metal contents. Disadvantages of land treatment are the large land-area requirements and the fact that this degradation process takes a relatively long time and may never be complete. Strong adsorption of hydrophobic chemicals, especially in weathered contaminants, can make them inaccessible to biodegradation and result in long-term persistence. Because regular tilling results in high rates of contact with the ambient atmosphere, high volatilization rates can be expected. Biologically recalcitrant or refractory VOCs are very likely to be emitted. To help control volatilization, prepared-bed systems are sometimes enclosed within greenhouse structures (Ryan et al., 1991), which have the added advantage of increasing the soil temperature. In such a case, the structure would have to be equipped with a ventilation system and an air emissions control system.

CASE STUDY

A laboratory study was conducted to simulate landfarming and to examine the effectiveness of such a method in the treatment of soil and sediments contaminated with pentachlorophenol (PCP) and creosote (Mueller et al., 1991). The effect of inorganic nutrient amendments on the treatment efficiency was also examined. The soil, which had a pH of 7.1, was contaminated with about 10,000 ppm (g/g soil) of creosote and PCP. The sediments, which had a pH of 10, were contaminated with about 70,000 ppm of the same waste. The experimental apparatus consisted of a funnel, lined with a filter paper, in which about 3 kg of the soil or sediment was placed. The funnel was seated on top of a beaker, in which leachate would be collected. The funnel/beaker setup was enclosed inside an amber chamber (to minimize photooxidation) and air was supplied through an oil-free compressor. Prior to entering the chamber, the air was saturated with water to minimize drying of the soil and the sediment. The air leaving the chamber was collected onto activated carbon to trap any volatile emissions. By monitoring for contaminants moving in the air and in the leachate, losses other than those due to biological activity can be taken into account.

Four chambers were prepared altogether, two for soil and two for sediments. One of each type of material was supplemented with an inorganic nutrient solution, while the other received distilled water. The experiment was run for 12 weeks, during which time the moisture content was maintained at a range of 8 to 12 percent, and the soil/sediment was tilled once a week. Samples were collected after tilling, at weeks 1, 2, 4, 8, and 12. Each time three samples were collected: one for moisture content, one for microbial population count, and one for contaminant concentration.

Considering that creosote is a complex mixture of chemicals with about 85 to 90 percent PAH content, it was not seen as practical to analyze for and monitor the disappearance of every chemical involved. Instead, 42 components of creosote, together with PCP, were monitored. These components were also grouped into 3 categories, based on molecular structure and biodegradability. Group 1, for example, contained 2-ring compounds, group 2 contained 3-ring compounds, and group 3 contained 4 and above ring compounds.

Results

In the soil samples, the addition of inorganic nutrients generally resulted in an increase in the rate and extent of biodegradation, especially with the low-molecular-weight PAHs. For example, by the end of the first week, 25 percent of the group 1 PAHs were degraded in the amended soil, compared to only 8 percent in the unamended soil. The effect of nutrient amendment was less pronounced in group 2 PAHs and even less so in group 3. Surprisingly, the extent of degradation (percent removal) of PCP in the unamended soil was higher than that in the amended one: 72 percent removal in the former compared to 55 percent in the latter.

The results observed in the sediment samples were somewhat different. The addition of inorganic nutrients did not seem to influence the rate of biodegradation of the different chemicals. The extent to which the group 1 PAHs were degraded, however, was higher in the amended sediment compared to the unamended one. The extent to which the other PAHs were degraded was about the same with or without amendments. In both sediments no PCP biodegradation was evident within the 12-week period.

These differences in observations between the soil and the sediment samples are thought to be somewhat related to the waste composition in each case. The creosote in the soil samples, having been exposed to the atmosphere, was thought to be more weathered. It contained a relatively lower concentration of the easily degradable low-molecular-weight PAHs. At the same time, the sediments contained several creosote phenolics which were not in the soil samples. The creosote phenolics were degraded before, and more rapidly than, the other contaminants in the sediments. A large drop in naphthalene and other 2-ring PAH concentrations was observed between weeks 8 and 12, after the creosote phenolics had disappeared.

Nutrient amendment did not seem to influence the number of total heterotrophic bacteria in the soil and in the sediments, when monitored over time. In the soil samples the number of total heterotrophic bacteria remained about the same throughout the duration of the experiment. In the sediments, on the other hand, bacteria seemed to acclimate over time and increase in number by about four orders of magnitude.

Questions to ponder

1. Why do you think the results for the sediment samples were so different from those for the soil samples?
2. What tests would you conduct to try to find out why the results were so different?
3. Do you think that the addition of distilled water to the unamended samples could have contributed to the results obtained?
4. Had the treatment period been longer, do you think that the results could have been different?

COMPOSTING

Composting is an aerobic biological process in which wet organic solids are oxidized to biologically stable forms such as humus. The most common applications of composting include treatment of agricultural residue, yard and kitchen waste, municipal solid waste, and sewage sludge. A considerable literature exists on composting, and readers are referred to other references such as Haug (1993) for additional information.

The high organic concentration and relatively low moisture content of compost result in substantial accumulation of heat released during biodegradation. Operating temperatures in compost piles often exceed 55°C. The high temperatures are useful in killing pathogenic organisms but also provide an environment for rapidly degrading a number of hazardous compounds.

The application of composting in the field of hazardous-waste treatment is relatively new (U.S. EPA, 1990; U.S. EPA, 1988). Hazardous materials are rarely present in concentrations necessary to support a composting operation. Instead, organics are added to contaminated soil in amounts that allow a composting environment to develop. Early attempts at hazardous-waste composting began in the 1960s when the process was used to biodegrade insecticides such as diazinon and parathion (Savage et al., 1985). Since then, composting has been used to biodegrade petroleum hydrocarbons, explosives, and chlorinated compounds in soils, sediments, and sludges.

Process Description

The basic principles involved in the composting of hazardous material are the same as for composting of nonhazardous waste. Four parameters need to be optimized in both cases: aeration, temperature, moisture content, and pH. In composting of hazardous material in contaminated soils, organic material is added in quantities needed to generate the necessary heat. While degradation of specific hazardous compounds is the focus of the process, design and operation is based on conventional composting criteria. An advantage, relative to composting of nonhazardous materials, is that organics used can be selected for ease of degradation and energy content.

Heat entrapment, or self-heating, is the most important distinguishing characteristic between land treatment and composting. While the waste is spread out in thin layers in land treatment, it is piled up in mounds or enclosed in reactors in composting. Pile dimensions are selected to ensure that the heat-generation rate resulting from microbial metabolism of organic material is greater than the rate of heat dissipated through the surface. Highly biodegradable material is usually added to a compost pile to increase metabolism and generate more heat.

In municipal waste composting, elevated temperatures are desired for pathogen destruction (pasteurization) so that the finished compost can be used as a soil amendment. For sewage sludge composting, a temperature of 55°C, for three continuous days, is required to ensure pathogen kills (Cookson, 1995). Pathogens are usually not of concern in hazardous materials. However, relatively high temperatures are desirable to enhance biodegradation rates and to increase solubility and mass transfer rates. Thermophilic microbial populations developing at high temperatures seem to have the ability to break down a wide range of chemical structures, and specific advantages may exist for high-temperature operation.

Composting is generally an aerobic process, although microscale anaerobic conditions surely occur in most systems. However, biodegradation of some hazardous materials, particularly halogenated compounds, is most readily initiated under anaerobic conditions. Examples include the initial steps in DDT and PCB biodegradation. Consideration should be given to design of systems in which the oxygen supply can be controlled.

TABLE 8.4
Types of amendments added in different processes

Process	Amendments added	Soil type	Reference
Windrow	Compost, 10% by volume	Variable	Sellers et al., 1993
Static pile	Mix of straw, wood chips, sawdust, and pine bark (inoculated with white rot fungus) 5% on dry-weight basis	Not specified	Holroyd and Caunt, 1995
Static pile	By percent volume: straw/manure: 47%, alfalfa: 38%, horse feed: 12%	Lagoon sediments	Williams and Myler, 1990
Windrow	Yard compost: 20% Turkey manure: 5%	Not specified	Biocycle, 1995
Windrow	Mix of dairy manure, wood chips, potato waste, and alfalfa: 70%	Not specified	Biocycle, 1995

Design Parameters

As was stated above, in hazardous-waste applications, the contaminated matrix (soil or sludge), on its own, does not provide the right environment for composting. For example, a compost pile needs to have a high enough concentration of readily degradable material to generate enough heat to raise the pile's temperature. Hazardous waste, for the most part, is not readily degradable. As such, the contaminated waste is usually mixed with highly degradable organic material that serves as a thermal source (LaGrega et al., 1994).

Compost piles usually need a bulking agent to improve soil structure, increase porosity, and allow for better air permeability. Addition of a bulking agent is particularly important with soils that are high in clay and silt, or with sludges and saturated soils where air permeability may be limited by the absence of air-filled pores. Selection of a bulking material capable of absorbing some of the excess moisture and creation of interconnected air-filled pores is important. For example, addition of two volumes of wood chips to one volume of sewage sludge can reduce the moisture content from 78 to 60 percent and provide satisfactory aeration capacity (Epstein and Alpert, 1980).

Often, the same material can serve as both a thermal source and a bulking agent. Sometimes a combination of different materials, collectively referred to as soil amendments, are used to create an optimum design. Examples of soil amendments used in different composting processes are given in Table 8.4. In most cases, the soil amendment used can also serve as a source of microbial seed to stimulate biodegradation, thus eliminating the need for a special inoculum. The types of materials used as soil amendments are discussed in the following sections.

Thermal source

Thermal sources are required where the contaminant concentration is not great enough to produce excess heat in the compost pile. The target contaminants are rarely present in high concentrations, and therefore most compost bioremediation

TABLE 8.5
Carbon-to-nitrogen ratio of some selected waste material

Waste	C:N
Sawdust	200–500
Straw, wheat	128–150
Straw, oats	48
Horse manure	25
Cow manure	18
Poultry manure	15
Grass clippings	12–15
Nonlegume vegetables	11–12
Activated sludge	6
Urine	0.8

Source: Diaz et al., 1993.

processes require addition of a thermal source. Material used as a thermal source should have a high, easily biodegradable organic content. Sources commonly used include livestock manure, such as chicken and horse manure, vegetation, such as tree and plant leaves, and food-processing wastes, such as molasses. Biodegradation of the organic material in the thermal source results in growth of high concentrations of microorganisms which usually include the species that degrade the target contaminants. In some cases the thermal source serves as a cometabolite needed for the degradation of a contaminant that does not serve as a carbon source for growth.

When adding the thermal source, the nutrient balance between carbon and nitrogen should be considered. The carbon-to-nitrogen ratio should not be higher than 20 or 25 to 1. Carbon-to-nitrogen ratios of selected waste materials are listed in Table 8.5. A waste that is too low in nitrogen may not be fully utilized by the microorganisms, in which case heat generation in the pile may be insufficient. Hence, if the thermal source used is too high in carbon, a nitrogen source may need to be added. The choice of what thermal source to use is often dictated by price and availability of appropriate materials.

Bulking agent

Addition of a bulking agent prevents compaction of soil and increases the porosity and availability of oxygen. The increased porosity allows greater drainage and may result in a decreased moisture content which can be detrimental to microbial activity. For this reason the use of an absorbent bulking agent which maintains a high moisture content while being resistant to compaction as well as degradation is desirable (Savage et al., 1985). Typical bulking agents in use include straw, hay (dried grass), rice hulls, and other fibrous vegetation, wood chips, and inert synthetic material. Shredded rubber tires can be used; they are nondegradable, but they do not absorb moisture and do not have much structure. Finished compost products can also be used. Paper is not a good choice, as it collapses (forms mats) when wetted.

Some bulking agents can be recycled and used in subsequent runs, decreasing waste production and providing acclimated microbial inoculum for subsequent contaminant degradation. Wood chips, finished compost, and shredded rubber tires are some examples of such bulking material. To recover the bulking material, however, the compost must be put through an energy- and time-consuming screening process which may result in aerosol emissions that could contain some hazardous constituents.

Inoculum

Satisfactory microbial inoculum for composting operations is usually present in the soil, thermal source, and bulking material. Petroleum hydrocarbons, nonhalogenated solvents, and many agricultural chemicals are metabolized by many microbial species found in soil. However, sewage sludge is often added to provide additional microorganisms and to decrease acclimation times. Recycling material from completed composting operations is also a good method of inoculating compost piles.

Specially developed inocula will need to be grown in the laboratory and added to the pile if there is reason to believe that the target contaminants can only be degraded by a limited group of microorganisms. For example, in a field application in Finland (Holroyd and Caunt, 1995), a fungal inoculum of *Phanerochaete chrysosporium* was needed to successfully biodegrade chlorophenol in a contaminated soil. This white rot fungus produces a special system of enzymes that is capable of oxidizing a very wide range of organic contaminants which are otherwise only partly biodegradable under anaerobic conditions. The fungus was first grown on a mix of straw, wood chips, sawdust, and pine bark to simulate the natural lignin environment. Once the fungus was well established, the mix was incorporated into the contaminated soil at 5 percent on a dry-weight basis. Over a period of 24 months, chlorophenol concentrations decreased from an average of 200 to 30 ppm. The requirement of a specific culture for degradation of chlorophenol is surprising because a number of bacterial species are capable of metabolizing the compound. However, particular environmental conditions may have been a factor in this case.

Pile composition

A key issue in designing a successful composting process is finding the right mix composition that will result in fast and extensive contaminant removal. Almost always laboratory or pilot-scale tests must be conducted to determine the optimum pile design. A relatively high percentage of amendments can result in higher porosity, better air distribution within the pile, more water-holding capacity, and hence better biodegradation. That is particularly true with soils having a high clay content. The more the amendments, however, the less soil that can be treated in one batch, and the more the land area that may be needed, or the longer the remediation period.

Stegmann et al. (1991) conducted laboratory studies on diesel-contaminated soil and showed that the higher the compost mass (in a soil-compost mix), the higher the microbial activity and the better the removal of the hydrocarbons. Best results were obtained with a soil:compost ratio of 2:1 (on a dry-weight basis) as compared to a ratio of 16:1, which had the worst results. Similar results were observed in bench-scale experiments on clay soil contaminated with the herbicide Dicamba (3,6-dichloro-2-methoxibenzoic acid) (Dooley et al., 1995). Compost mixes containing 41 percent amendment by weight (35 percent wood chips and 6 percent cow manure) had shorter lag periods, faster rate of degradation, greater extent of reduction, as well as greater extent of mineralization compared to compost mixes with just 10.8 percent amendment by weight (4.5 percent wood chips and 6.3 percent cow manure).

Hayes et al. (1995) reported on pilot-scale experiments in which they observed that soil amendment and nutrient addition had little effect on improving contaminant degradability. Selected amendments including shredded tree waste, wheat straw, and refinery sludge, at two concentration levels in each case, were used to treat soil contaminated with refinery waste and crude oil dating from 1920. Fifteen piles were constructed with varying configurations, and operated for 45 weeks. The results showed an average TPH removal of about 55 percent for all piles, with no significant difference between the different treatments. The main reason for this result is believed to be decreased bioavailability of the contaminants due to weathering. The limiting factor in biodegradability was not microbial activity in this case but rather the fact that the contaminants were either inaccessible to the microorganisms or were recalcitrant. Decreased bioavailability is caused by strong sorption onto clay and silt surfaces, binding of the chemicals to soil organic matter and humus, and entrapment in micropores.

Moisture content

As with landfarming, maintaining a moisture content that enhances microbial growth is essential for a successful composting process. In landfarming, optimum moisture content that ensures that enough pore space is filled with air to allow for aerobic activity is measured as a percentage of field capacity. In composting, moisture content is measured as a percentage of the water-holding capacity of the pile mix. Because of the added amendments the water-holding capacity of the compost mix is usually higher than the field capacity of unamended soil.

A moisture content equivalent to approximately 60 percent of water-holding capacity was optimum for microbial activity in soil compost mixtures (Stegmann et al., 1991). Higher moisture content results in decreased activity due to a reduction in air-filled pores. A lower moisture content produced a similar result because of reduced bioavailability. While the optimum moisture content is case-specific and depends on the pile composition and the kind of environment needed by the degrading microorganisms, a range of 50 to 80 percent of water-holding capacity seems to be generally acceptable.

Heat generation in compost piles

As microorganisms begin biodegrading the waste, the heat generated by metabolic activity is produced at a rate faster than the dissipation rate. Temperature profiles typically begin with a short lag phase associated with microbial acclimation. The temperature then increases at an exponential rate until a maximum is

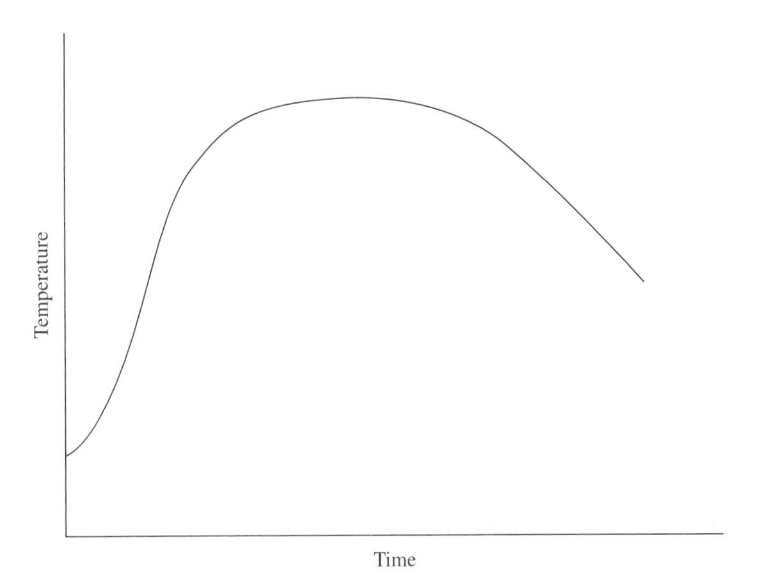

FIGURE 8.3
Change in temperature with time during composting.

reached. If the temperature enters the thermophilic range (> 45°C), major changes will take place within the microbial population. Microorganisms unable to tolerate high temperatures either die or sporulate, while thermophilic bacteria become favored and predominate. If temperatures are allowed to rise higher than 55 or 60°C, the thermophilic bacteria become affected and activity will decrease. Normally, compost piles are operated within either the mesophilic range (30 to 40°C) or the thermophilic range (50 to 60°C). Selection of the operating range is based on pilot-scale investigations and availability of thermal source material at an acceptable cost.

A typical temperature profile can be seen in Figure 8.3. Note that under controlled conditions, the temperature begins to decrease after a certain time period. That drop is associated with diminishing food supplies and a corresponding drop in microbial activity. This rise and fall in temperature is often used to monitor the performance of a compost pile (especially with sewage sludge). As the pile cools off and temperatures within the pile approach ambient temperature, the period of active composting is considered complete.

Another characteristic of successful composting is the change in texture and in smell of the pile at the end of treatment. At start-up, offensive odors may emanate from the sewage or manure added as a thermal source while at completion compost smells much like garden soil. Mixture texture also changes from rough and fibrous to a more homogeneous one. Changes in odor and texture are both the result of biodegradation of the organic matter. This mass conversion from solid to vapor form also results in reduction in pile size. Depending on the amount of organic material mixed in with the soil, the pile's mass can be expected to be reduced by up to 40 percent (Hart, 1991).

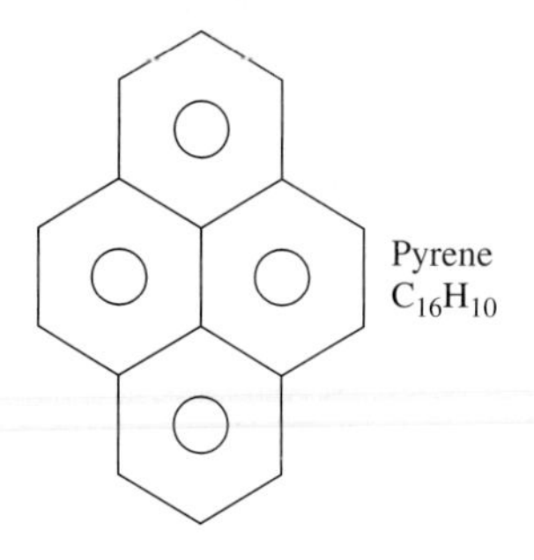

EXAMPLE 8.3. PERFORMANCE OF A COMPOSTING SYSTEM. Sludge contaminated with pyrene (20,000 ppm) is to be treated in a compost pile that weighs approximately 2,500 kg. The pile is prepared such that the sludge makes up 25 percent by mass of the compost mixture. Treatment will be in two stages, an active composting stage, for 40 days, and a curing stage (during which the pile is turned periodically), for 90 days. If the half-life for pyrene during the active stage is 30 days, and during the curing stage is 55 days, and if the compost mass is expected to be reduced by 30 percent during the active stage, but hardly at all during the curing stage, determine the expected final concentration of pyrene in the pile.

Solution

1. Find the concentration of pyrene initially available in the compost pile:

$$C_{\text{pyrene}} = \frac{20{,}000 \text{ mg pyrene}}{\text{kg sludge}} \times \frac{0.25 \text{ kg sludge}}{\text{kg compost}}$$

$$= 5{,}000 \text{ mg pyrene/kg compost} = 5{,}000 \text{ ppm}$$

$$M_{\text{pyrene}} = \frac{5{,}000 \text{ mg pyrene}}{\text{kg compost}} \times 2{,}500 \text{ kg compost}$$

$$= 1.25 \times 10^7 \text{ mg pyrene}$$

2. Find the drop in pyrene concentration based on degradation rates in the active and curing stages:
   In the active stage:

$$\ln(0.5) = -k \text{ (30 days)}$$

$$k = 0.023 \text{ per day}$$

After 40 days, pyrene concentration is

$$\frac{C}{C_o} = e^{-kt}$$

$$\frac{C}{C_o} = e^{-(0.92)} = 0.4$$

$$C_{40} = 0.4C_o$$

Curing stage:

$$\ln(0.5) = -k \text{ (55 days)}$$

$$k = 0.013 \text{ per day}$$

After 130 days, pyrene concentration is

$$\frac{C_{130}}{C_{40}} = e^{-kt}$$

$$\frac{C_{130}}{C_{40}} = e^{-(1.17)} = 0.31$$

$$C_{130} = 0.31(0.4)C_o = 0.124\ C_o$$

3. To find final concentration in the pile:

$$M_{\text{compost}} = (1 - 0.3)(2{,}500 \text{ kg}) = 1{,}750 \text{ kg}$$

$$M_{130} = 0.124 \times \frac{5{,}000 \text{ mg pyrene}}{\text{kg compost}} \times 2{,}500 \text{ kg compost}$$

$$= 1.55 \times 10^6 \text{ mg}$$

$$C_{130} = \frac{1.55 \times 10^6 \text{ mg pyrene}}{1{,}750 \text{ kg compost}} = 886 \text{ mg/kg} = 886 \text{ ppm}$$

As can be seen, the actual reduction in pyrene concentration is not very great.

Types of Composting Systems

Three types of composting systems, windrow, static piles, and closed reactors, are in widespread use. Windrow and static piles, which are sometimes referred to as open systems, are more commonly used than closed reactors. In open systems, the material to be composted is piled on an impervious platform, such as concrete or asphalt. Often a polyethylene liner is used as an extra precaution to ensure that no contaminants leach to the soil through cracks, should they exist. The mode of aeration is what differentiates between a windrow and a static pile. In a windrow operation, the pile is aerated by turning the compost mixture, either manually or mechanically. In static piles forced aeration is utilized. A system of perforated pipes is laid out at the base of the pile (Figure 8.4) and aeration can be achieved in a positive mode (pushing air through the pile) or in a negative mode (by applying vacuum on the pile). In closed system (also known as an in-vessel system) the compost mixture is placed in an enclosed reactor and mixing and aeration are achieved by turning and forced aeration.

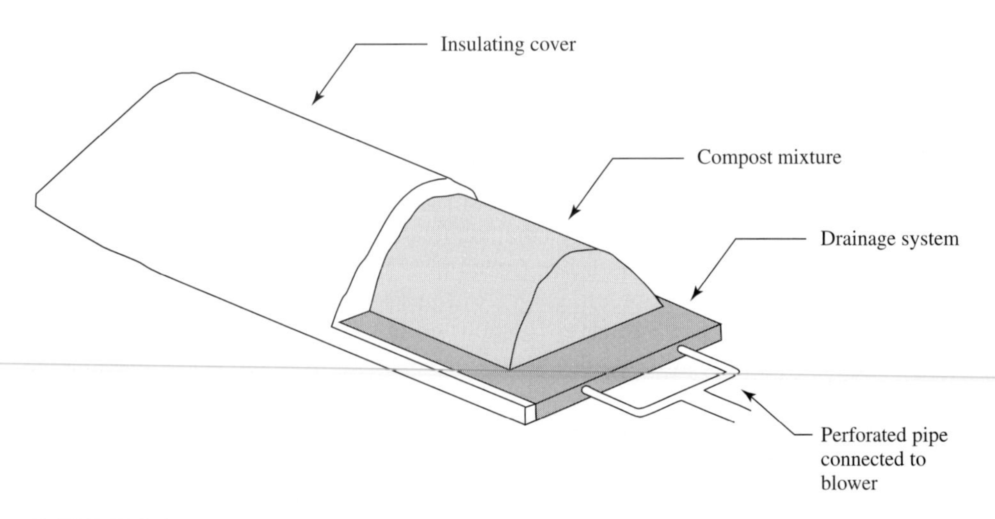

FIGURE 8.4
Schematic diagram of an aerated static pile.

Setting the ground for a biopile to treat soil contaminated with petroleum hydrocarbons. A network of embedded piping is installed to deliver oxygen to the degrading microorganisms. *(Courtesy of Groundwater Technology, Inc.)*

Windrows

Windrows are the simplest of the three compost-system designs. The name is given because of the way the material to be composted is piled on a platform in long mounds, or windrows. Windrow size significantly affects the performance of a compost pile. To maintain high temperatures within the windrow, the compost pile should be large enough to allow heat generated by metabolic processes to exceed the heat lost at the exposed surfaces. The larger the cross-sectional area of the windrow, the smaller the surface-to-volume ratio and hence the more heat will be retained in the pile (Hay and Kuchenrither, 1990). Temperature of the windrow can be controlled by turning the pile, which also provides aeration. Pile width is normally between 3 and 4 m (10 and 12 ft), whereas the height can be up to 1.2 to 1.5 m (4 to 5 ft) (Cookson, 1995). Some windrows are as wide as 6 m (Biocycle, 1995).

To provide better homogeneity within the mix and hence better biodegradation, the material to be composted is usually mixed together before forming the pile or windrow. Premixing is especially important if nutrients and other amendments are to be added. Soluble amendments and nutrients can be added with the makeup water to ensure good distribution within the pile. If added later with the irrigation water, nutrient mobility and bioavailability may be limited by the movement of liquid in the compost matrix. In some cases the material to be composted is laid out in layers and then mixed in to form the windrow. For example, at Seymour Johnson Air Force Base in North Carolina a windrow system was set up in five layers. A compost layer (composted tree trimmings and grass) was set at the bottom to protect the sealed base from damage from the heavy machinery. A layer of contaminated soil was added followed by a layer of compost and then a layer of turkey manure as a thermal source. The compost layer made up approximately 10 percent of the pile mass, while the turkey manure made up only about 5 percent. Two passes with a turner were then made to ensure thorough mixing (Biocycle, 1995).

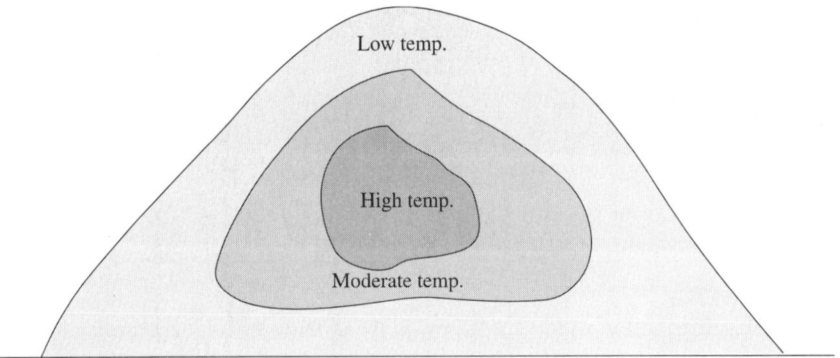

FIGURE 8.5
Typical temperature profile distribution within a windrow.

Windrow aeration. Sizing of the windrow depends to some extent on the aeration method used. Mechanical mixing is typically done using either a front-end loader or a turner. A front-end loader is less expensive, but the efficiency of mixing depends on the time spent by the operator to thoroughly turn the pile. A turner, on the other hand, mounts the windrow and turns and mixes the soil as it drives over the windrow, and hence results in better mixing and aeration.

The frequency of turning depends on the objectives to be achieved. Turning is primarily done to aerate the pile and/or release heat and decrease temperatures. A typical temperature profile distribution can be seen in Figure 8.5. The difference in temperature between the zones and the size of each zone depends to some extent on the frequency of turning. Turning helps redistribute the temperature profile so that the outer layers which are at lower temperature become exposed to the higher temperature within. An added advantage to turning is mixing of the contaminants with the soil amendments, which results in better distribution and increased bioavailability. Windrows are turned as often as once a day (U.S. EPA, 1995), to as little as once a month (Sellers et al., 1993), to never at all through the duration of the treatment period.

Piles which are not turned depend on what is sometimes referred to as passive aeration to maintain aerobic conditions. Passive aeration is the outcome of temperature buildup within the pile and the resulting temperature gradient between the inside of the pile and ambient atmosphere which causes convective airflow in and out of the pile. Such aeration is limited by the porosity of the matrix (air-filled space) and the depth of the pile as indicated in Figure 8.6. The outer layer which is exposed to the atmosphere would have the higher oxygen concentration whereas the layer deep within the pile would be the most oxygen-deficient. If the pile dimensions are large enough, the oxygen diffusing through the outside layers will be used up before it can reach deep inside the windrow.

An example of the limited ability of oxygen to diffuse inside a windrow was observed in a field application (Banazon et al., 1995) where a pile 12 m wide, 2.5 m high, and 26 m long was constructed to treat soil contaminated with ethylbenzene, styrene, and other petroleum hydrocarbons. To enhance passive aeration, four pipes were installed at the base of the pile and three more were embedded at a depth of

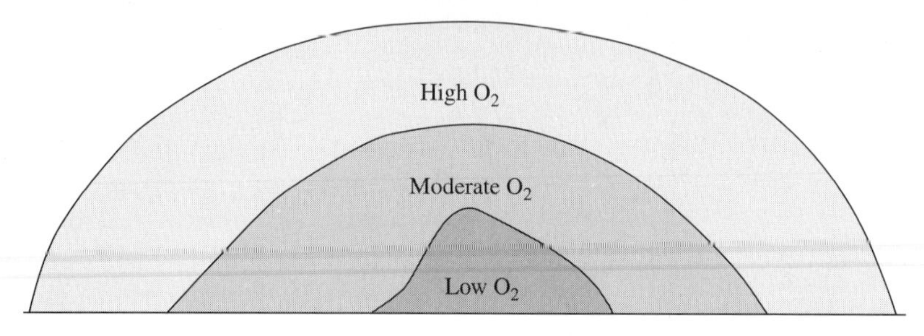

FIGURE 8.6
Oxygen distribution within a windrow in the absence of frequent turning.

1.5 m below the surface of the pile. The pile was then covered with a 30-cm layer of wood chips and a 20-mil polyethylene cover. After 168 days of treatment both the ethylbenzene and the styrene concentrations had dropped from a high of 2,190 and 365 ppm, respectively, to less than 1 ppm, within the top 80-cm layer. Total petroleum hydrocarbon concentrations dropped from a high of 30,000 ppm to close to 1,000 ppm within the same depth. At depths greater than 0.8 m, however, the contaminant concentrations had changed insignificantly. The average temperature of the pile during the cold winter months was close to 15°C at the time that ambient temperatures were less than –10°C. The results seemed to indicate that oxygen diffusion was limited to the top layer.

Covering windrows. In some cases windrows are constructed in warehouses or similar structures. In such a protected environment there is usually no need for a cover. However, windrow composting is usually carried out in open environments and covers are necessary to maintain temperatures within the pile, minimize wind erosion, and prevent saturation by rain. Rain is a particular problem because drainage from saturated piles will contain contaminants. Covers are also needed if there is a potential for hazardous VOC emissions. The covering material used is typically either synthetic such as high-density polyethylene or a layer of organic material such as wood chips or compost.

Static piles

In this configuration, the material to be composted is laid out on a system of perforated pipes connected to a blower or a vacuum pump. Aeration of static piles can be achieved either through a positive mode (forced aeration) or through a negative mode (vacuum applied). Vacuum-induced aeration is generally preferred because emissions of volatile compounds are minimized and off-gases can be treated separately using biofiltration (see Chapter 10) or catalytic oxidation. Off-gas recycling may also be effective, a process in which the static pile itself becomes a form of biofilter. Negative aeration may not be the best choice in climates with severe cold temperatures. Cold air being sucked into the pile would cause a drop in temperatures, especially at the outer layers. Piles with vacuum-induced aeration are sometimes referred to as vacuum-heap treatment. Positive-mode aeration results in heating of the pile because the blower discharge will be heated by compression.

Airflow is used to control both the temperature and the oxygen content within the static pile. The layout of the perforated pipes and the rate of aeration used are

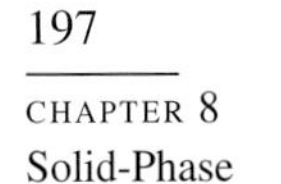

Ventilation system for biopiles at a petroleum refinery. Protective sheeting is provided to prevent runoff and to trap heat during cold weather. *(Courtesy of Groundwater Technology, Inc.)*

TABLE 8.6
Pile characteristics in some static pile operations

Pile dimensions, m			Cover	Mode of aeration	Comments	Reference
Width	Height	Length				
30	2	50	HDPE liner	Positive	Pile inoculated with white rot fungus	Holroyd and Caunt, 1995
16.5	2.4	16.5	9 mil HDPE	Negative	Covers supported by PVC frame	Peterson et al., 1995
9.1	3.0	24.4	3 layers of 6 mil plastic	Negative	Piping embedded at 1.2 and 2.1 m. Slotted PVC pipes to provide passive airflow, laid next to vapor extraction pipes	Sellers et al., 1993
5.5	1.6	9.1	Sawdust, wood	Negative	Blower cycling controlled by timer and temperature feedback system	Williams and Myler, 1990
12	2.5	26	0.3-m layer of wood chips, and 20-mil polyethylene cover	Negative	75% of the off-gas was recycled into the pile	Benazon et al., 1995

essential design parameters in static pile treatment. Pipes at the base are always embedded in a layer of highly permeable material such as gravel, sand, wood chips, or compost. The ability to aerate a static pile without disturbing the compost mixture allows use of larger pile dimensions than what is typical for windrows. Static pile heights up to 3 m are common (Table 8.6), but piles as high as 6 m have been used (Cookson, 1995). Piles several meters high are sometimes referred to as biopiles. In deep piles pipe networks may be placed at several elevations for aeration and the delivery of moisture and nutrients. Smaller static piles are usually watered by means of soaker hoses laid out at the outer surface of the pile.

Static pile aeration. For a pile design to be successful, the rate of aeration used must correspond to the microbial activity taking place. When treatment is initiated and microbial activity accelerates, the need for oxygen is high, the temperature builds up relatively quickly, and a high airflow rate is required. As remediation proceeds, organic concentrations diminish, microbial activity decreases, the need for oxygen decreases, the temperature drops, and a lower airflow rate is required. Three modes of aeration are possible, depending on the aeration rate (Leton and Stentiford, 1990):

1. Fixed rate: The rate of aeration is fixed and control is achieved by turning the airflow off and on, as required. For example, aeration may be on for 6 min and off for 18 min. The main disadvantage of fixed-rate aeration is that the pile may be overaerated (cooled) at the beginning of the treatment period when microbial activity is not as high, and underaerated later in the process when microbial activity and hence temperature are at their highest.
2. Variable rate: This is the least used mode of aeration. In this case, the rate of aeration starts out high and is gradually decreased (step function) with time. This method is hard to implement because day-to-day monitoring is required.
3. Automated aeration: Computer programming is used to automatically regulate aeration in response to fluctuations in temperature readings.

Biopiles for the treatment of approximately 20,000 yd^3 of soil contaminated with petroleum hydrocarbons from underground storage tanks. *(Courtesy of Groundwater Technology, Inc.)*

Static pile cover. If the pile is covered, especially with an impermeable layer such as plastic sheeting or HDPE liner, measures must be taken to allow for air to penetrate the pile to compensate for what is being pulled out. The simplest technique would be to create openings or slots in the cover. However, short circuiting in the open areas and no airflow in the covered parts may result. To overcome this problem, Peterson et al. (1995) used a PVC framing system to keep the liner 6 to 24 in above the soil surface. The support, however, was relatively weak. It resulted in sagging, water ponding, and tears in the liner.

An alternative approach is to allow for passive airflow into the pile by using embedded, slotted PVC pipes that are open to the atmosphere, or by placing a layer of gravel between lifts to create an air-permeable layer. A third alternative is to redirect part of the outlet stream back into the pile, while supplementing with fresh air. The degree of efficiency and problems associated with each of these techniques are not well documented.

Closed reactors

In-vessel composting, although capital-intensive, provides the most process control and has several advantages over open systems. Closed reactors are generally equipped with blending devices, designed to allow for frequent, if not continuous, mixing of the waste (rotating drums, mixed tanks or chambers). Mixing improves distribution of the target contaminants within the compost mass and allows for more contact between the microorganisms and the chemicals, thus increasing the potential for biodegradation (Hart, 1991). Closed reactors also offer better control of air emissions. Volatile organic compounds, as well as offensive odors, are contained and can be recycled or treated separately. Since the pile is not exposed to the atmosphere, heat dissipation is minimized and temperature control and oxygenation can be achieved through forced aeration. At the same time, the closed environment allows for maintaining optimum moisture content, as well as eliminating leaching, and hence soil and groundwater contamination in the treatment area.

CASE STUDY

A field study was conducted at the Louisiana Army Ammunition Plant to investigate the effectiveness of composting, as a technology, in remediating lagoon sediments contaminated with explosives and propellants (Williams and Myler, 1990; Williams et al., 1992). The contaminants included TNT (2,4,6-trinitrotoluene), HMX (octahydro-1,3,5,7-tetranitri-1,3,5,7-tetraazocine), RDX (hexahydro-1,3,5-trinitro-1,3,5-triazine), and tetryl (*N*-methyl-*N*,2,4,6-tetranitroaniline).

The first step in preparing the compost mixture was to homogenize the contaminated sediment and analyze for explosives and propellant concentrations. The sediments were found to contain 56,800 mg/kg of TNT, 17,900 mg/kg of RDX, 2,390 mg/kg of HMX, and 650 mg/kg of tetryl. Bulking agents and thermal sources were then added and mixed with the sediments. The composition of the final mix, on a weight basis, was 24 percent contaminated sediments, 10 percent alfalfa, 25 percent soiled bedding (straw/manure), and 41 percent horse feed. A small amount of fertilizer was also added to provide nitrogen and phosphorus for microbial growth. A carbon-to-nitrogen ratio of 30:1 was used as a design criterion. Each pile was about 26.6 m^3 in volume and 4,400 kg in weight. Sawdust, wood chips, and baled straw were used as an insulating cover for the pile, as well as a base.

Two aerated static piles were constructed on top of concrete test pads lined with drainage channels. The drained water was reapplied to the compost pile whenever moisture adjustments were needed. Each compost pile was individually covered by an open-sided structure to protect against precipitation and leaching. An explosionproof radial-blade blower, connected to a system of perforated polyethylene tubing, was used to pull air through the compost pile. A programmable timer, together with a temperature feedback system, was used to control blower cycling and temperature within the pile. One of the piles was kept at approximately 35°C (mesophilic range), while the other was kept at approximately 55°C (thermophilic range). Two landfill thermocouple probes were also used to monitor temperature within the pile. One was placed near the blower end of the pile and the other was placed in the center of the pile, next to the thermistor controlling the cycling of the blower.

The piles were operated over a period of 153 days during which samples were taken at 9 different times and analyzed for contaminant concentration. The samples were obtained from the central region of the pile, using a soil auger, from at least three different locations along the length of the pile at each sampling time. The moisture content in the piles ranged between 25 and 56 weight percent. It was noticed that whenever the moisture level fell below 35 percent, the temperature in the pile dropped significantly. As a result, water was added whenever the moisture dropped below 40 percent. Through the length of the experiment, the compost piles were dismantled, re-moistened, and remixed three different times.

Results

Williams et al. (1992) observed greater reductions in the contaminants concentrations in the thermophilic compared to the mesophilic pile. Total explosives in the thermophilic pile dropped from an average of 17,870 to 74 mg/kg, while in the mesophilic pile the decrease was from 16,460 to 326 mg/kg. One should keep in mind, however, that these values only reflect the concentrations of the extractable fraction of the contaminants. The fraction that is sorbed onto the compost matrix or bound to humic substances is not included in these numbers. Volatilization was not considered to be a removal pathway owing to the low vapor pressures of the contaminants.

Aside from monitoring the concentrations of the parent compounds, the authors also monitored the concentrations of specific TNT transformation products. Transformation products increased in concentration up to sixfold over the first several weeks, and then decreased to very low levels by the end of the test period.

Visual examination of the collected soil samples indicated a larger amount of fungal growth in the mesophilic pile compared to the thermophilic pile. The appearance and smell of the compost pile changed considerably over the duration of the experiment. When the pile was first mixed, it had a rough-textured fibrous look and it smelled of manure. At the end of the experiment the pile had the look and smell of loamy soil.

Questions to ponder

1. Why do you think temperature in the pile dropped as moisture content dropped?
2. Why do you think contaminant removal in the thermophilic pile was higher than in the mesophilic pile?
3. Comment on the results of the transformation products.

Advantages and Disadvantages

Advantages of composting include relatively low energy demands, low sludge and brine production, applicability to most organic compounds, and tolerance to relatively high metal concentrations (U.S. EPA, 1985). Retention time, i.e., degradation time for similar compounds, is much shorter in composting than it is in in situ or land treatment, for example, weeks rather than months (Savage et al., 1985). Land requirements are also less for composting than they are for landfarming, and water-contamination problems are minimized (U.S. EPA, 1990). Compared to nonbiological treatment processes, such as incineration, composting is much cheaper and a much simpler technology to apply. Disadvantages of composting include high maintenance requirements and high air emissions because of the high temperatures involved. Air emissions are easier to control, however, in composting than in landfarming. Because of the high temperatures and the periodic aeration, the moisture content needs to be monitored closely to maintain optimum microbial activity.

PROBLEMS AND DISCUSSION QUESTIONS

8.1. A land treatment unit has a surface area of 4 acres and a soil depth of 9 in, with a soil field capacity of 32 percent and a dry bulk density of 1.78 g/cm^3. If the soil currently has a moisture content of 12 percent, then how much water (in gallons) should be added to raise the moisture content to 80 percent of field capacity?

8.2. A soil has a field capacity of 27 percent (g water per g dry soil). If the moisture content is raised from 60 percent to 90 percent of field capacity, then what is the drop in air porosity? Assume that the soil is saturated at field capacity.

8.3. A land treatment unit (LTU) has an area of 2 acres and an average soil depth of 70 cm. Soil-water analysis shows that the top 20 cm has a moisture content of 11 percent whereas the remaining 50 cm has a moisture content of 14 percent. How much water (in liters) should be added to the LTU to bring the moisture content to 16 percent? The soil has a dry bulk density of 1.85 g/cm^3.

8.4. In a laboratory-scale experiment, 20 mg of ammonium nitrate (NH_4NO_3) and 3 mg of potassium phosphate (KH_2PO_4) are added to a flask containing 10 g of soil. Find the N:P ratio of the inorganic nutrients added.

8.5. Soil from a contaminated site has an organic carbon content of 1.5 percent by weight, a nitrogen content of 10 ppm, and a phosphorus content of 4 ppm. Determine the mass of diammonium phosphate [$(NH_4)_2HPO_4$] that must be added to the soil to raise the C:N:P ratio to 100:10:1.

8.6. In a composting process, 20 kg of straw (oats) are mixed together with 10 kg of horse manure and 975 kg of soil contaminated with crude oil, to form one cubic meter of compost mix.

a. If the soil has an organic carbon content of 8,000 ppm and a nitrogen content of 150 ppm, then using Table 8.1 find the C:N ratio of the mix.

b. How much activated carbon would need to be added to raise the ratio to 20:1? Comment on your answer.

8.7. Clay soils generally have a higher porosity compared to sandy soils. Given that fact, Equation (8.1) would suggest that oxygen can diffuse deeper in a clay soil compared to a sandy soil. If that is the case, then why do you think it is that the EPA recommends shallower depth for clay soils in land treatment units compared to sandy soils?

8.8. Using Equation (8.1), find the depth to which oxygen can diffuse in a ponded soil that has a porosity of 0.35. Assume that oxygen solubility in water is 8 mg/L and that oxygen diffusivity in water is 2.5×10^{-5} cm²/s.

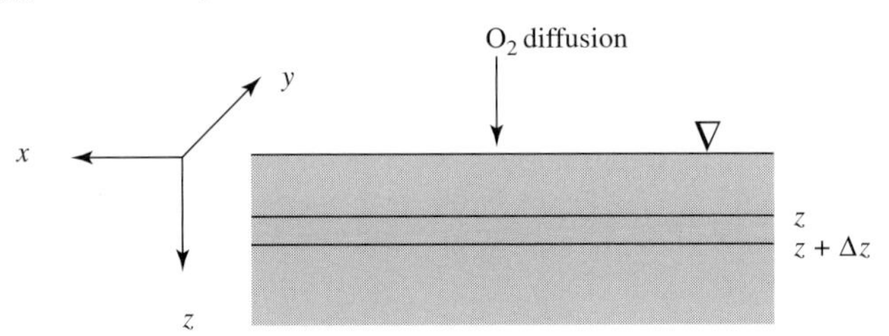

8.9. Soil in a land treatment unit is 12 in deep; it has a bulk density of 1.55 g/cm³ and a moisture content of 14 percent by weight.
a. Find the air porosity of the soil.
b. How fast will oxygen be consumed if the rate of oxygen uptake is 150 g/m³ · h? Assume a soil temperature of 20°C.

8.10. A laboratory experiment was conducted to simulate land treatment of diesel-fuel–contaminated soil. Results monitored over 46 days showed a drop in concentration from 900 to 300 ppm. Assuming first-order rate kinetics ($C = C_o e^{-kt}$):
a. Find the degradation-rate constant k.
b. Find the half-life ($t_{1/2}$).
c. Considering that the experiment was conducted under room temperature (22°C), you are asked to estimate the degradation-rate constant at a field temperature of 27°C. Use the Arrhenius equation: $k_T = k_{22°C}\,\Theta^{(T-22°C)}$ (use $\Theta = 1.02$).
d. What is the half-life for diesel degradation at field temperatures?

8.11. A proposal has been made to treat an organic sludge composed of semivolatile hydrocarbons by landfarming. The tentative plan is to inject the sludge into soil to a depth of 15 cm at a concentration of 1 g TOC/kg of soil. Pilot-scale experiments have been conducted and the degradation rate of contaminants in soil has been determined to be

$$r_{TOC} = -2\,s \text{ mg TOC/kg-day}$$

where s = soil concentration, g TOC/kg soil

The z direction flux of contaminants is

$$J_{z\,TOC} = -D_g \frac{\partial C_{g\,TOC}}{\partial z} \text{ g/m}^2 \cdot \text{day}$$

Assume local phase equilibrium.

$$K_{SD} = \frac{s}{C_{L\,TOC}}$$

$$HC_{L\,TOC} = C_{g\,TOC}$$

a. Write a gas-phase TOC mass balance for the horizontal slice of the 15-cm layer shown below. Use the sorbed concentration s as the dependent variable, convert the liquid-phase concentration ($C_{L\text{TOC}}$) and the gas-phase concentration ($C_{g\text{ TOC}}$) to s. State your assumptions for the boundary conditions.

b. Briefly suggest appropriate regulatory constraints on the use of landfarming for this type of waste.

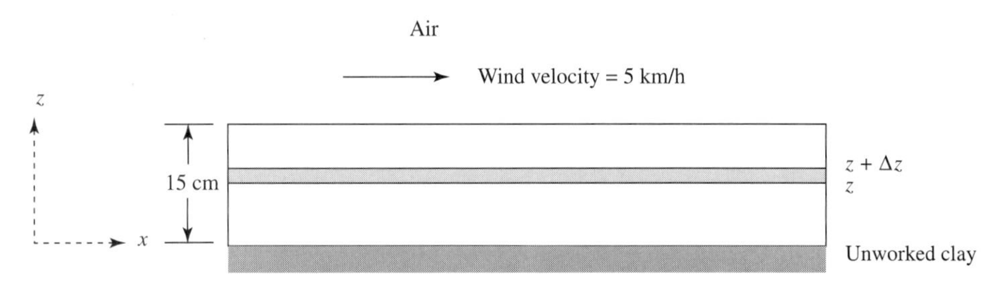

8.12. In a field-scale land treatment operation, the rate of removal of diesel was monitored over two depths, 0 to 15 cm and 15 to 30 cm. A first-order rate constant was estimated for each depth and found to be 0.018/day for the top layer and 0.022/day for the bottom one.

a. Find the half-life in days for each layer.

b. Are these the results you would expect? Why?

c. What factors could possibly contribute to these results?

8.13. Leaking underground gasoline storage tanks in California have contaminated a large number of sites with fuel containing approximately 15% MTBE by volume. Assume that gasoline can be characterized by three compounds, MTBE, benzene, and octane. Discuss the applicability of excavation and landfarming for bioremediation of the soils.

Compound	H	$\log K_{OW}$	Solubility, mg/L
MTBE	0.03	0.08	40,000
Benzene	0.20	2.12	1,780
Octane	0.14	5.18	72

REFERENCES

Benazon, N. D., W. Belanger, D. B. Scheurlen, and M. J. Lesky (1995): "Bioremediation of Ethylbenzene- and Styrene-Contaminated Soil Using Biopiles," in *Biological Unit Processes for Hazardous Waste Treatment* edited by R. E. Hinchee, R. S. Skeen, and G. D. Sayles, Battelle Press, Columbus, Ohio.

Benedict, A. H., E. Epstein, and J. Alpert (1988): *Composting Municipal Sludge—A Technology Evaluation,* Noyes Data Corporation, Park Ridge, NJ.

Biocycle (1995): "Cleaning Carolina Soil, Successful Bioremediation with Compost," *Biocycle,* vol. 36(2), pp. 57–59.

Blackburn, J. W., and W. R. Hafker (1993): "The Impact of Biochemistry, Bioavailability and Bioactivity on the Selection of Bioremediation Techniques," *Trends in Biotechnology,* vol. 11.

Bleckmann, C. A., E. J. Wilson, K. W. Hayes, and N. L. Hercyk (1995): "Land Treatment of Produced Oily Sand," in *Applied Bioremediation of Petroleum Hydrocarbons* edited by R. E. Hinchee, J. A. Kittel, and H. J. Reisinger, Battelle Press, Columbus, Ohio.

Bogart, J. D., and J. R. League (1988): Biological Remediation of Underground Storage Facilities. Air Pollution Control Association, 81st Annual Meeting of APCA, paper 88-7.2.

Calabrese, V. G., R. E. Elsavage, G. K. Bissonnette, and A. J. Sexstone (1993): "Mobility and Enhanced Biodegradation of a Dilute Waste Oil Emulsion During Land Treatment," *Journal of Industrial Microbiology,* vol. 12, pp. 13–20.

Cookson, J. T. (1995): *Bioremediation Engineering Design and Application,* McGraw-Hill, New York.

Diaz, L. F., G. M. Savage, L. L. Eggerth, and C. G. Golueke (1993): *Composting and Recycling Municipal Solid Waste,* Lewis Publishers, Boca Raton, FL.

Dooley, M. A., K. Taylor, and B. Allen (1995): *Composting of Herbicide-Contaminated Soil in Bioremediation of Recalcitrant Organics,* R. E. Hinchee, D. B. Anderson, and R. E. Hoeppel, eds., Battelle Press, Columbus, Ohio.

Dupont, R. R., R. C. Sims, J. L. Sims, and D. L. Sorensen (1988): *In Situ* Biological Treatment of Hazardous Waste-Contaminated Soils, in *Bioremediation Systems,* vol. II, edited by Donald L. Wise, CRC Press, Florida.

Ellis, E. (1994): "Reclaiming Contaminated Land: In Situ/Ex Situ Remediation of Creosote- and Petroleum Hydrocarbon-Contaminated Sites," in *Bioremediation Field Experience,* P. E. Flathman, D. E. Jerger, and J. H. Exner, eds., Lewis Publishers, Chelsea, MI, pp. 107–143.

Epstein, E., and J. E. Alpert (1980): "Composting of Industrial Wastes," *New Ultimate Disposal Options,* vol. 4, chap. 15, R. B. Pojasek ed., Ann Arbor Science, Michigan, pp. 243–252.

Fiorenza, S., K. L. Duston, and C. H. Ward (1991): "Decision Making—Is Bioremediation a Viable Option?," *Journal of Hazardous Materials,* vol. 28, pp. 171–183.

Flathman, P. E., B. J. Krupp, J. R. Trausch, J. H. Carson, R. Yao, G. J. Laird, P. M. Woodhull, D. E. Jerger, P. R. Lear, and P. Zottola (1995): Biological Solid-Phase Treatment of Vinyl Acetate-Contaminated Soil: An Emergency Response Action, *Proceedings of the Air and Waste Management Association 88th Annual Meeting and Exhibition,* San Antonio, TX, paper 95-FA163.02.

Fogel, S. (1994): "Full-Scale Bioremediation of No. 6 Fuel Oil-Contaminated Soil: 6 Months of Active and 3 Years of Passive Treatment," in *Bioremediation Field Experience,* P. E. Flathman, D. E. Jerger, and J. H. Exner, eds., Lewis Publishers, Chelsea, MI, pp. 161–175.

Genes, B. R., and C. C. Cosentini (1993): "Bioremediation of Polynuclear Aromatic Hydrocarbon Contaminated Soils at Three Sites," in *Hydrocarbon Contaminated Soils,* vol. III, E. J. Calabrese and P. T. Kostecki, eds., Lewis Publishers, Chelsea, MI, pp. 323–331.

Hanstveit, A. O., W. J. Th. van Gemert, D. B. Janssen, W. H. Rulkens, and H. J. van Veen (1988): "Literature Study on the Feasibility of Microbiological Decontamination of Polluted Soils," in *Biotreatment Systems,* vol. I, ed. by Donald L. Wise, CRC Press, Inc., Florida.

Harmsen, J. (1991): "Possibilities and Limitations of Landfarming for Cleaning Contaminated Soils," in *On-Site Bioreclamation Processes for Xenobiotic and Hydrocarbon Treatment,* R. E. Hinchee and R. F. Olfenbuttel, eds., Butterworth-Heinemann, Massachusetts, pp. 255–272.

Hart, S. A. (1991): "Composting Potentials for Hazardous Waste Management," in *Biological Processes, Innovative Hazardous Waste Treatment Technology Series,* vol. 3, ed. by H. M. Freeman and P. R. Sferra, Technomic, Lancaster, PA.

Haug, R. T. (1980): *Compost Engineering: Principles and Practice,* Ann Arbor Science, Ann Arbor, MI.

Haug, R. T. (1993): *The Practical Handbook of Compost Engineering,* Lewis, CRC Press, Boca Raton, Florida.

Hay, J. C., and R. D. Kuchenrither (1990): "Fundamentals and Application of Windrow Composting," *Journal of Environmental Engineering,* vol. 116, pp. 746–763, July/August.

Hayes, K. W., J. D. Meyers, and R. L. Huddleston (1995): "Biopile Treatability, Bioavailability, and Toxicity Evaluation of a Hydrocarbon-Impacted Soil," in *Applied Bioremediation of Petroleum Hydrocarbons,* ed. by R. E. Hinchee, J. A. Kittel, and H. J. Reisinger, Battelle Press, Columbus, Ohio.

Hinchee, R. E., and M. Arthur (1991): "Bench Scale Studies of the Soil Aeration Process for Bioremediation of Petroleum Hydrocarbons," *Applied Biochemistry and Biotechnology,* vol. 28/29, pp. 901–906.

Holroyd, M. L., and P. Caunt (1995): "Large-Scale Soil Bioremediation Using White-Rot Fungi in Bioaugmentation for Site Remediation," R. E. Hinchee, J. Fredrickson, and B. C. Alleman, eds., Battelle Press, Columbus, Ohio.

Huling, S. G., D. F. Pope, J. E. Mathews, J. L. Sims, R. C. Sims, and D. L. Sorenson (1995): "Wood Preserving Waste-Contaminated Soil: Treatment and Toxicity Response," in *Bioremediation of Recalcitrant Organics,* R. E. Hinchee, D. B. Anderson, and R. E. Hoeppel, eds., Battelle Press, Columbus, Ohio, pp. 101–109.

Jerger, D. E., P. M. Woodhull, P. E. Flathman, and J. H. Exner (1994): "Solid-Phase Bioremediation of Petroleum Hydrocarbon-Contaminated Soil: Laboratory Treatability Study through Site Closure," in *Bioremediation Field Experience,* P. E. Flathman, D. E. Jerger, and J. H. Exner, eds., Lewis Publishers, Chelsea, MI.

King, R. B., G. M. Long, and J. K. Sheldon (1992): *Practical Environmental Bioremediation,* Lewis Publishers, Boca Raton, FL.

Leavitt, M. E. (1992): "In Situ Bioremediation of Diesel Fuel in Soil," in *Bioremediation: The State of Practice in Hazardous Waste Remediation Operations,* Seminar sponsored by Air and Waste Management Association and HWAC, January, 1992.

LaGrega, M. D., P. L. Buckingham, and J. C. Evans (1994): *Hazardous Waste Management,* McGraw-Hill, New York.

Leton, T. G., and E. I. Stentiford (1990): "Control of Aeration in Static Pile Composting," *Waste Management and Research,* vol. 8, pp. 299–306.

McGinnis, G. D., H. Borazjani, M. Hannigan, F. Hendrix, L. McFarland, D. Pope, D. Strobel, and J. Wagner (1991): "Bioremediation Studies at a Northern California Superfund Site," *Journal of Hazardous Material,* vol. 28, pp. 145–158.

Nyer, E. K. (1992): "Treatment for Organic Contaminants, Physical/Chemical Methods," in *Bioremediation: The State of Practice in Hazardous Waste Remediation Operations,* Seminar Sponsored by Air and Waste Management Association and HWAC, January, 1992.

Peterson, M. A., S. S. Huismann, and R. J. Jardine (1995): A Demonstration of Enhanced Bioremediation of Chlorinated Solvent and Diesel Contaminated Soils at Fort Gillem, Georgia, *Proceedings of the Air and Waste Management Association 88th Annual Meeting and Exhibition,* San Antonio, TX, paper 95-FA163.05.

Portier, R. J., and J. A. Christiansen (1994): "Closure of an RCRA Surface Impoundment by Employing a Modified Biological Treatment Approach," in *Bioremediation Field Experience,* P. E. Flathman, D. E. Jerger, and J. H. Exner, eds., Lewis Publishers, Chelsea, MI.

Reynolds, C. M. (1993): "Field Bioremediation Rates in a Cold Region Landfarm: Spatial Variability," in *Hydrocarbon Contaminated Soils,* vol. III, E. J. Calabrese and P. T. Kostecki, eds., Lewis Publishers, Chelsea, MI, pp. 487–499.

Ryan, J. R., R. C. Loehr, and E. Rucker (1991): "Bioremediation of Organic Contaminated Soils," *Journal of Hazardous Materials,* vol. 28, pp. 159–169.

Savage, G. M., L. F. Diaz, and C. G. Goluekc (1985): "Disposing of Organic Hazardous Wastes by Composting," *BioCycle,* January/February, pp. 31–34.

Sellers, K. L., T. A. Pedersen, and C. Fan (1993): "Review of Soil Mound Technologies for the Bioremediation of Hydrocarbon Contaminated Soil," in *Hydrocarbon Contaminated Soil,* vol. III, E. J. Calabrese, and P. T. Kostecki, eds., Lewis Publishers, Chelsea, MI.

Sims, J. L., R. C. Sims and J. E. Mathews (1990): "Approach to Bioremediation of Contaminated Soil," *Hazardous Waste and Hazardous Materials,* vol. 7, no. 4, pp. 117–149.

Stegmann, R., S. Lotter, and J. Heerenklage (1991): "Biological Treatment of Oil-Contaminated Soils in Bioreactors," in *On-Site Bioreclamation Processes for Xenobiotic and Hydrocarbon Treatment,* R. E. Hinchee and R. F. Olfenbuttel, eds., Butterworth-Heinemann, Massachusetts.

Troy, M. A., S. W. Berry, and D. E. Jerger (1994): "Biological Land Treatment of Diesel Fuel–Contaminated Soil: Emergency Response through Closure," in *Bioremediation Field Experience,* P. E. Flathman, D. E. Jerger, and J. H. Exner, eds., Lewis Publishers, Chelsea, MI, pp. 145–160.

U.S. Environmental Protection Agency (1983): *EPA Guide for Identifying Clean up Alternatives at Hazardous Waste Sites and Spills: Biological Treatment,* EPA 600/3-83/063.

U.S. Environmental Protection Agency (1988): *Technology Screening Guide for Treatment of CERCLA Soils and Sludges,* EPA 540/2-88/004.

U.S. Environmental Protection Agency (1990): *Available Models for Estimating Emissions Resulting from Bioremediation Processes: A Review,* EPA 600/3-90/031.

U.S. Environmental Protection Agency (1993): *Bioremediation Using the Land Treatment Concept,* EPA 600/R-93/164.

U.S. Environmental Protection Agency Member Agencies of the Federal Remediation Technologies Roundtable (1995): Remediation Case Studies: Bioremediation, EPA-542-R-95-002.

Williams, R. T., and C. A. Myler (1990): "Promising Research Results, Bioremediation Using Composting," *Biocycle,* pp. 78–82, November.

Williams, R. T., P. S. Ziegenfuss, and W. E. Sisk (1992): "Composting of Explosives and Propellant Contaminated Soils under Thermophilic and Mesophilic Conditions," *Journal of Industrial Microbiology,* vol. 9, pp. 137–144.

Ziegenfuss, P. S., R. T. Williams, and G. A. Myler (1991): "Hazardous Materials Composting," *Journal of Hazardous Materials,* vol. 28, pp. 91–99.

CHAPTER 9

Slurry-Phase Bioremediation

Slurry-phase treatment is used for bioremediation of contaminated soil. The contaminated material is usually excavated and brought to the slurry-phase reactor. In some cases sludges derived from long-term storage of contaminated material are treated in-place under slurry conditions. Alternative names for slurry-phase treatment that will be found in the literature include *bioreactor system, bioslurry treatment,* or *liquid-solid contactor.* The distinguishing characteristic of slurry-phase treatment is suspension of the contaminated soil or sludge in an aqueous medium; that is, treatment is conducted under saturated conditions. The slurry is formed by adding water or wastewater to the contaminated soil to form a desired slurry density. Quantities being treated and contaminant concentration are the principal factors in determining slurry density. The amount of mixing energy required increases sharply with density. When large quantities of material are to be treated, lower slurry densities are preferable. Similarly, high concentrations of contaminants result in the need for high oxygen transfer rates and dilution may be required. In most applications slurry-phase reactors are well mixed to keep the solids in suspension.

As the slurry is mixed, contact between the microorganisms and the contaminant compounds is increased, resulting in increased mass transfer and reaction rates. Biodegradation is usually optimized by adding nutrients (inorganic and/or organic) and by controlling the pH and temperature to meet microbial growth requirements. Continuous mixing and aeration help break up clumps of soil and help dissolve contaminants. Compared to other treatment processes, slurry-phase reactors generally provide the most contact between contaminants, microorganisms, oxygen, water, and nutrients. Because of these features, slurry-phase treatment may be particularly applicable to the bioremediation of soils contaminated with oily wastes and wastes with tarlike consistency (Lewis, 1992).

Slurry-phase treatment can be carried out in situ or on site. The distinction between the two is somewhat vague depending on the materials handling and operation. Generally in situ refers to the treatment of sludge in the impoundment or lagoon from which it originated, without much site disturbance. On-site, on the other hand, refers to dredging and excavation of the contaminated sludge and soil for treatment to be carried out in especially designed bioreactors. In true in situ slurry treatment, a floating mixer and/or aerator may be set in place and treatment of the entire lagoon is carried out in a single batch operation. Once the treatment goal is achieved, the solids in the lagoon are allowed to settle and the liquid is pumped out. Should the treated sludge be destined to remain in place, as would be the case if the impoundment's liner were intact and no soil contamination had occurred, or if there was no liner but soil contamination was not considered a problem, then the solids can be either dried, or stabilized with lime or fly ash addition, before capping or backfilling.

For impoundments that are greater than about 8,000 m^2 (2 acres) in area, treatment in a single operation is not recommended because of difficulty in providing homogeneity within the impoundment (U.S. EPA, 1993). Large impoundments should be divided into cells using sheet piles or earthen berms before remediation is initiated.

Reactors used for on-site slurry-phase treatment can be purchased or constructed depending on the size and materials selected. Commercially available bioreactors range in size from 3 to 15 m (10 to 50 ft) in diameter and from 4.5 to 7.6 m (15 to 25 ft) in depth (King et al., 1992). Portable tanks can hold about 75 m^3 (20,000 gal) of liquid, whereas constructed in situ reactors typically hold between 300 and 1,325 m^3 (80,000 and 350,000 gal) (League, 1991), although much larger units have been used. In situ reactors are usually built where contaminated soil has been excavated. To protect against groundwater contamination, the soil may be compacted and lined with clay to form a low-permeability layer. A thick liner such as high-density polyethylene is then used to contain the liquid and keep it from leaking underground.

PROCESS DESCRIPTION

Slurry-phase bioremediation systems are typically operated in a batch or semibatch mode because of the nature and quantities of materials involved. Continuous-flow operation is possible but is not commonly used. In batch operation, a single reactor vessel is used. Contaminated soils or sludges are deposited into the reactor; nutrients, water, and microbial cultures are added; and the slurry is mixed and aerated until the conversions of the targeted compounds attain a satisfactory level. Mixing and aeration are then stopped and the solids are allowed to separate from the fluid by sedimentation (soils) or flotation (some sludges). The solids are removed and, if appropriate, returned to their original location while the liquid may be sent to a wastewater-treatment plant, allowed to evaporate or recycled to the next soil batch. A portion of the slurry is retained in the reactor to be used as seed for the sequential treatment runs. Such a process is termed a sequencing batch reactor, or SBR.

In semibatch operation the treatment steps are carried out in separate tanks. The primary tank is usually used to mix the solids with the makeup water as well as to add the necessary nutrients, inoculum, and pH adjustments. The mixture is then transferred to the treatment tank (or tanks) where most of the biodegradation occurs

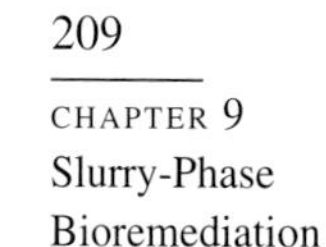

Full-scale slurry-phase biological treatment of wood-preserving wastes in four 200,000-gal reactors at the southwestern wood-preserving Superfund site, Mississippi. This project was the first ex situ biological slurry-phase project completed under the Superfund program. *(Courtesy of OHM Remediation Services Corp.)*

with continuous mixing and aeration. The slurry is then pumped to the last tank, where the liquid is separated from the solids. Some biodegradation may continue to occur during the separation process, which may bring the contaminants to the desired level of treatment. The semibatch system may be more efficient in terms of volume utilization because the water-addition and solids-separation steps can be carried out in smaller tanks. However, the additional complexity and costs associated with processing may make the single-tank SBR systems more economical.

REACTOR CONFIGURATIONS AND MODELING

The two configurations are shown schematically in Figures 9.1 and 9.2. Note that the SBR configuration shown in Figure 9.1 is a time sequence and that each of the steps is for the same physical unit.

The time to dilute the contaminated soil suspension to the desired consistency and to add nutrients can be determined from physical constraints such as size of pumps and the quantity of material being processed. Settling time will be a function of the soil-particle properties and the density of the suspension. Settling velocities decrease with particle size and particle concentration. Reaction time is a function of the biodegradability of the contaminants, the concentration of microorganisms present, and the partitioning of the contaminants between the liquid and solid phases. As noted in Chapters 4 and 5, bacteria obtain all of their nutrients from the liquid phase and thus partitioning between phases and/or mass transfer rate are extremely important factors in slurry-phase reactor design.

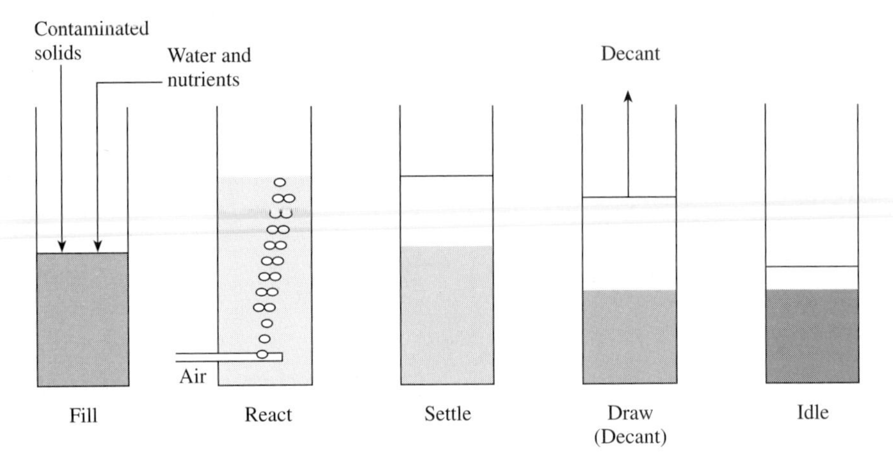

FIGURE 9.1
Schematic diagram of steps in time sequence for a sequencing batch reactor. The time required for each step is specific to the material being treated.

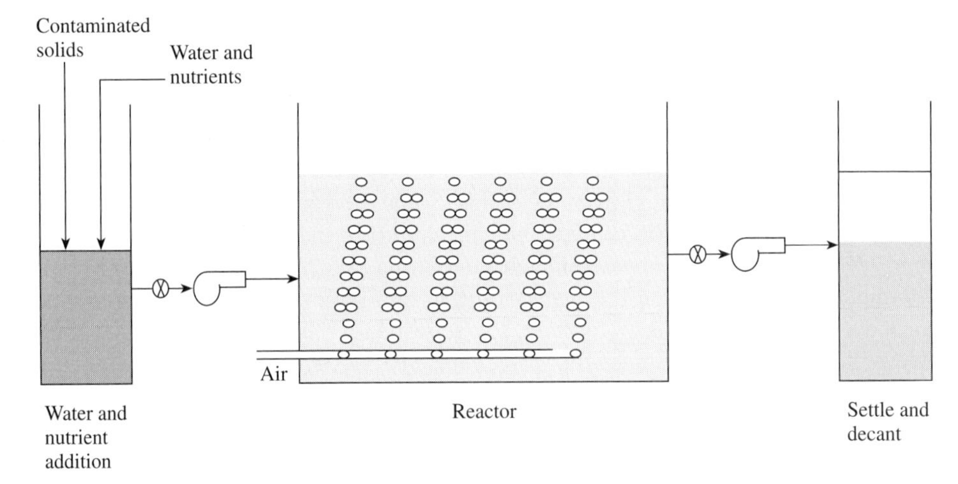

FIGURE 9.2
Schematic diagram for semibatch reactor system in which water addition, reaction, and solids separation are carried out in separate units.

Batch Reactor Modeling

Although full-scale slurry-phase reactors are often quite difficult to mix, ideal mixing is nearly always assumed. Mass balances on ideal batch reactors reduce to the accumulation term being equal to the reaction rate, as shown below.

$$\frac{dM}{dt} = Vr_o \tag{9.1}$$

where M = mass of contaminants in reactor, kg
t = time, day
V = reactor volume, m^3
r_o = rate of contaminant biodegradation, $kg/m^3 \cdot day$

In the discussion that follows the volume of liquid is assumed to be equal to the total volume V ($V_L + V_{solids}$). The volume of solids is negligible in comparison to the liquid volume in virtually all cases. Equations 9.1–9.5 will need to be modified in cases where the approximation is invalid.

The mass of contaminants present includes both the material sorbed onto the solids and the material in solution:

$$M = s\, X_S V + CV \tag{9.2}$$

where s = mass of contaminant sorbed per unit mass of solids present in reactor, kg/kg
X_S = mass concentration of solids (assumed to be contaminated soil), kg/m^3
C = mass concentration of soluble contaminant, kg/m^3

When the reaction is slow, equilibrium conditions are approximated and the soluble and sorbed materials are related by the soil distribution coefficient, K_{SD} (see Chapter 3 for a complete discussion of the soil distribution coefficient):

$$K_{SD} = \frac{s}{C_S} \tag{9.3}$$

Fast reaction kinetics result in nonequilibrium conditions, and separate mass balances must be made on the sorbed and liquid phases.

For the sorbed phase:

$$\frac{dM_{soil}}{dt} = -VK_L a(C_S - C) \tag{9.4a}$$

$$X_S \frac{ds}{dt} = -K_L a\left(\frac{s}{K_{SD}} - C\right) \tag{9.4b}$$

For the liquid phase:

$$V\frac{dC}{dt} = VK_L a(C_S - C) + Vr_o \tag{9.5a}$$

$$\frac{dC}{dt} = K_L a(C_S - C) + r_o \tag{9.5b}$$

where K_L = mass transfer rate coefficient, m/day
a = interfacial area per unit volume, m^{-1}
C_S = concentration in the liquid at equilibrium with the solid phase, as defined by Equation (9.3), kg/m^3

In a large fraction of the cases, equilibrium conditions are approximated and Equation (9.1) can be written as

$$(K_{SD} X_S V + V)\frac{dC}{dt} = Vr_o \tag{9.6a}$$

$$\frac{dC}{dt} = \frac{r_o}{K_{SD} X_S + 1} \tag{9.6b}$$

EXAMPLE 9.1. BIODEGRADATION IN A SLURRY REACTOR. A contaminated soil is to be remediated in a slurry reactor. The soil contaminant concentration, measured as COD, is 800 mg/kg dry soil and the allowable concentration (because of toxicity) is 40 mg/kg. Based on laboratory studies, $K_{SD} = 0.2$ m^3/kg and biodegradation can be described as a first-order function of the dissolved COD concentration. A value of the first-order rate coefficient of 0.05 per day has been determined. Adequate mixing can be achieved at a solids concentration of 10 kg/m^3. Determine the time required to remediate the soil. Density of the soil is 2,600 kg/m^3.

Solution

1. Determine the volume of water required per kg of soil to produce a slurry concentration of 10 kg/m^3.

$$V_{\text{soil}} = \frac{1 \text{ m}^3}{2{,}600 \text{ kg}} = 3.85 \times 10^{-4} \text{ m}^3/\text{kg}$$

$$10 \text{ kg/m}^3 = \frac{1 \text{ kg}}{3.85 \times 10^{-4} \text{ m}^3 + V_w}$$

$$V_w = 0.1 \text{ m}^3/\text{kg of soil}$$

2. Determine the initial and final concentrations of the contaminant in the liquid phase (based on 1 kg of soil).

$$C = \frac{s}{K_{SD}}$$

$$M = 8 \times 10^{-4} \text{ kg} = s(1 \text{ kg}) + VC$$

$$= (K_{SD} + V)C = (0.2 + 0.1)C$$

At $t = 0$
$$C_o = \frac{8 \times 10^{-4} \text{ kg}}{0.3 \text{ m}^3} = 2.67 \times 10^{-3} \text{ kg/m}^3$$

At $t = t_f$
$$C_f = \frac{4 \times 10^{-5} \text{ kg}}{0.3 \text{ m}^3} = 1.33 \times 10^{-4} \text{ kg/m}^3$$

3. Apply Equation (9.6*b*) and determine the time required for remediation.

$$\frac{dC}{dt} = \frac{r_o}{K_{SD}X_S + 1}$$

$$= -\frac{kC}{(0.2 \text{ m}^3/\text{kg})(10 \text{ kg/m}^3) + 1}$$

$$\frac{dC}{C} = -\frac{0.05 \text{ per day}}{3}\, dt$$

Integrating from C_o to C_f and from 0 to t

$$\ln \frac{C_f}{C_o} = -0.0167\, t$$

$$t = -59.9 \ln 0.05 = 179 \text{ days}$$

Oxygen Supply

Oxygen is required for aerobic biodegradation of contaminants, as was discussed in Chapter 5. Under certain conditions NO_3^-, NO_2^-, or SO_4^- can substitute for O_2 as the terminal electron acceptor. However, in most situations oxygen is the electron acceptor of choice because of the range of and completeness of biodegradation that can be achieved. Oxygen can be supplied by diffused aeration, turbine spargers, or surface aeration (Metcalf and Eddy, 1991). In diffused aeration oxygen is supplied by forcing air through porous devices such as ceramic diffusers located at the bottom of the tank. As the bubbles rise, oxygen diffuses across the gas-liquid interface of the bubbles. Efficiency of transfer is a function of the bubble size and the contact time. The contact time is a function of the liquid depth, and most diffused aeration tanks are 5 m or more in depth. Turbine spargers are combined mechanical and diffused aeration systems in which air is released below a flat-bladed turbine located at the bottom of the reactor. Turbulence, caused by the turbine, breaks the air into small bubbles and results in good transfer and mixing characteristics. In deep tanks, mixing cells may form and an intermediate-depth turbine may be required to maintain complete mixing. Surface aeration is provided by a turbine located at the liquid surface. When the surface turbine rotates, water is drawn upward and thrown outward in small droplets. The droplets provide a large surface for oxygen transfer and the turbulence resulting from the surface turbine provides mixing. The mixing depth is limited, although draft tubes can be installed to increase the effective depth. Definition sketches of the three types of oxygen transfer devices are presented in Figure 9.3. For dilute suspension, such as those used in municipal wastewater treatment, oxygen transfer rates are usually between 1.2 and 2.5 kg O_2·kW·h for all three systems. However, oxygen transfer rates decrease with increasing suspended solids concentrations and care

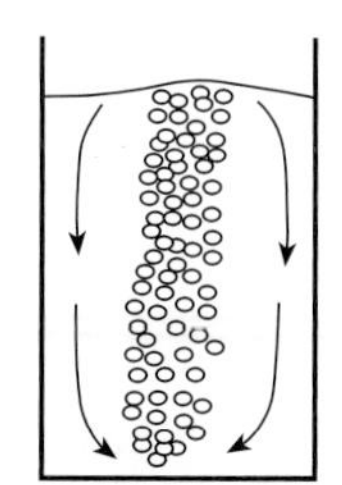

a. Diffused aeration

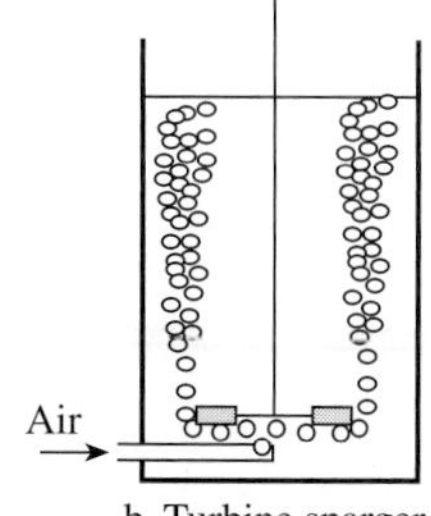

b. Turbine sparger

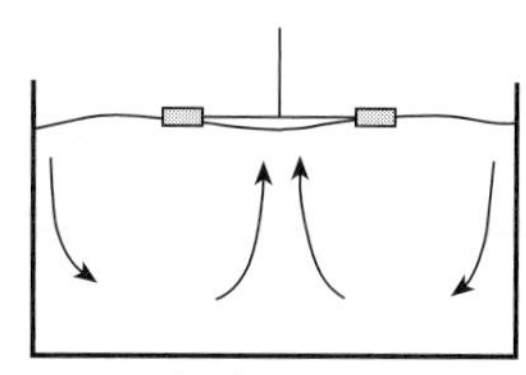

c. Surface aerator

FIGURE 9.3
Definition sketches of oxygen transfer systems used in slurry-phase bioremediation.

must be taken in selecting devices for slurry-phase bioremediation processes. Oxygen transfer rates well below 1 kg O_2·kW·h can be expected as solid concentrations approach 10 percent.

The oxygen transfer rate required is a function of the organic degradation rate r_o and the microbial growth rate r_g. Oxygen uptake rates, and hence required oxygen transfer rates, are less than the rate of COD removal because a portion of the organic material is used to form new bacteria rather than being oxidized. Unfortunately, bacterial mass is exceedingly difficult to determine in a slurry-phase system. An approximate relationship for the oxygen uptake rate which is based on wastewater-treatment design (Metcalf and Eddy, 1991) is given in Equation (9.7).

$$r_{O_2} = r_o\left(1 - \frac{0.6}{1 + 0.05\tau}\right) \tag{9.7}$$

where r_{O_2} = oxygen uptake rate, mg/L · h
τ = solids residence time, days

$$\tau = \left(1 + \frac{V_R}{V}\right)t \tag{9.8}$$

where V_R = volume of settled solids retained in tank, m^3
t = reaction time, days

EXAMPLE 9.2. OXYGEN TRANSFER REQUIREMENT IN A SLURRY-PHASE REACTOR. Determine the oxygen transfer requirement for the bioremediation system of Example 9.1 if the volume fraction retained in each operation is 10 percent.

Solution

1. Determine the solids residence time τ.

$$\tau = 1.1\ t = 1.1\ (18 \text{ days}) = 19.8 \text{ days}$$

2. Determine r_{O_2}

$$r_{O_2} = r_o\left(1 - \frac{0.6}{1 + 0.05\tau}\right)$$

$$= -kC\left(1 - \frac{0.6}{1 + 0.05\tau}\right) = -0.7\ kC$$

$$C = C_o e^{-kt} = C_o e^{-0.0167t}$$

$$r_{O_2} = -0.7(0.05)C_o e^{-0.0167t} = -0.035 C_o e^{-0.0167t}$$

$$C_o = 2.67 \times 10^{-3} \text{ kg/m}^3$$

$$r_{O_2} = -9.35 \times 10^{-5} e^{-0.0167t} \text{ kg/m}^3 \cdot \text{day}$$

3. The oxygen uptake rate (absolute value) will decrease as the COD decreases (Figure 9.4)—that is, as bioremediation progresses. The oxygen transfer rate required is equal to the oxygen uptake rate. Although the oxygen transfer rate required decreases to a very low level, the mixing requirement will remain the same to keep the soil in suspension. For this reason turbine speeds cannot be decreased as oxygen demand decreases.

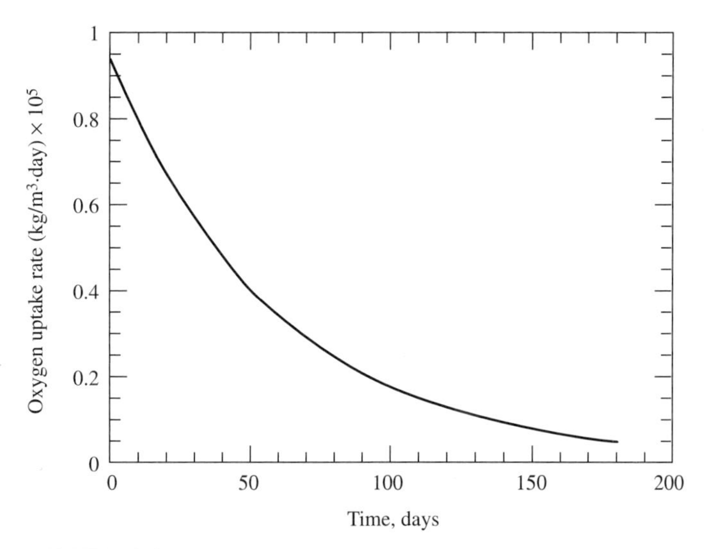

FIGURE 9.4
Oxygen transfer rate as a function of reaction time in Examples 9.1 and 9.2.

Mixing

Mixing results in greater homogeneity throughout the fluid. When concentration or temperature gradients exist within a slurry reactor, the reaction rates will vary spatially. In some cases spatial variation is desirable because optimal use of reactor volume results (Tchobanoglous and Schroeder, 1985). In slurry-phase reactors, a completely mixed fluid having minimal concentration gradients is usually desired to maximize desorption rates and minimize possible toxicity to microorganisms. Mixing theory has been developed for relatively small vessels (< 50 m^3) while slurry-phase treatment systems are often quite large (> 500 m^3). Selection of mixing equipment is relatively empirical under any conditions, but where large, unbaffled tanks are used a combination of experience, manufacturer's recommendations, and experimentation is usually involved. Because soil settles rapidly the mixing regime should be turbulent. The radius of influence of a turbine mixer used with soil suspensions is usually between 2 and 3 times the impeller diameter. Thus to obtain complete mixing in a square tank 60 m on a side with a 5-m-diameter impeller between 16 and 36 units would be required.

The degree of turbulence generated can be characterized by the impeller Reynolds number N_{Re_i}.

$$N_{\mathrm{Re}_i} = \frac{N D_i^2 \rho}{\mu} \tag{9.9}$$

where N = rotational speed of impeller, rps
D_i = impeller diameter, m
ρ = density of the suspension, kg/m^3
μ = dynamic viscosity of suspension, $kg/m \cdot s$

Maintenance of turbulent conditions in the mixing zone, or zone of influence, will require N_{Re_i} values greater than 10,000. For a 5-m impeller and a suspension having approximately the same characteristics as water ($\rho_w \approx 1{,}000$ kg/m^3, $\mu_w \approx 0.001$ kg/m·s), the rotational speed N will only need to be greater than 0.024 min^{-1}. However, typical operating N values are 40 to 60 rpm or 0.7 to 1 rps and operating N_{Re_i} values are between 1 and 3×10^7.

Power requirements for mixing are determined empirically and can be estimated from manufacturer's equipment specifications. Typical power requirements for complete mixing are in the 20 to 50 kW/1,000 m^3 range for moderately thick suspensions. Thick suspensions may require 100 to 250 kW/1,000 m^3 and the mixer radius of influence will be substantially reduced. For this reason thick suspensions are more easily treated in a deep, small-diameter tank than in a shallow, large-diameter tank.

Nutrient Requirements

Nutrient addition to slurry-phase reactors should be based on laboratory studies of the contaminant degradation. As noted above, microbial growth is difficult to measure in situations where a large amount of inorganic particulate material is present, and for this reason estimating the amount of nitrogen, phosphorus, and other elements required is also difficult.

An approach similar to that used for estimating oxygen requirements is possible. Nitrogen composes approximately 10 to 14 percent of microbial cell mass and phosphorus composes 0.5 to 2 percent in most cases. The rate of nitrogen uptake, as N, is a function of the microbial cell growth rate, which in turn is a function of the organic removal rate. If the rate of organic removal is measured in units of oxygen demand (COD or BOD_U) the rate of nitrogen uptake can be approximated by Equation (9.10).

$$r_N \approx \frac{0.06}{1 + 0.05\tau} r_o \tag{9.10}$$

Nitrogen can be added as NH_4^+ or NO_3^- and the quantities of salt added to supply a required amount of N can be quite large. For example, adding $NaNO_3$ as a source requires 6.1 g of the salt to supply 1 g of N. One result is a large increase in the total dissolved solids concentration.

An expression similar to Equation (9.10) can be used to estimate the rate of phosphorus (P) uptake. Note that P is usually supplied as orthophosphate, PO_4^{-3}, and approximately 3 g of orthophosphate must be supplied to provide 1 g of P.

$$r_P \approx \frac{0.01}{1 + 0.05\tau} r_o \tag{9.11}$$

Experimental studies at laboratory scale are advised in determining nutrient requirements because the stoichiometry of microbial reaction processes is quite variable. Degradation of complex compounds, such as those that are commonly found in contaminated soils, often results in surprisingly low cell yields, and hence low nutrient requirements. Laboratory studies can be conducted in which an ex-

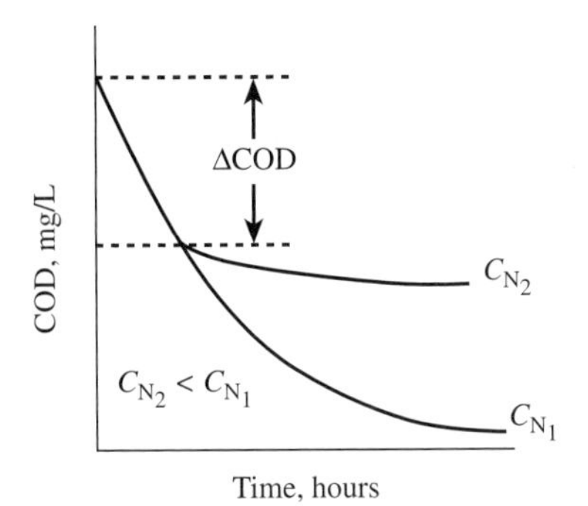

FIGURE 9.5
Change in COD removal rate resulting from depletion of nitrogen. The approximate nitrogen requirement is $C_{N_2}/\Delta COD$.

cess concentration of a single organic contaminant is provided and biodegradation is followed for a range of nutrient concentrations. In depletion studies a sharp decrease in the rate of organic removal will be seen when the nutrient becomes limiting. The stoichiometric requirement can then be estimated as the ratio of the mass of nutrient added to the change in the mass of organic, as indicated in Figure 9.5.

PRETREATMENT

Pretreatment of contaminated soil is often necessary to remove uncontaminated or less contaminated materials and to remove materials that are not compatible with slurry-phase treatment. Contaminants are usually somewhat hydrophobic and will be associated with the finer fraction of the soil which has the larger surface to volume ratio. Rocks, gravel, and sand are generally relatively clean and can be separated from the soil which will be treated in the slurry reactor. These same materials are difficult to keep in suspension and are likely to settle in reactor corners or other regions of lower turbulence, and their removal is desirable for that reason, also. Additionally, nonsoil debris (sticks, plastic materials, machine parts, construction material, etc.) should be removed prior to slurry-phase treatment. Soil fractionation and soil washing are two methods commonly used to pretreat contaminated soil.

Soil Fractionation

In soil fractionation the soil is screened and separated based on particle size. Large debris such as wood, plastics, and asphalt are removed while the coarse soil particles (gravel and sand) are separated from the finer particles (silt, clay, and humus). The purpose of separating the soil fines from the larger particles is to concentrate

TABLE 9.1
Distribution of COD by soil component

Component	%mass	%COD
Coarse	29	2.9
Fines	55	88.8
Tars	0.3	8.8
Water	15.7	0

Source: Black et al., 1991.

the waste and reduce the amount of soil that would have to be treated in a slurry-phase reactor. The finer particles have a higher surface-to-volume ratio as well as a higher surface activity, and as such the particles adsorb more of the contaminants, on a mass per mass basis, than do the larger particles. A typical distribution of organics by soil class is given in Table 9.1. The data in Table 9.1 were developed from studies of soil at a chemical manufacturing facility and substantial variation would be expected as a function of both the contaminant and the soil characteristics, although the general pattern of distribution would be expected to be consistent with that given. The tars reported are essentially NAPLs, and small mass fractions of such materials are often present in contaminated soils. If possible NAPLs should be removed for separate disposal.

Soil Washing

The objective of soil washing, like that of fractionation, is to separate highly contaminated fine particles from the less contaminated large particles. However, because of the washing and scrubbing involved and the resulting solubilization of contaminants, the larger particles may be cleaned to the point that would allow for low-cost disposal. As such, soil washing is sometimes used as a separate physical treatment technology aimed at significantly reducing the volume of contaminated soil. Soil washing is a highly mechanized, expensive process and is generally uneconomical unless greater than 70 percent of the feed mass can be recovered as clean material (Boyle, 1993).

Soil washing is relatively new to the United States, with the first applications being developed in the early 1980s (Griffiths, 1995). Several vendors have developed modifications of the basic concept, almost all of which utilize mechanisms and principles employed in mining and ore processing. Most soil-washing processes start with coarse screening of the soil to remove large debris and other unwanted materials, such as plastic sheeting and large stones. Water is then added to form a slurry. Cleaning of the larger particles is then carried out in an attrition cell (Boyle, 1993). Rotating impellers are used to generate shear and mechanical stress to break up clumps of soil while achieving vigorous agitation. Intensive scrubbing of the slurry results in separation between the fine and large soil particles and the cleaning of the large particles by surface abrasion. Particle washing may also be achieved through countercurrent flow.

Following washing, partial classification of the particles by size is then carried out using trommels (rotary screens), hydrocyclones, or vibrating screens. Particles larger than 200 mesh (74 μm in diameter) are often clean enough to be discarded (Griffiths, 1995). Solids remaining, those passing the screen, are then separated from the washwater through settling of the heavier particles or through air flotation of the lighter ones. Coagulants and flocculants are often needed to help with the separation process. The resulting sludge can then be treated biologically. Metals concentrations need to be monitored as the concentration steps may result in toxicity problems. The washwater can be recycled and used as makeup at the inlet or can be treated before discharge to a municipal sewer.

Some variations among soil-washing processes include warming the water to achieve greater solubilization of the contaminants. Surfactants or other chemicals may be added to help desorb and dissolve hydrophobic compounds. Acids may be added to precipitate metals, and in some cases caustics are added to improve organic acid extraction, as with pentachlorophenol (Griffiths, 1995). Note that additives can complicate further treatment of both the soil and the washwater and should be used with care.

From an economical standpoint, soil washing is more appropriate for sandy, coarse soils as compared to clay soils. With the former, a larger volume of soil may be recovered as clean since the larger portion of the mass is made up of bigger particles. Feasibility studies should be conducted to determine whether soil washing is economically justifiable. For the purposes of slurry-phase treatment, fractionation achieves a similar objective to soil washing, which is to eliminate the larger particles prior to treating the highly contaminated finer particles which are easier to suspend and mix. Fractionation may be more applicable where low contaminant concentrations prevail and the bulk of the contamination is in the finer particles. However, treatment of the larger soil particles may still be needed.

DESIGN CONSIDERATIONS

As discussed above, slurry-phase treatment can be carried out in batch, semibatch, or continuous-flow mode. Relatively small in situ operations are usually carried out in batch mode. Large impoundments of contaminated sludges or discarded soils may be treated in situ by dividing the site into cells with sheet piling. The cells can be treated in sequence using portable mixers and aeration devices if the time to complete remediation is not limiting. On-site, tank-based operations with separate tanks for chemical addition and solids separation (see Figure 9.1) are suitable for small operations or where time is not limiting. As noted above, continuous-flow mode is seldom used because of the large hydraulic residence times required and the difficulty of moving soils and sludges (Brown et al., 1995).

Microbial Inoculum

Three considerations are particularly important in developing and adding a microbial inoculum for slurry-phase reactors: (1) development of an appropriate microbial population, (2) provision of a suitable concentration of microorganisms, and (3) control of toxicity.

Microbial population

Each bacterial species, or strain within a species, has a limited number of reaction capabilities. In most cases a species capable of initiating aerobic degradation of a compound can oxidize the carbon molecules to CO_2. A portion of the carbon skeleton of complex compounds may be returned to the liquid phase because a particular bond structure cannot be broken, and in such cases other species may be able to grow on the "table scraps." At times the nondegraded fraction may be toxic to the bacteria initiating degradation, and without a secondary culture consuming the scraps, toxicity would accumulate to the point that the system would shut down. Generally, a number of species will carry out each of the reaction steps required and the relative numbers of each species or group present will be a function of environmental conditions (temperature, pH, presence of particular metals, . . .). Mixed populations or mixed cultures are nearly always more stable than pure cultures in terms of process performance. The actual species present may change significantly with time due to local or large-scale environmental changes. However, because a number of species can carry out the reactions the effect on process performance is minimal when mixed cultures are used.

Inoculum development

The steps involved in developing a microbial culture for bioremediation purposes are illustrated in Figure 9.6 and Table 9.2. Development of microbial cultures is usually started by collecting samples from a number of sources known to contain a broad spectrum of bacterial species such as activated-sludge wastewater-treatment plants and agricultural soils. The contaminated soil or sludge being remediated is a possible source but the question must be asked, "If the contaminated zone contains appropriate microorganisms why does the contamination still exist?" A mixture of material from each of the sources can be used to start a culture in the laboratory. Initially, a number of small cultures should be started using: organisms from each of the sources separately, mixtures of organisms, and controls. If toxicity characteristics of the contaminated material are unknown, a range of concentrations should be used. The contaminants are usually nonvolatile if slurry-phase treatment is a viable option, and diffused aeration or shaking can be used to aerate the samples. If the possibility of toxic fume production exists, off-gases must be passed through activated carbon and final concentration must be monitored. Contaminant concentration can be measured using a surrogate parameter such as COD. In some cases COD is an inappropriate parameter because a specific contaminant present in low relative concentration must be removed regardless of the overall organic removal.

The data from the first phase of microbial population development will result in a culture of suitable microorganisms, and initial information on concentrations that can be treated. In many cases cultures can be developed which are capable of treating significantly higher concentrations than observed in the first set of experiments. The reason may be slow development of secondary populations which degrade toxic by-products. In any event further study of limiting concentrations is appropriate.

Cultures found suitable should be grown in enrichment cultures by a process of adding contaminated material, aerating and mixing until the contaminants are degraded, allowing the solids to be separated, removing the supernatant liquor, and adding more contaminated slurry. Monitoring the cell counts by standard

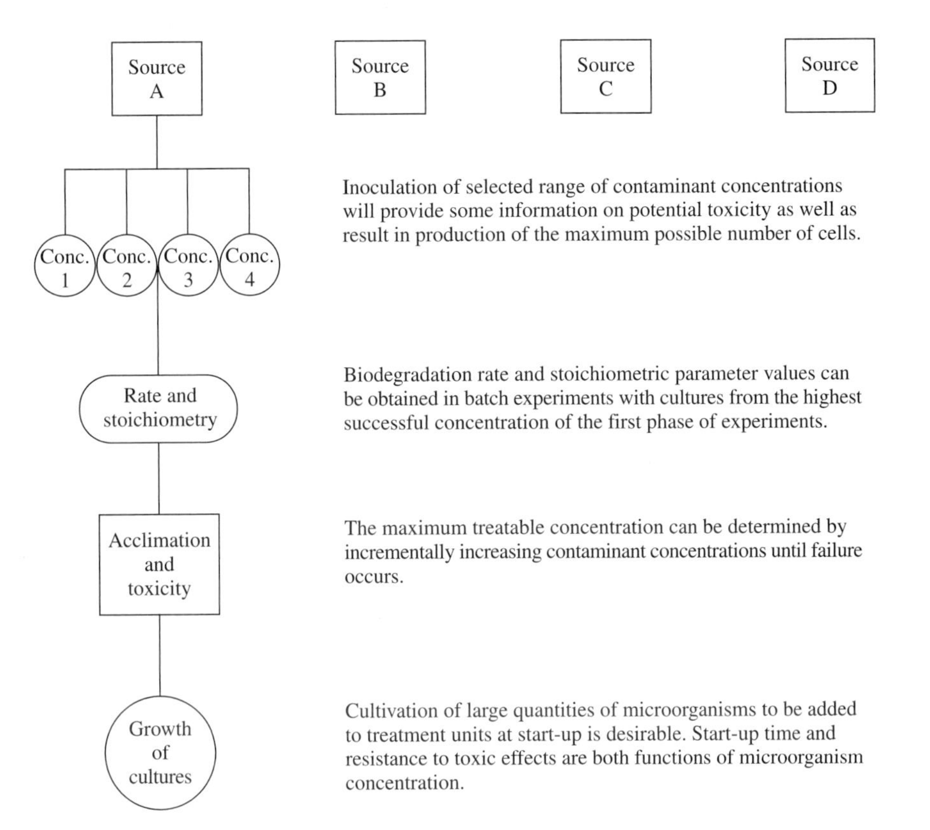

FIGURE 9.6
Sequence to be followed in developing microbial cultures for use in slurry-phase reactors. The steps shown for source *A* should be followed for other sources. Cultures that behave well can be mixed in the final step.

TABLE 9.2
Steps in microbial culture development for slurry reactors

Step	Objective	Comment
1. Selection of microbial sources	To find microbial groups or species capable of degrading the target contaminants	More than one source should be used. Soil and activated sludge should always be included
2. Growth of cultures in laboratory	To select the best sources of microorganisms for future use	Mixing sources may be the best choice
3. Determination of kinetic and stoichiometric parameters	To develop design and operating parameters	Design of oxygen transfer system and nutrient addition requirements is dependent on this step
4. Determination of toxicity limitations	To determine concentration limits on operation	Toxicity to microorganisms is a common problem
5. Growth of dense cultures	To provide microorganisms for treatment processes	In general, the higher the population density the better

microbiological techniques (e.g., plate counts) is a good idea. Microbial concentrations should be greater than 10^8 per mL in the reactor at start-up.

Reactor Selection

Batch reactors are appropriate when the volume of soil to be treated is relatively small and the time to treat it is not restricted. Holding tanks may be necessary for preparing a nutrient mix or maintaining a microbial inoculum to add to the slurry. A batch operation is useful if the degradation kinetics are first-order with respect to the contaminant concentration, in which case degradation proceeds most rapidly initially, when the contaminant concentration is highest. There is usually a lag period though, since in most cases the inoculum comes from a liquid culture, and microorganisms, though they may be indigenous, take some time to acclimate to the new environment of the slurry reactor.

EXAMPLE 9.3. DESORPTION-RATE LIMITATION. A contaminated soil slurry having a solids concentration of 10 kg/m^3 is to be remediated in a slurry reactor. The soil contaminant concentration s measured as COD is 320 mg/kg dry soil and the allowable liquid concentration (because of toxicity) is 50 mg/L. Based on laboratory studies, $K_{SD} = 2 \times 10^{-2}$ m^3/kg, the desorption rate coefficient $K_La = 2$ per day, and biodegradation can be described as a first-order function of the dissolved COD concentration with a rate coefficient of $k = 0.05$ per day. Determine the time required to remove 90 percent of the contaminants from the soil. Assume that homogeneous mixing can be achieved and that removals can be described on a unit volume basis.

Solution

1. Write mass-balance equations on the soil and on the liquid (Equations 9.4*a* and 9.5*b*).

$$\frac{dM_{\text{soil}}}{dt} = -VK_La(C_S - C)$$

$$\frac{dC}{dt} = K_La(C_S - C) - kC$$

2. Substitute Equation (9.3).

$$M_{\text{soil}} = (1\text{ m}^3)(10\text{ kg/m}^3)(s\text{ kg/kg})$$

$$= (10\text{ kg})(K_{SD}\text{ m}^3/\text{kg})(C_S\text{ kg/m}^3)$$

$$= (0.2\text{ m}^3)(C_S\text{ kg/m}^3) = (200\text{ m}^3)(C_S\text{ mg/L})$$

$$(200\text{ m}^3)\frac{dC_S}{dt} = (1\text{ m}^3)K_La(C_S - C)$$

3. Determine the initial conditions on a unit volume basis.

$$s_o = 3.2 \times 10^{-4}\text{ kg COD/kg soil}$$

$$C_o = C_{S_o} = \frac{s_o}{K_{SD}} = \frac{3.2 \times 10^{-4}\text{ kg COD/kg soil}}{2 \times 10^{-2}\text{ m}^3/\text{kg}} = 1.6 \times 10^{-2}\text{ kg/m}^3 = 16\text{ mg/L}$$

4. The coupled ordinary differential equations can be solved numerically. The time required will be approximately 250 days as shown in Figure 9.7.

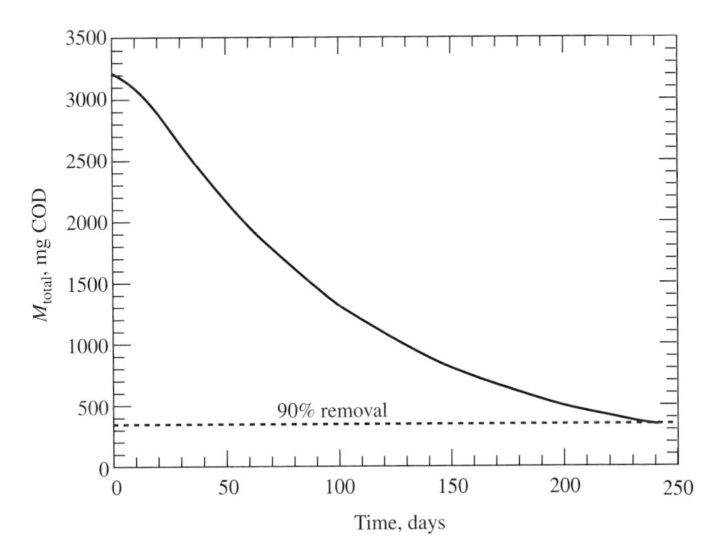

FIGURE 9.7
Total mass of contaminant on solid and in liquid as a function of time in Example Problem 9.3.

In continuous-flow operations there is usually no lag period. A recycle stream from the effluent supplies a high concentration of active, acclimated microorganisms to the influent. This is particularly helpful if the feed stream contains volatile compounds. If there is no lag period, degradation of volatile compounds begins immediately on entry to the reactor and the mass volatilized is decreased. Maintaining a high concentration of microorganisms through recycling may also be necessary if the degradation rate is dependent on the number of active microorganisms. Another aspect of continuous-flow operation is the instantaneous dilution of the feed stream. High concentrations of contaminants or of toxicants such as metals are immediately diluted, thus reducing possible inhibitory effects on the microorganisms.

A continuous-flow mode allows for treatment of a certain amount of soil in a shorter time period, compared to a batch mode since soil is continuously fed to the reactor and there is no wait period between batches. However, because of the continuous pumping involved in a continuous-flow operation, the cost may be too high, especially considering that slurry-phase processes have residence times on the order of days and weeks compared to activated-sludge processes which operate on the same principle but have much shorter residence times.

A semibatch process offers a medium solution. Compared to continuous-flow, semibatch cuts down on energy costs since pumping is only intermittent. Compared to batch, semibatch can process soil faster since there is no wait between batches. With reactors set in series, the first unit can be used for preparing the slurry; the second would be used for aerating and mixing and the third for settling and separating liquid from solids. A recycle stream from the last reactor to the first can provide acclimated, active microorganisms. At the same time, since degradation can proceed in all three reactors, the effective residence time can be increased.

The size of the reactor (or reactors) depends on the hydraulic retention time (HRT) needed for treatment

$$V = Q \text{ HRT} \tag{9.12}$$

where V = volume of the reactor and Q = slurry flow rate. The retention time itself is dependent on the biodegradability of the contaminants, the initial concentrations, and the level of treatment expected. Often, degradation kinetics are simplified into first-order:

$$C = C_o e^{-kt} \tag{9.13}$$

where C = concentration, kg/m^3, at time t, days
C_o = initial concentration, kg/m^3
k = degradation rate constant, per day

EXAMPLE 9.4. DESIGN OF SLURRY-PHASE REACTOR. Soil from an underground storage tank facility is shown to contain approximately 650 ppm total petroleum hydrocarbons (TPH) as diesel. In laboratory-scale slurry-phase experiments on the soil, degradation of the petroleum hydrocarbons could be approximated by a first-order rate model with a half-life of 15 days. If a continuous-flow slurry-phase reactor is to be used to treat the soil, then:

a. Determine the hydraulic residence time required to bring the concentration level of the contaminants to less than or equal to 100 ppm.
b. What volume reactor is needed given that the slurry flow rate is 6 m^3/day?
c. How long would it take to treat 400 m^3 of the soil, given that 1 m^3 of soil makes up 4 m^3 of slurry?

Solution

1. Given a half-life of 15 days, find the degradation rate constant:

$$C = C_o e^{-kt}$$

where

$$\frac{C}{C_o} = 0.5 = e^{-k(15)}$$

$$k = \frac{\ln 0.5}{-15} = 0.046 \text{ per day}$$

2. The hydraulic residence time can then be calculated by substituting the values of C_o, C, and k in Equation (9.13):

$$100 = 650\ e^{-0.046t}$$

$$t = \text{HRT} = 41 \text{ days}$$

Keep in mind that when extrapolating from lab studies to large-scale applications, the actual residence time should be expected to be larger. Degradation rates in larger systems are expected to be lower especially since a first-order rate model may not satisfactorily apply over the entire concentration range, and since large-scale operations do not offer the same homogeneous controlled environment existing under laboratory conditions. The result is that prototype reactor volumes should be larger than those calculated on the basis of laboratory data.

3. Reactor volume

$$V_{\text{liquid}} = Q \text{ HRT} = (6 \text{ m}^3/\text{day})(41 \text{ days}) = 246 \text{ m}^3$$

Head space is estimated as a fraction f of the total volume. A value of $f = 0.05$ is often used.

$$0.05 = \frac{fV_{\text{total}}}{fV_{\text{total}} + 246}$$

$$fV_{\text{total}} = 13 \text{ m}^3$$

$$V_{\text{total}} = 259 \text{ m}^3$$

Reactor dimensions can be approximately 4 m deep and 9.5 m in diameter (283 m^3). Head space depth will be 500 mm, which may be necessary if foaming is expected to be a problem.

4. To find the time needed to treat 400 m^3 of soil, first find:

$$\text{Soil processing rate} = \left(\frac{1 \text{ m}^3 \text{ soil}}{4 \text{ m}^3 \text{ slurry}}\right)\left(\frac{6 \text{ m}^3 \text{ slurry}}{\text{day}}\right) = 1.5 \text{ m}^3 \text{ soil/day}$$

$$\text{Time to treat} = \frac{400 \text{ m}^3}{1.5 \text{ m}^3\text{/day}} = 267 \text{ days} = \text{approximately 9 months}$$

OPERATING PARAMETERS AND PROCESS CONTROL

As mentioned earlier, slurry-phase remediation is a highly mechanized, energy-intensive process. Several different parameters have to be controlled if treatment is to proceed successfully and cost-effectively. The most important is the solids concentration of the slurry, and dissolved oxygen. Other parameters include temperature, pH, and microbial biomass. At the same time, operational problems, such as volatile air emissions, foam production, and contaminant sorption, have to be kept under control. The following is a brief discussion of some of these issues.

Solids Concentration

The recommended concentration of solids in the slurry depends primarily on the reactor design, the type of soil, and the contaminant concentration. For in situ operations, a solids content of between 5 and 20 percent is typical (U.S. EPA, 1993). So, if the sludge layer is thick and complete suspension of the solids is not possible, it may be necessary to divide the impoundment into cells in order to control the solids loading in the treatment unit. On-site processes, on the other hand, are generally better able to handle slurries with higher solids concentrations. A steel reactor, for example, with a tapered bottom, equipped with aerators and powerful mixers can be designed to handle a slurry with a solids content as high as 50 percent. A good design would mean that solids are kept in suspension and are not allowed to settle. A high solids content means a shorter treatment period, or alternatively a smaller-sized reactor, which is why it is desirable. A high concentration of contaminants, however, may be toxic to the microorganisms, in which case a lower solids content may be necessary to create a more dilute waste.

Another problem encountered with a high solids concentration is reduced oxygen mass transfer rate. The importance of this parameter stems from the fact that oxygen has to transfer to the liquid and dissolve in it before it can be available to the microorganisms. Two factors that directly influence oxygen transfer are solids concentration and airflow velocity.

Experimental results reported by Andrews (1990) show that, per unit mass of slurry, the maximum oxygen transfer rate increases with increased airflow velocity but decreases with increased solids content. For solids content above 40 percent the oxygen transfer rate becomes very small and almost independent of airflow velocity. The maximum oxygen transfer rate is measured as

$$r_{O_2\max} = \frac{K_L a C(1-\omega)}{\rho} \tag{9.14}$$

where K_L = mass transfer coefficient, m/s
a = specific surface area, m^2/m^3
C = oxygen concentration, g/m^3
ω = fractional slurry solids content, kg solids/kg water
ρ = liquid density, g/m^3

Based on visual observations, the reason for the form of Equation (9.14) is increased bubble coalescence associated with increased solids concentration (Andrews, 1990; Joosten et al., 1977). Andrews noted that for water with no solids, an increase in airflow velocity results in an increase in the number of bubbles, but almost no change in bubble size. The result is an increase in the interfacial area (a) between the gas and the liquid, thus increasing mass transfer. For a liquid with a high concentration of solids, however, an increase in airflow velocity seems to result in an increase in the size of the bubbles but not much change in the number of bubbles, i.e., coalescence. With a concentrated slurry those bubbles can be seen bursting at the surface. An increase in the bubble size, for the same number of bubbles, results in a minor increase in interfacial area, and so minimal change in the mass transfer rate.

As a general rule, slurry-phase bioremediation is most easily carried out with relatively small batches of contaminated soil because of the difficulties inherent in mixing and aeration. For most operations, the solids content typically ranges between 10 and 40 percent (Ross, 1990). Values for solids contents as low as 5 percent and as high as 50 percent, however, have been reported (LaGrega et al., 1994).

EXAMPLE 9.5. SLURRY DENSITY DETERMINATION. Three hundred cubic yards of soil have been excavated to be treated in a slurry-phase reactor. The soil has a moisture content (w) of 0.19 g/g soil, a particle density of 2.65 g/mL, and a bulk density of 1.2 g/mL. If the soil is to be treated in 20 m^3 batches, determine the quantity of water that must be added, per batch, to produce a slurry that is 25 percent solids by weight. What is the resulting slurry density?

Solution

1. The mass of solids M_S in 20 m^3 of soil can be obtained from the bulk density:

$$\rho_b = \frac{\text{mass solids}}{\text{total volume}} = 1.2 \text{ g/mL} = 1{,}200 \text{ kg/m}^3$$

$$M_S = \rho_b V_t = (1{,}200 \text{ kg/m}^3)(20 \text{ m}^3) = 24{,}000 \text{ kg}$$

2. Determine the mass of water in the slurry.

$$0.25 = \frac{\text{mass solids}}{\text{mass slurry}} = \frac{M_S}{M_S + M_W}$$

$$0.25M_W = (1 - 0.25)M_S$$

$$M_W = 3\ M_S$$

$$= 3\ (24{,}000\ \text{kg}) = 72{,}000\ \text{kg}$$

3. Determine the mass of water originally present in the soil.

$$M_{SW} = w\ M_S = (0.19)\ 24{,}000\ \text{kg} = 4{,}560\ \text{kg}$$

Note that approximately 6 percent of the water needed to form the slurry comes from the water originally present in the soil.

4. Determine the mass and volume of water that must be added.

$$M_{W\text{added}} = 72{,}000\ \text{kg} - 4{,}560\ \text{kg} = 67{,}440\ \text{kg}$$

$$V_W = \frac{M_{W\text{added}}}{\rho_W} = \frac{67{,}440\ \text{kg}}{1{,}000\ \text{kg/m}^3} = 67.4\ \text{m}^3$$

5. Determine the slurry density ρ_{sl}.

$$\rho_{sl} = \frac{\text{mass solids} + \text{mass water}}{\text{vol solids} + \text{vol water}}$$

$$= \frac{M_S + M_W}{V_S + V_W}$$

$$V_S = \frac{M_S}{\rho_S} = \frac{24{,}000\ \text{kg}}{2.65\ \text{kg/m}^3} = 9.06\ \text{m}^3$$

$$\rho_{sl} = \frac{24{,}000\ \text{kg} + 72{,}000\ \text{kg}}{9.06\ \text{m}^3 + 72\ \text{m}^3}$$

$$= 1{,}184\ \text{kg/m}^3 = 1.18\ \text{kg/L}$$

Aeration and Oxygen Demand

Slurry-phase bioremediation, as described in this chapter, is an aerobic process, and one of the most important operating parameters is dissolved oxygen (DO) concentration. To ensure aerobic conditions within the soil particles, a minimum liquid-phase DO concentration of 2 mg/L is desirable. Aeration devices in common use were described earlier in this chapter. Unlike conventional wastewater-treatment processes, mixing energy requirements may be greater than aeration energy requirements in slurry-phase bioremediation reactors. Use of standard wastewater-treatment design parameters is generally inappropriate. For thick slurries, where propeller-type mixers are used, separate design of aeration and mixing systems may be necessary. Mixing helps dissolve and distribute oxygen, as well as increase the interfacial area among all three phases (solid, liquid, and gas). Often thick suspensions require axial mixing, as well as radial mixing, to achieve full suspension of the solids.

Temperature and pH

Because water is a poor heat conductor, small variations in ambient temperature will not produce noticeable differences in slurry temperature. However, because of the long residence times required for slurry-phase bioremediation significant temperature variation can be expected on a seasonal basis. For example, in a field-scale study carried out in Canton, Mississippi, on PAH-contaminated soil, Jerger and Woodhull (1994) observed an increase in the rate and extent of degradation of the PAHs in the warm season as compared to the colder season. The reactor temperature ranged between 25 and 40°C in the spring, summer, and fall months as compared to 15 and 21°C in the winter. Temperature control may be necessary if systems are to be operated under extreme weather conditions where temperature might become the limiting factor in microbial activity.

Surfactants and Other Additives

Surfactants are sometimes used to help desorb and solubilize the contaminants in order to enhance biodegradation. The usefulness of surfactants in bioremediation has not been well documented. In a pilot-scale study on PAH-contaminated soil, Lewis (1992) observed that the addition of the surfactant Tween 80 did not result in any significant improvement in degradation. In another study, Melcer et al. (1995) argued that the reason surfactant addition did not enhance biodegradation is that the surfactant added was below the critical micelle concentration (CMC). However, at a higher concentration, the nonionic surfactant may have inhibited biodegradation. Other researchers believe that biosurfactants are better designed to enhance biodegradation as compared to synthetically produced surfactants. For example, Castaldi and Ford (1992) suggest that maintaining a high microbial biomass may result in a higher production of microbial surfactants. These biosurfactants are believed to act as emulsifiers that help desorb some of the more hydrophobic contaminants and bring them into the aqueous phase.

In some cases oxidizing agents are added to the slurry in order to oxidize highly stable molecules of contaminants and help make them more biodegradable. An example of such a chemical is Fenton's reagent, which has been used recently in degrading high-molecular-weight PAHs (Brown et al., 1995). Fenton's reagent, a solution of hydrogen peroxide and iron salts, is used to carry out a chemical oxidation step that results in the hydroxylation of the large PAH molecules. As mentioned in earlier chapters, hydroxylation of the ring is the first step in biodegrading PAHs. For those with a high number of rings, hydroxylation is often the rate-limiting step.

VOC Emissions

Because of the continuous mixing and agitation, the release of volatile organic compounds is usually a concern in slurry-phase treatment. Portable reactors can be covered with a hood to collect and treat air emissions. Covering in situ reactors is generally more difficult because they tend to be larger in size. Most VOC emissions occur during the first few days of operation, particularly during loading and start-up. The mass of organics volatilized depends on the compound properties as well as the concentration. Cost effectiveness of covering the reactors is questionable, especially if the contaminants are mostly semivolatile. For example, Lewis (1992) estimated that the concentrations of volatile emissions released from the slurry-phase treatment of PAH-contaminated soils were equivalent to what would be expected when loading other physical/chemical treatment processes such as incinerators or thermal desorption units.

Foam Production

A problem often encountered in slurry-phase treatment is foam production. The presence of naturally occurring organics in certain soils seems to promote formation of stable foams that result in operational problems (Glasser et al., 1994). Preliminary treatability studies should be conducted to evaluate the probability of foam formation. Foaming problems tend to be magnified when high slurry densities are used. Reducing the mixing speed or adding an antifoam reagent can help control foam production.

CASE STUDY

A bench-scale study was performed at the U.S. EPA Test and Evaluation Facility in Ohio to investigate the effects of selected design criteria on the treatment of creosote-contaminated soil (Glasser et al., 1994). The design parameters used included solids concentration, mixing rate, and the addition of a dispersant. As part of pretreatment, the soil was first sieved to remove particles greater than 0.25 in in diameter. Water was then added, and the concentrated slurry was passed through a hydrocyclone to remove the sand/gravel fraction.

The reactors used were glass, with 8-L capacity and covered to control volatile emissions. Each reactor had a variable-speed mixer. The mixing shaft was equipped with two impellers which provided radial mixing (lower impeller), and downward thrust mixing (upper impeller). Air was introduced just below the lower impeller to optimize the spread of oxygen through the liquid. Airflow into the reactor was adjusted to maintain a dissolved oxygen concentration of at least 2 mg/L.

The reactors were operated for 10 weeks and were tested for two solids concentrations: 10 percent and 30 percent. Initial total PAH concentration in the soil averaged 1,750 ppm in the 10 percent solids slurry and 2,047 ppm in the 30 percent slurry. For each solids concentration two mixing speeds were tested: a high rate for complete off-bottom suspension, and a low rate, arbitrarily set at 200 rpm less. The high rate for the 10 percent solids was 650 rpm, and for the 30 percent solids it was 900 rpm. The addition of a dispersant (Westvaco Reax 100M) was also tested to evaluate its ability to reduce foam production.

The concentration of solids in the slurry seemed to have a significant effect on the removal rate and the final concentration levels of PAHs in the soil at the end of the 10-week period. Most of the removal occurred in the first 7 days of treatment. In the 10 percent solids slurry, total PAH concentrations were reduced by about 74 percent, compared to 82 percent reduction in the 30 percent solids slurry. In both cases, higher removals were observed in the 2- to 3-ring PAHs (90 percent range), compared to the 4- through 6-ring compounds (64 to 75 percent range).

In the 30 percent solids slurry all of the PAH removal occurred in the first 21 days. Further treatment of the soil after that period did not result in any significant reduction in PAH concentrations. At the end of the 10-week period, the total PAH concentration of approximately 300 ppm persisted. However, in the 10 percent solids slurry, PAH removal continued throughout the length of the experiment. The total PAH concentration dropped by about 50 percent between days 21 and 70. The final concentration was approximately 170 ppm. The mixing speeds did not appear to have any effect on PAH removal. The addition of the dispersant did not reduce foam formation, nor did it increase PAH removals. Foam production was better controlled by reducing the mixing speed or through the addition of an antifoam reagent.

Questions to ponder

1. Why do you think that degradation in the 30 percent slurry stopped after 21 days?
2. How can the composition of the waste matrix influence degradation?
3. How can mixing speed improve biodegradation?
4. What do you think happened to the cell biomass through the duration of the experiment?

Field demonstration of an integrated chemical and biological slurry-phase treatment process for contaminated soils from a former manufactured gas plant process, New Jersey. *(Courtesy of OHM Remediation Services Corp.)*

FIELD-SCALE APPLICATIONS

Although there are numerous reports in the literature of bench- and small pilot-scale applications of slurry-phase treatment, there have been very few reports of full-scale operations. In one example cited by Coover et al. (1994), liquid/solid treatment was

TABLE 9.3
Operating parameters of pilot-scale slurry reactor bioremediation process

C:N:P	Temperature, °C	pH	O_2 uptake rate, mg/L · min	Microbial cell count, CFU/mL
100:2:0.2	20–25	5.8	0.2	10^7–10^8

Source: Coover et al., 1994.

used as part of a treatment train to remediate three impoundments containing petroleum sludges at a former refinery site in Sugar Creek, Montana. One of the impoundments, 19,000 m^3 (5 million gal capacity), was used as a liquid/solid reactor while another impoundment was used as a prepared-bed land treatment unit. The in situ slurry-phase reactor was equipped with float-mounted aerators and mixers and was operated in a batch mode. Each batch took between less than 60 and up to 90 days to achieve a 66 percent reduction in oil and grease concentrations. Once that level of treatment was achieved, the solids were allowed to settle, after which they were pumped to the land treatment unit for continued biodegradation. No detailed results were available as of the time of the writing of the aforementioned report.

Another example of a large-scale slurry-phase treatment was cited by Coover et al. (1994). The 1 million gal pilot-scale demonstration was conducted on petroleum sludge from an impoundment at a refinery site located on the Gulf Coast. An abandoned concrete clarifier, 47 m in diameter, was used as the bioreactor. Sidewall and center depths were 2.7 and 4.3 m, respectively. Sludge was dredged from the impoundment and loaded into the reactor over a period of 5 days. The solids content of the slurry was set at about 10 percent. Float-mounted mixers and aerators were then installed, but aeration was started incrementally, over a period of 3 days, to minimize VOC stripping. Once mixing started, the reactor was seeded with 83 m^3 (22,000 gal) of activated sludge from the refinery wastewater-treatment plant. The reactor was operated as a single batch for a period of 8 weeks. The operating parameters are listed in Table 9.3.

Operational problems reported included the formation of a foam layer (10 to 25 cm thick) that persisted throughout the experiment, and the settling of material at the bottom of the tank. At the end of treatment, approximately 25 percent of the solids in the tank were settled material. Most of the biodegradation occurred within the first 2 weeks.

ADVANTAGES AND DISADVANTAGES

The treatment of hazardous waste in slurry-phase reactors is generally faster and requires less land area than landfarming systems. Comparative rate studies suggest that clean-up times may be an order of magnitude shorter in many cases (Mueller et al., 1991). The ability to control slurry-phase treatment systems is also much greater than for other soil treatment methods, and hence slurry-phase treatment may be the most effective bioremediation technology available. However, because the vigorous mixing and forced aeration enhance the release of air emissions, slurry-phase may not be a good choice for soils contaminated with volatile organics.

Slurry-phase combines pretreatment, vigorous mixing, and forced aeration, which makes it a highly mechanized process. Hence the most obvious disadvantages of this treatment process are the high capital, operation, and maintenance costs. Cost reductions may be possible through use of slurry-phase treatment as a preliminary step before land disposal or by operation in a semicontinuous mode.

PROBLEMS AND DISCUSSION QUESTIONS

9.1. A soil contaminated with methylene chloride, CH_2Cl_2, was being bioremediated using closed batch slurry-phase treatment. The reactor is closed and a portion of the air is recycled after CO_2 stripping. After a period of time the system failed, as indicated in the curve below. The compound is known to be rapidly degraded by a limited number of bacterial species, most of which are common soil bacteria. The dimensionless Henry's law coefficient of methylene chloride is 0.094 at 20°C and the log K_{OW} is 1.3.

a. Why is the reactor covered?
b. Why is CO_2 stripping a good idea?
c. Explain the probable cause of the failure.
d. Suggest a method of operation of the slurry-phase systems that is likely to be successful.

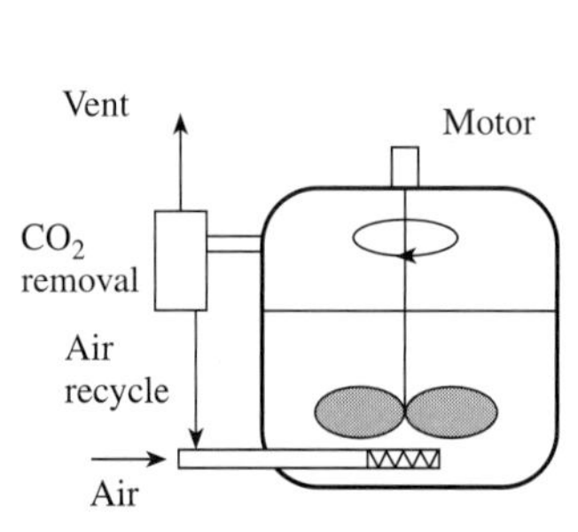

Closed slurry-phase reactor.

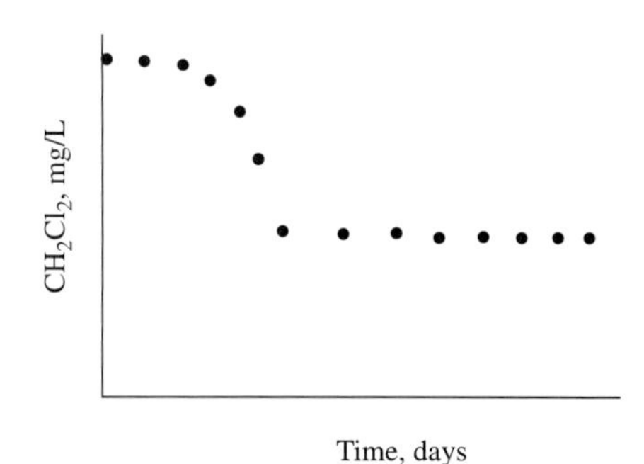

Liquid-phase CH_2Cl_2 concentration versus time.

9.2. Mass balances on microbial cells in ideal stirred-tank reactors operated at steady state result in the equations below:

$$r_g = \frac{X}{\theta_H}$$

where r_g = rate of microbial growth, mg/L · day
X = cell mass concentration, mg/L
$\theta_H = \frac{V}{Q}$ = hydraulic residence time, days

For the classic Monod rate model (Chapter 4):

$$r_g = Y\frac{kCX}{K+C} - k_d X$$

A plot of C vs. θ_H results in four asymptotes.

a. Derive the asymptote values.

b. Sketch the curves and determine which are real and which are not real. Explain your answers.

9.3. In a bench-scale experiment, a continuous-flow slurry-phase system is used to treat creosote-contaminated soil. The first reactor, where initial mixing occurs, has a volume of 50 L. Fenton's reagent is added at a rate of 1 mL/min to the second reactor, which has a volume of 10 L. The slurry then flows into the third reactor where final polishing is achieved. If the slurry leaves reactor 1 at a rate of 6 L/day, and if the total hydraulic residence time (HRT) of the slurry, not including the Fenton's reagent, is to be 14 days or greater, determine the minimum volume necessary for reactor 3.

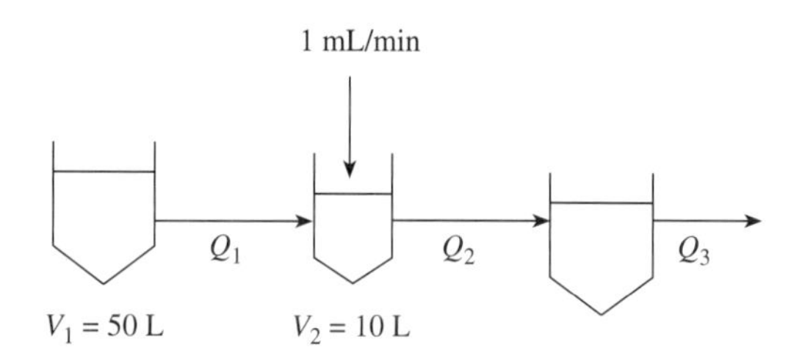

9.4. Determine the total volume of tank(s) needed to treat soil in a slurry-phase reactor system, at an average rate of 8 metric tons per hour, a solids concentration of 20 percent by weight, and a hydraulic residence time of 21 days. Assume a soil particle density of 2.65 g/cm^3, and a moisture content (w) of 0.1 g/g soil.

9.5. The results listed in the table below were obtained during a bench-scale experiment with two identical slurry-phase reactors operated in series. Hydraulic residence times given are based on the individual reactors. Slurry solids concentration is 100 g/L and particle density is 2.65 kg/L. Using the data provided, estimate the volume required for a full-scale system treating 1,000 m^3/day if the final soil TOC concentration must be less than 200 mg/kg. Note that the concentrations C_i, C_1, and C_2 given are for the liquid phase while the concentration C_{Ti} includes both the soluble and sorbed TOC. Assume first-order rate kinetics based on liquid-phase concentrations. What problems might you expect when extrapolating to a field design?

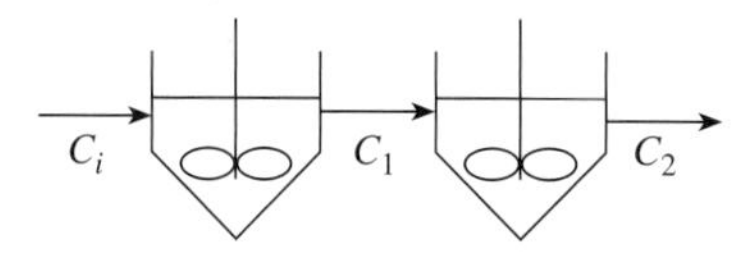

θ_H, days	C_{Ti}, mg/L	C_i, mg/L	C_1, mg/L	C_2, mg/L
10	1,200	200	180	161
20	1,200	200	162	131
40	1,200	200	136	92

9.6. The results listed in the table below were obtained during a bench-scale experiment with two slurry-phase reactors operated in series, with a total hydraulic residence time of 15 days. Using this data estimate the time needed to reach a concentration of 200 ppm or less for all constituents. Assume first-order rate kinetics. What problems might you expect when extrapolating to a field design?

Contaminant	Initial concentration, ppm	Final concentration, ppm
Anthracene	8,000	150
Phenanthrene	8,000	100
Fluoranthrene	6,500	210
Chrysene	4,500	450
Pyrene	6,000	300

9.7. A slurry-phase reactor is being operated in a batch mode to treat soil contaminated with 2- and 3-ring PAHs. Total PAH concentration in the soil is 4,300 ppm, and the soil makes up 15 percent (by weight) of the solid's content in the slurry. If on the average 3 g of oxygen are consumed per g of PAH degraded, and assuming no other oxygen is needed for microbial activity, then what is the minimum oxygen aeration rate needed to maintain a dissolved oxygen concentration of 2 mg/L? What is the corresponding airflow rate needed? Assume degradation follows first-order rate kinetics with $k = 0.09$/day. Also assume a saturation oxygen concentration of 8 mg/L, which is also the initial concentration of oxygen. The K_La value is 0.12 per day.

9.8. A soil slurry containing 350,000 kg of soil and having a solids weight fraction ω of 0.1 is contaminated with a nonvolatile, biodegradable contaminant with a K_{SD} of 0.2 m^3/kg. The dry density of the soil is 2,700 kg/m^3. Laboratory studies on biodegradation of the materials have resulted in the conclusion that the biodegradation kinetics will be 0 order with k = 10 mg/L·day (based on liquid-phase concentration changes). If the initial liquid-phase concentration of contaminant C_L is 50 mg/L, determine the time necessary in a batch process to reduce the concentration to 1 mg/L.

9.9. A well-sealed pond has been used for disposal of wastes from an industrial process utilizing a large number of organic chemicals for a 20-year period ending in 1985. Records of materials that have been placed in the pond are very incomplete but compounds commonly used in the plant include substituted phenols (e.g., 2-chlorophenol and 2,6-dichlorophenol, and pentachlorophenol), unchlorinated solvents (e.g., acetone, toluene), and chlorinated solvents (1,1,1-trichloroethane, trichloroethene, and dichloromethane). Periodic evaluation of the pond contents over the last 10 years has resulted in the data below. A small mixer was installed on 2/28/88.

Date	BOD_5, mg/L	COD, mg/L
2/25/85	6,850	15,850
3/15/86	6,200	14,900
2/28/87	5,400	13,640
2/14/88	4,900	12,900
3/3/89	2,300	10,400
3/1/90	2,250	10,250
2/15/91	2,150	10,150
2/12/92	2,090	10,020

The mixer was installed in 1988 because the decreasing contaminant concentration values were noted and it was believed that biodegradation would be enhanced.

a. What is the basis for assuming that biodegradation might be occurring?

b. Is biodegradation the only explanation for the decrease in COD and BOD?

c. What explanation might be given for the fact that the BOD_5 remains at 2,090 mg/L after 7 years?

d. What tests might be run to verify your answer to *c*?

REFERENCES

Andrews, Graham (1990): "Large-Scale Bioprocessing of Solids," *Biotechnology Progress,* vol. 6, pp. 225–230.

Black, W. V., R. C. Ahlert, D. S. Kosson, and J. E. Brugger (1991): "Slurry-Based Biotreatment of Contaminants Sorbed onto Soil Constituents," in *On-Site Bioreclamation Processes for Xenobiotic and Hydrocarbon Treatment,* edited by R. E. Hinchee and R. F. Olfenbuttel, Butterworth-Heinemann, Massachusetts.

Bogart, J. D., and J. R. League (1988): Biological Remediation of Underground Storage Facilities. Air Pollution Control Association, 81st Annual Meeting of APCA, paper 88-7.2.

Boyle, C. (1993): "Soils Washing," in *Remedial Processes for Contaminated Land,* edited by M. Pratt, Institution of Chemical Engineers, Warwickshire, UK.

Brown, K. L., B. Davilla, and J. Sanseverino (1995): Combined Chemical and Biological Oxidation of Slurry-phase Polycyclic Aromatic Hydrocarbons, presented at the 88th Annual Meeting and Exhibition of Air and Waste Management Association, San Antonio, TX, paper 95-RA127.06.

Castaldi, F. J., and D. L. Ford (1992): "Slurry Bioremediation of Petrochemical Waste Sludges," *Water Science and Technology,* vol. 25, no. 3, pp. 207–212.

Coover, M. P., R. M. Kabrick, H. F. Stroo, and D. F. Sherman (1994): "In Situ Liquid/Solids Treatment of Petroleum Impoundment Sludges: Engineering Aspects and Field Applications," in *Bioremediation Field Experience,* edited by P. L. Flathman, D. E. Jerger, and J. H. Exner, Lewis Publishers, Boca Raton, FL.

Glasser, J. A., J. S. Platt, M. A. Dosani, P. T. McCauley, and E. R. Krishnan (1994): Evaluation of Bench-Scale Slurry-Phase Bioreactors for Treatment of Contaminated Soils, *Proceedings of the 87th Annual Meeting of the Air and Waste Management Association,* Ohio.

Griffiths, R. A. (1995): "Soil-Washing Technology and Practice," *Journal of Hazardous Material,* vol. 40, pp. 175–189.

Jerger, D. E., and P. M. Woodhull (1994): Slurry-Phase Biological Treatment of Polycyclic Aromatic Hydrocarbons in Wood Preserving Wastes, *Proceedings of the 87th Annual Meeting of the Air and Waste Management Association,* Ohio.

Joosten, G. E., J. G. Schilder, and J. J. Janssen (1977): "The Influence of Suspended Solid Material on the Gas-Liquid Mass Transfer in Stirred Gas-Liquid Contactors," *Chemical Engineering Science,* vol. 32, pp. 563–566.

King, R. B., G. M. Long, and J. K. Sheldon (1992): *Practical Environmental Bioremediation,* Lewis Publishers, Boca Raton, FL.

LaGrega, M. D., P. L. Buckingham, and J. C. Evans (1994): *Hazardous Waste Management,* McGraw-Hill, New York.

League, Jim (1991): "Liquid Solids Contact (LSC) or How to Bug Hazardous Waste to Death," in *Biological Processes, Innovative Hazardous Waste Treatment Technology Series,* vol. 3, edited by H. M. Freeman and P. R. Sferra, Technomic, Lancaster, PA.

Lewis, R. F. (1992): SITE Demonstration of Slurry-Phase Biodegradation of PAH Contaminated Soil, *Proceedings of the 85th Annual Meeting of the Air and Waste Management Association,* Missouri.

Melcer, H., and C. E. Aziz (1994): Bioremediation of Coal Tar Contaminated Soil in a Slurry Bioreactor, presented at the 87th Annual Meeting and Exhibition of Air and Waste Management Association, Cincinnati, Ohio, paper 94-TA45A.04.

Melcer, H., C. E. Aziz, and J. Anderson (1995): Slurry Bioremediation of Coal Tar Contaminated Soil, presented at the 88th Annual Meeting and Exhibition of Air and Waste Management Association, San Antonio, TX, Paper 95-FA163.01.

Metcalf and Eddy, Inc. (1991): *Wastewater Engineering,* 3d ed., McGraw-Hill, New York.

Mueller, J. G., S. E. Lantz, B. O. Blattmann, and P. J. Chapman (1991): "Bench-Scale Evaluation of Alternative Biological Treatment Processes for the Remediation of Pentachlorophenol- and Creosote-Contaminated Materials: Slurry-Phase Bioremediation," *Environmental Science and Technology,* vol. 25, pp. 1055–1061.

Ross, Derek (1990): "Slurry-Phase Bioremediation: Case Studies and Cost Comparisons," *Remediation,* vol. 1, pp. 61–74.

Tchobanoglous, G., and E. D. Schroeder (1985): *Water Quality; Characteristics, Modeling, Modification,* Addison-Wesley, Reading, MA.

U.S. Environmental Protection Agency (1993): Pilot Scale Demonstration of a Slurry-Phase Biological Reactor for Creosote Contaminated Soil, Cincinnati, OH, EPA 540/A5-91/009.

Yare, B. S. (1991): "A Comparison of Soil-Phase and Slurry-Phase Bioremediation of PNA-Containing Soils," in *On-Site Bioreclamation Processes for Xenobiotic and Hydrocarbon Treatment,* edited by R. E. Hinchee and R. F. Olfenbuttel, Butterworth-Heinemann, Massachusetts.

CHAPTER 10

Vapor-Phase Biological Treatment

Vapor-phase contaminants are found in off-gases from soil and groundwater remediation operations, from industrial processes, and from wastewater-treatment systems. Compounds commonly found in air that are amenable to biological treatment include petroleum hydrocarbons, halogenated and unhalogenated solvents, sulfides (e.g., H_2S), and ammonia. The contaminants must be transferred to the liquid phase to be available for microbial metabolism. Thus vapor-phase biological treatment involves three steps; gas-liquid transfer, liquid-phase transport to the microorganisms, and microbial transformation of the contaminants. Two general process configurations exist, suspended growth/diffused aeration and packed beds. Suspended-growth applications have been almost entirely associated with using contaminated air for aeration of activated-sludge processes. In such cases the treatment of vapor-phase contaminants is a fortuitous artifact rather than an engineered system. Processes designed for biological treatment of vapor-phase contaminants have been almost entirely packed beds. The first engineered packed-bed systems were used for control of odors at wastewater-treatment plants (Pomeroy, 1957, 1982; Carlson and Leiser, 1966). Packing material used in the first systems was soil and volumetric gas fluxes ($m^3/m^2 \cdot s$) were relatively low. The systems were given the name *soil filters* and when alternative packings began to be used the term *biofilter* came into use. Conceptually, biofilters are very similar to packed-bed systems used for wastewater treatment. The principal differences are the presence of vapor-phase rather than liquid-phase contaminants and the lack of a moving liquid phase. The latter difference has been modified by the introduction of *biotrickling filters,* vapor-phase systems to which nutrients are supplied by a continuously recycled liquid stream added as a spray at the top of the column (Diks and Ottengraf, 1991; Sorial et al., 1995). A modification of the basic biofilter technology is the *bioscrubber,* a process in which a suspended culture is sprayed over the packing, collected at the bottom, and recycled to a suspended growth reactor (Hammervold et al., 1995).

TABLE 10.1
Comparison of vapor-phase pollutant control technologies

Treatment technology	Residuals/by-products	Energy costs	Comments
Adsorption	Spent activated carbon (regenerable systems usually combined with condensation or incineration)	Moderate to high	Limited to low- to moderate-concentration emissions and molecular weights between approximately 45 and 130
Absorption	Wastewater, chemical sludges	Moderate	Limited to soluble compounds (e.g., H_2S, acetone, methanol)
Thermal oxidation (incineration)	NO_x, CO, HCl, potentially toxic organic compounds	High	Stable performance with sufficient time, temperature, and turbulence
Catalytic oxidation (catalytic incineration)	NO_x, CO, HCl, potentially toxic organic compounds	Moderate to high	H_2S, HCl, or particulate matter can destroy catalyst
Condensation	Compound not destroyed; however, potential for product recovery	High	Low range of compounds at high concentrations
Biofiltration	Compost media changed every 2–5 years	Low	Low- to moderate-concentration biodegradable emissions
Biotrickling filter	Synthetic media, low-flow-rate cell waste stream	Low to moderate	Moderate- to high-concentration biodegradable emissions

Important advantages of biofiltration systems over other air-pollution-control alternatives include low capital and operating costs, low energy requirements, and the absence of residuals and by-products requiring further treatment or disposal. Although the intent of conventional systems for VOC removal from gaseous waste streams is gas-phase pollution control, each produces a waste stream which must be either treated or disposed. A summary of existing VOC control technologies, process residuals and by-products, energy costs, and process limitations is shown in Table 10.1.

BIOFILTERS

Biofilters are closed packed-bed reactors through which contaminated air is either blown or drawn, as indicated in Figure 10.1. Microbial communities grow as biofilms on the packing surface. Biofilms are composed of microbial cells (principally bacteria), extracellular polysaccharides, and bound water. A liquid film must exist around the microorganisms because they extract all of their nutrients from the liquid phase. Whether a layer of water in addition to that in the biofilm is required to maintain a satisfactory environment is unclear. However, liquid films would be very thin in conventional biofilters operated at 50 to 60 percent moisture by weight. Based on typical compost packing-specific surface areas of 6 to 10 m^2/g, the liquid films would be 0.5 to 5 μm in thickness.

Drawing air through the packing is generally believed to provide better flow distribution. However, pressure drops are generally below 10 mm H_2O per m of packing, and head loss through the bed itself is not usually a problem unless biomass accumulation as the result of high organic loadings clogs the interstices between the packing particles. The low pressure drop can lead to airflow distribution problems. Pressure losses in treatment units upstream and downstream of the biofilter can result in significant pressure differentials and negative pressure units have collapsed as a result of large vacuums. Air flux values used depend on the contaminants being removed but typical values are between 1 and 2 $m^3/m^2 \cdot min$ based on total cross-sectional area. Most full-scale biofilter applications have been for odor control but an increasing number of systems are being used for VOC removal. Sulfide concentrations treated range from less than a ppm_v to several hundred ppm_v. Similar VOC concentration ranges have been successfully treated by biofiltration. Organic loading rates of up to 4 g carbon/$m^2 \cdot min$ (approximately 10 times the loading rates applied to high-rate rock trickling filters used in wastewater treatment) can be achieved but satisfactory process loading constraints have not been developed at this time.

A biofilter system is typically composed of a blower, humidification system, biofilter unit, and in some cases, a granular activated-carbon backup or polishing unit as indicated in Figure 10.2. In addition to the humidification system, a method of adding water directly to the packing through a manually controlled spray is usually included. An alternative to using direct water addition would be to introduce a cooling coil to lower the outlet gas temperature and condense vapor-phase moisture back into the bed.

Biofilters have been in use for odor control at wastewater-treatment plants, composting plants, and industrial processes since the mid-1950s (Carlson and

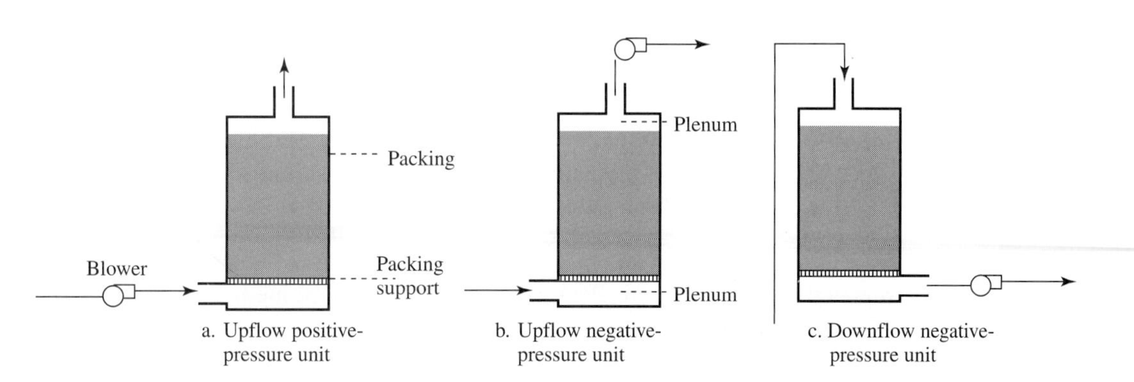

FIGURE 10.1
Typical biofilter configurations.

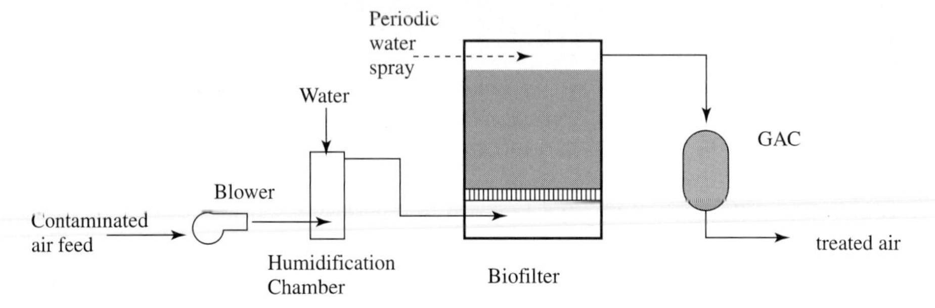

FIGURE 10.2
Biofilter system including blower, humidification chamber, biofilter unit, and backup GAC adsorption bed.

Leiser, 1966). The application of biofiltration systems to the control of VOC emissions, however, is an innovative technology which has been gaining acceptance in the last decade. Contaminant gas streams that have low concentrations of biodegradable and relatively soluble VOCs are well suited to biofiltration. Biofiltration may be combined with soil vapor extraction or air stripping of groundwater to bioremediate a contaminated site. A summary of recent field and laboratory studies of biofiltration systems is given in Table 10.2. Other potential biofiltration applications include chemical, pharmaceutical, paper, textile, polymer, and plastics manufacturing; printing industry; surface coating processes; and petroleum production and refining.

Contaminant Characteristics

Contaminants can be classified in three groups: inorganic, hydrophilic organics, and hydrophobic organics. The principal inorganic contaminant is H_2S, a compound that is quite soluble in water and easily absorbed into the liquid phase. The principal product of H_2S biodegradation is H_2SO_4. Extremely low pH values develop in the reaction zones of biofilters treating H_2S (pH values of approximately 1 are not uncommon) and severe corrosion problems can develop. Sulfide-oxidizing bacteria function well at low pH values but long-term accumulation of acid will eventually limit process performance. Modifications such as biotrickling filters and bioscrubbers in which acid is removed from the treatment unit in a liquid stream can be used to control pH.

Ammonia, NO, and N_2O can be removed by biofiltration. As in the case of H_2S, ammonia oxidation results in the formation of an acid. Unlike the sulfide-oxidizing bacteria, nitrifying bacteria do not function well below pH 6.5. Consequently pH control will be necessary in ammonia treatment systems. Oxidation of NO_x in a laboratory biofilter has been demonstrated but removal rates have been very low (Davidova et al., 1997). The most promising method for biological NO_x (NO and N_2O) removal from air streams is denitrification (du Plessis et al., 1996; Apel et al., 1995; Barnes et al., 1995). Reduction of NO_x requires locally anaerobic conditions, a carbon source, and an energy source. Alkalinity is produced in denitrification and the pH of the medium will rise.

TABLE 10.2
Summary of research relating to biofiltration of VOCs

Reference	Media	Compounds	Inlet concentration, ppm$_v$	Residence time, min	Removal efficiency, %
Ottengraf and van den Oever (1983)	Compost	TOL Butylacetate Ethylacetate Butanol	73–1,400 10–290 20–400 200	0.68–1.2	40–99
Kampbell et al. (1987)	Soil	Propane *n*-Butane Isobutane	2,000 as total HC	15	>85
Utgikar et al. (1991)	GAC	TOL DCM	50–1,000 1,000–3,500	1.8	75–99 20–60
Togna and Folsom (1992)	Peat	Styrene	200–300	0.5–1	90–95
Deshusses and Hamer (1992)	Compost and polystyrene spheres	MEK MIK	560–2,000	1–1.5	98
Leson et al. (1993)	Bark and compost	Ethanol Methane	500–2,200 1,000	0.2–0.5	50–90
Shareefdeen et al. (1993)	Peat moss and perlite	Methanol	4,600	5.6	90
Peters et al. (1993)	Compost and perlite	Kerosene Gasoline	60–1,000 as carbon	3	>90
Ergas et al. (1994)	Compost and perlite	DCM TOL	3–50	0.5–1	>96 >99
Ergas et al. (1995)	Compost and perlite	BZ TOL XYL DCM TCM	0.01–10	0.5–2	83–95 88–97 88–93 20–40 20–40
Wright et al. (1997)	Compost and perlite	Gasoline	100–1,000	2	>95

CM—chloromethane, EtBZ—ethylbenzene, MEK—methyl ethyl ketone, MIK—methyl isobutyl ketone, DCM—dichloromethane, BZ—benzene, TOL—toluene, XYL—xylene, TCE—trichloroethene, TCM—trichloromethane.

Hydrophilic organic compounds usually have high solubilities and may accumulate in the liquid phase of biofilter packing. Consequently significant removals of hydrophilic organic compounds may be due to absorption rather than biodegradation. If drainage is significant, as in the case of biotrickling filters, the liquid effluent may contain high concentrations of contaminants. Hydrophobic compounds, such as the BTEX group, do not accumulate in significant concentrations in the packing, and biodegradation rates are approximately the same as the gas-liquid mass transfer rates.

Packing media should have a high specific surface area and high permeability and should provide a good source of nutrients for microbial growth. Materials used have included natural materials such as soil, compost, peat, and wood chips and synthetic materials such as ceramic saddles, polyethylene Pall rings, and extruded diatomaceous earth pellets. Natural materials such as gravel and rock could be used but the surface-to-volume ratios are low, which results in low volumetric reaction rates.

Compost

Compost has become the most commonly used natural packing. A wide range of compost types have been used (yard waste, wastewater biosolids, manures). Although compost has been used effectively in many biofiltration systems, its properties can be highly variable. This variability can be a source of undesirable problems. For instance, some composts may have low levels of available nutrients and/or low microbial population densities. The microbial population in compost is generally broad, and addition of microorganisms is unnecessary. System start-up, however, may be more rapid if a microbial inoculum is added. Active composts have more available nitrogen than finished composts, and temperatures in active composts may exceed 60°C.

Moisture content of compost packing should be 50 to 60 percent on a wet weight basis. High moisture contents decrease effective porosity and airflow and may result in locally anaerobic conditions. Low moisture contents result in decreased microbial activity and channelization. Moist or wet compost compacts with time and bulking agents must be added to maintain packing porosities and structure and decrease pressure drops. Bulking agents used include porous ceramics, perlite, wood chips, bark, and Styrofoam pellets. Compost and bulking agents are mixed in approximately equal volumes. To prevent drying of the compost during operation, contaminated air fed to compost biofilters should be saturated with water. In many cases the contaminated air will need to be humidified.

Heat produced in biodegradation, or due to exposure to sunlight, can result in drying of the media, and a method of water addition or humidity control is necessary. Humidity control methods are discussed below. Monitoring moisture content is difficult. Available moisture sensors have not proved successful for monitoring moisture content. Periodic sampling of the packing is a possibility but requires sampling ports at a number of locations. Care in sampling is required to ensure that low-density regions are not produced that result in channelization of flow. One commercially available biofilter system incorporates load cells to monitor moisture content (van Lith, 1989).

Nutrients and buffers can be added with the water during the initial system setup. Addition of water and nutrients during operation can be in combination with periodic water addition.

Synthetic packing

Synthetic packing materials including plastic Pall rings, Raschig rings, and granular activated carbon (GAC), and extruded diatomaceous-earth pellets have been used in laboratory and pilot-scale experiments. Granular activated carbon has been the most extensively used synthetic packing material. Performance of systems with synthetic packing materials has been essentially indistinguishable from that of systems with natural packing. Conceptual advantages of synthetic packing include lower head losses and less biofouling due to larger interstices between packing granules or pieces, larger specific surface areas, and in the case of GAC, solid-phase adsorption of contaminants. However, head losses in compost systems run for periods of several years remain in the range of 1 to 3 cm, and of the commonly used synthetic packings only GAC has a specific surface area larger than that of compost. Because GAC pores are smaller than bacterial cells, the high specific surface areas do not provide increased reaction capacity. Further, for continuous-flow systems the GAC quickly comes to equilibrium with the vapor phase. If the contaminants are biodegradable GAC may provide a buffer for transient loadings, but adequate supporting data for this hypothesis does not exist.

As in the case of compost biofilters, humidified air must be used in synthetic-packing medium units. Nutrients can be supplied with the humidification system or the packing can be periodically soaked in a nutrient solution (Ergas et al., 1994). Kinney (1996) developed an aerosol moisture delivery system for a laboratory biofilter system in which the packing was composed of diatomaceous-earth pellets (Celite R-635). The aerosol particles were created from a nutrient solution using a nebulizer. The resulting system had a very small excess water stream, low head losses, and very stable performance in treating toluene and methylene chloride.

Gas Distribution

Biofilters may be operated in an upflow or downflow mode. Although they are commonly operated in upflow mode, downflow operation eliminates the problem of liquid hold-up during spray humidification. Four types of gas-distribution systems are currently in use in biofiltration systems: (1) perforated pipes, (2) pressure-chamber systems, (3) sinter-block systems, and (4) plenums. In perforated-pipe systems, the bed is underlain with a network of perforated pipes set in a gravel bed. In pressure-chamber filter systems, air supply and distribution is performed using a large pressure chamber at the bottom or top of the bed. This type of system is limited to smaller filter beds owing to instability of large pressure chambers and the weight of the filter material in large filters (Hartenstein and Allen, 1986). In the sinter-block system, air distribution is accomplished through prefabricated slotted concrete blocks which provide both aeration and drainage systems. The blocks have sufficient mechanical strength to provide access to the filter material by a front-end loader. Plenums are used in small biofilters where the packing weight can be supported by expanded metal grids. Typical plenum heights are 150 to 300 mm.

Humidification

Moisture content of the biofilter media is critical to successful operation. Too little moisture in the media results in dry zones, loss of microbial activity, and cracking of the bed. Too much moisture may inhibit gas transport and results in the development of anaerobic zones. Generally, the moisture content of the biofilter media should be maintained between 50 and 65 percent by weight:

$$\text{Moisture content} = \frac{\text{mass of water}}{\text{mass of water} + \text{mass of dry material}} \tag{10.1}$$

Moisture is usually provided to the bed by spraying water over the surface of the bed and by humidifying the influent gases. Humidification of the influent gases is essential except in applications with already humid gas streams, such as the control of emissions from wastewater-treatment processes. Relative humidity of air entering the biofilter should be approximately 100 percent at the bed temperature. Relative humidity is defined as the ratio of the partial pressure of the vapor present in the gas to the vapor pressure of pure liquid water at the gas temperature:

$$H_r = 100\frac{P_{H_2O}}{P_{vH_2O}} \tag{10.2}$$

where H_r = relative humidity, percent
P_{H_2O} = partial pressure of the vapor actually present in the gas, atm
P_{vH_2O} = vapor pressure of liquid water, atm

A typical humidification chamber is shown in Figure 10.3. The chamber has a mist generator located on the top and air blown through is humidified both in the chamber and as a result of carrying mist particles downstream. Residence time of the air in the chamber is usually less than 2 s. Humidification typically reduces gas temperature because of the latent heat of evaporation, and supplemental heat input may be necessary.

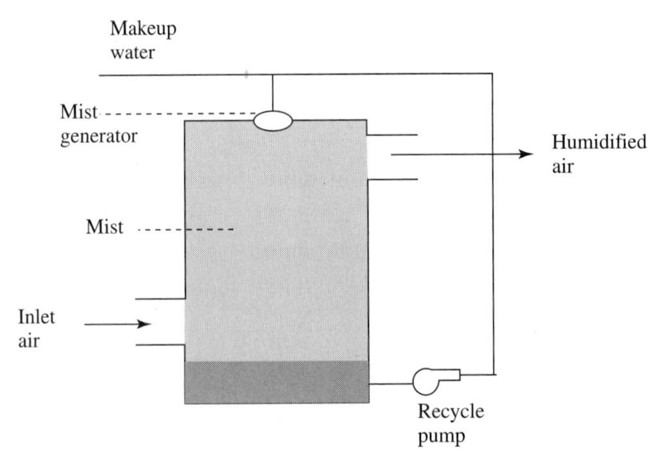

FIGURE 10.3
Definition sketch of humidification chamber for biofilter system.

pH Control

Biological metabolism of reduced sulfur compounds and chlorinated organic compounds produces acid by-products with a corresponding decrease in pH of the filter media. Highly acidic conditions generally inhibit microbial activity since most bacteria favor neutral conditions. Sulfur-oxidizing bacteria, however, are well adapted to acidic conditions. An obligate aerobic chemoautotrophic genus, they produce sulfuric acid through oxidation of H_2S and function well at pH values as low as 1. In general, however, for VOC removal, the pH of the filter material should be maintained between 7 and 8.5.

Agricultural liming agents may be added to the filter bed to control acidification of the filter material. Materials traditionally used for liming soils are oxides, hydroxides, carbonates, and silicates of calcium or calcium and magnesium. The reaction of all liming materials involves the neutralization of H^+ ions in solution by OH^- or $SiO_2(OH)_2^{2-}$ ions furnished by the buffer, as with the reaction of calcium carbonate:

$$CaCO_3 + H_2O \rightarrow Ca^{2+} + HCO_3^- + OH^- \qquad (10.3)$$

$$H^+ + OH^- \rightarrow H_2O \qquad (10.4)$$

Agricultural liming materials include slaked lime [$Ca(OH)_2$], unslaked lime (CaO), limestone ($CaCO_3$), crushed oyster shell ($CaCO_3$), dolomite [$CaMg(CO_3)_2$], marl (earth + $CaCO_3$), and slags ($CaSiO_3$). Differences in the neutralizing capacity and speed of reaction of these materials depend upon molecular composition, material purity, and fineness of particles. In addition to liming, Yang and Allen (1994) found that periodic washing of the biofilter bed with distilled water or a sodium bicarbonate solution was successful in removing hydrogen ions and extending the life of the bed. As mentioned previously, bioscrubber and biotrickling filter systems have been used with continuous pH buffer addition to the nutrient medium.

Temperature Control

The rate of biochemical reactions is governed by temperature as well as other factors. In general an increase in temperature increases the rate of reaction up to some optimal temperature above which there is a decrease in reaction rate. Each microorganism has an optimal temperature range for growth. In general, the microorganisms common in biofilters are mesophiles that can grow between 15 and 45°C and have optimal growth in the range of 25 to 35°C. Biofilters have been used successfully in cold climates such as Wisconsin (Kampbell et al., 1987) and Finland (Lehtomaki et al., 1992). However, in very cold climates, biofilters require insulation and heating of inlet gases. Inlet gases coming from warm process streams may need to be cooled prior to treating them using biofiltration. Operation in the thermophilic region, 50 to 65°C, appears to be practical if a stable supply of warm air is available.

An acclimation period or lag period during start-up of biofilter operations is virtually always reported. The observed lag period is the result of the acclimation of the microbial population to the contaminants as a carbon and energy source and/or the growth of significant microbial communities from an initially small number of organisms. Ottengraf and Van Den Oever (1983) reported that a 10-day period was required before steady-state conditions were observed in their systems treating lacquer thinner emissions. Peters et al. (1993) observed a 1-week acclimation period before steady-state degradation of kerosene and unleaded gasoline vapors occurred. A 1-year period was required for initial acclimation of a compost-based biofilter treating methyl tert-butyl ether (Eweis et al., 1997*a*). In later experiments with the bacterial culture, isolated from the compost-based biofilter, the acclimation period was reduced to 3 weeks in a biofilter packed with synthetic material (Eweis et al., 1997*b*). Acclimation periods have also been reported in biofiltration systems after periods of shutdown, when concentrations have been increased, or due to shock loads.

Concentration data from the start-up of a laboratory-scale biofiltration column is shown in Figure 10.4. The compost-filled column was receiving an airstream containing 40 ppm_v toluene and had a residence time of 1 min. During the initial start-up period, the outlet concentration was equal to the inlet concentration. After

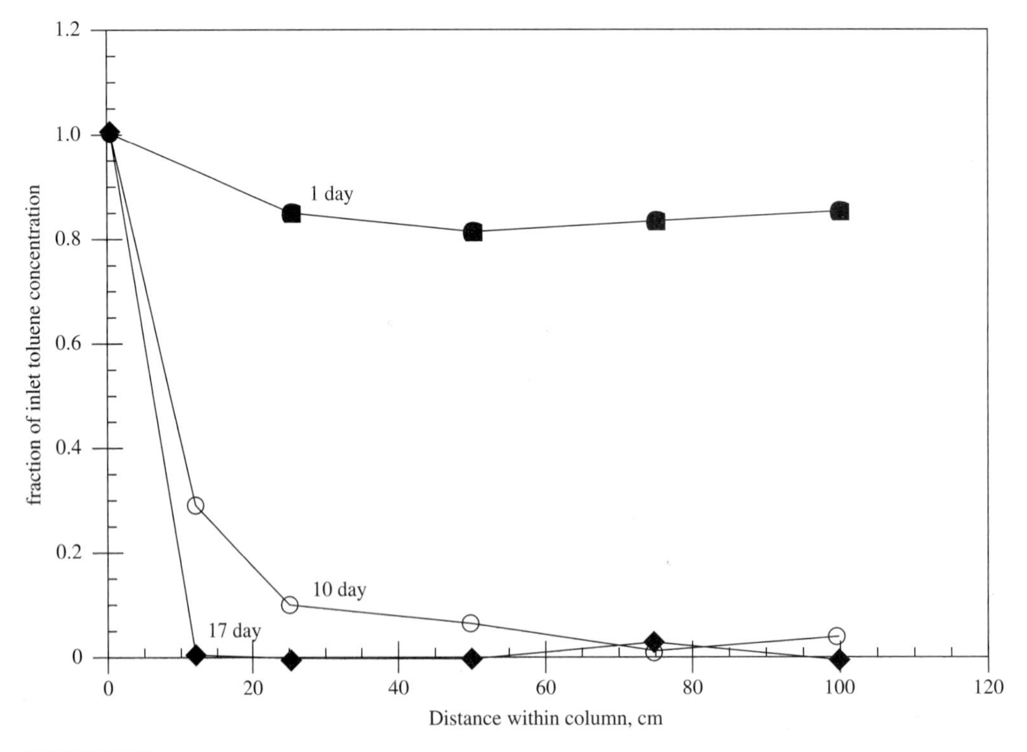

FIGURE 10.4
Acclimation period of a compost biofilter degrading toluene (Ergas, 1993).

the initial lag period, biodegradation began to occur, but at a low rate. After 17 days, the biofilter reached a steady removal rate.

Acclimation is a concern in biofilter operations because reduced removal efficiency may occur at start-up, after periods of shutdown, or due to variations in inlet concentration. Explanations put forth to explain the acclimation period include: (1) microbial enzymes may be induced only after exposure to the chemical or (2) initially small populations of microorganisms capable of degrading the pollutant may be present and time is required to allow them to grow up and cover sufficient surface to the point where significant degradation can occur.

BIOTRICKLING FILTERS

Biotrickling filters are conceptually identical processes to biofilters, as can be seen in Figure 10.5. The only physical difference is that moisture is provided by spraying water continually over the top of the packing rather than providing water-saturated inlet air through humidification. However, the liquid spray results in several practical differences between biofilters and biotrickling filters:

- The liquid spray results in a mobile liquid film flowing over the packing. Gas-liquid contaminant transport remains a key step in the process but the liquid film is thicker than in biofilters and transport through the liquid phase will be slower.
- Compost packing is unsuitable for use in biotrickling filters because water will accumulate within the compost, effective porosity will be decreased, anaerobic conditions will develop, and head losses will increase.

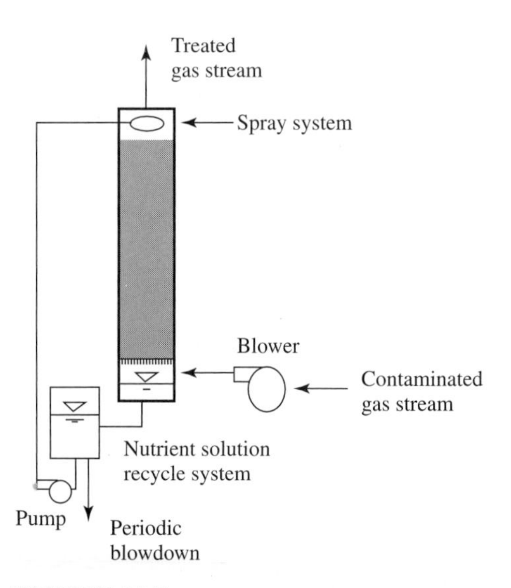

FIGURE 10.5
Definition sketch of biotrickling filter system. The airflow direction is optional and a makeup solution will need to be provided because of liquid loss in the outlet gas stream.

- Size of packing material will need to be increased to allow combined airflow and liquid flow. Laboratory studies with extruded diatomaceous-earth pellets having approximate dimensions of 5 mm in diameter and 10 mm in length have resulted in good performance but significant biofouling problems (Sorial et al., 1995*a,b;* Smith et al., 1995).
- Both makeup water and blowdown stream disposal must be included in the process design. Water is lost in the outlet gas stream requiring makeup water be added. Because of the accumulation of salts in the recycled spray stream a blowdown stream is required.

Liquid spray rates used are about 1 $m^3/m^2 \cdot$ day (Sorial et al., 1995*a*), a value which is typical of loading rates used for low-rate wastewater trickling filters. As in the case of low-rate trickling filters, the biofilm is not removed from the packing by hydraulic shear but instead accumulates to the point of increasing head losses significantly and causing channelization of flow. Severe clogging has been reported to develop in biotrickling filter systems with organic loading rates as low as 1.2 kg organic carbon/$m^3 \cdot$ day (Sorial et al., 1995*a;* Smith et al., 1995). Similar problems are rarely observed for compost biofilters, even at organic loading rates as high as 9 kg carbon/$m^3 \cdot$ day (Morgenroth, 1994). Working with an extruded diatomaceous-earth pellet packing, Kinney (1996) observed mild biofouling and decreased removals after approximately 60 days of operation at organic loading rates of 1 kg carbon/$m^3 \cdot$ day and developed a strategy in which the airflow direction was reversed at periods of 1 to 3 days. The result was a more even distribution of biofilm through the column and stable operation without increased head losses or decreased performance due to biofouling.

Solutions to the biofouling problem of biotrickling filters include:

- Increasing the packing size to provide larger interstices. This approach will result in a lower specific surface area and lower removal rates.
- Periodic backwashing of the packing (Smith et al., 1995; Sorial et al., 1995*b*).
- Use of aerosol sprays to provide moisture and directional switching provides more uniform volumetric loading (Kinney, 1996).
- Nutrient limitation to control cell production (Holubar et al., 1995).

DESIGN AND OPERATING PARAMETERS

Design and operation of biofilters (and biotrickling filters) is based on four loading-rate parameters; the empty-bed residence time, the gas flux, the contaminant mass loading rate, and the elimination capacity.

The empty-bed residence time t_r represents the theoretical average time that a gas molecule would spend inside an empty filter bed.

$$t_r = \frac{V}{Q} \tag{10.5}$$

where V = empty-bed volume, m^3
Q = volumetric gas flow rate, m^3/s

Empty-bed residence times used in VOC removal studies vary between 0.3 and 12 min with higher residence times necessary for less degradable or less soluble compounds. Compounds such as hexane, benzene, toluene, and xylene have relatively low solubilities (11 to 1,780 mg/L) but require t_r values of less than 1 min for complete removal. Apparently the mass transfer rate is rarely limiting in biofilters.

The gas flux or superficial velocity v (m/s) is a measure of the average fluid velocity through the empty bed:

$$v = \frac{Q}{A} \tag{10.6}$$

where A = bed cross-sectional area, m^2

Empty-bed residence time and gas flux are related by

$$v = \frac{h}{t_r} \tag{10.7}$$

where h = height of the filter bed, m

Biofilter heights are commonly around 1 m, although biotowers of 1.5 m or more have been reported (Hartenstein and Allen, 1986). Concentration profiles through the bed are usually very steep and often most of the removal takes place in the first 250 mm. Increased bed depths will be necessary only if reaction or mass transfer rates are low or if interference between reactions exists. Using deep beds for treatment of high contaminant concentrations will result in increased microbial growth at the inlet end and increased clogging problems. Thus construction of deep beds may be counterproductive.

The theoretical velocity of the gas through the pores of the filter material (v_{pore}) or the Darcy velocity is calculated by dividing the gas flux by the void fraction f (dimensionless).

$$v_{\text{pore}} = \frac{v}{f} \tag{10.8}$$

The contaminant mass loading rate, R_m (g/m^3 · s) can be determined as

$$R_m = \frac{QC_i}{V} \tag{10.9}$$

where C_i = inlet gas contaminant concentration, g/m^3

Elimination capacity EC (g/m^3 · s) of a biofilter bed is the overall contaminant-removal rate and can be calculated from the following:

$$EC = \frac{Q(C_i - C_o)}{V} \tag{10.10}$$

where C_o = outlet gas contaminant concentration, g/m^3

Data from pilot biofilter studies are often plotted as EC vs. contaminant loading rate. An example from a pilot system used to control H_2S emissions is shown

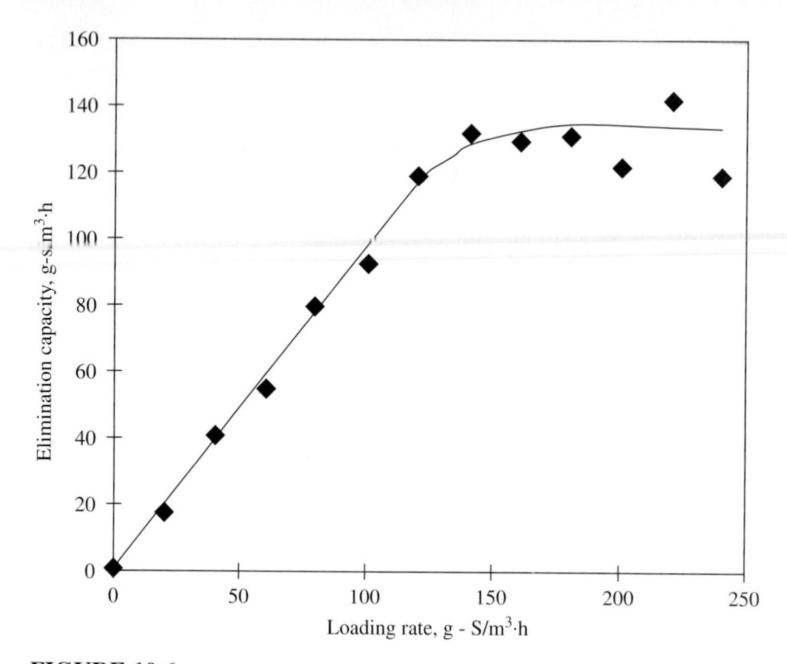

FIGURE 10.6
Determination of maximum elimination capacity of biofilter media based on mass of sulfur (S) loaded. (*From Yang and Allen, 1994.*)

in Figure 10.6 (Yang and Allen, 1994). At low loading rates (low inlet concentration or gas flow rate), the elimination capacity of the bed increases as the loading rate increases. At high loading rates, the elimination capacity of the bed remains constant with increased loading.

EXAMPLE 10.1. PACKING VOLUME DETERMINATION. Using the data of Figure 10.6, determine the volume of packing required for a biofiltration system with an inlet H_2S concentration of 100 ppm$_v$ and a flow rate of 2,000 ft³/min (assume P = 1 atm, T = 25°C).

Solution

1. Calculate total sulfur mass flow rate M_S.

$$M_S = \frac{100(10^{-6})\text{L H}_2\text{S}}{\text{L air}} \frac{1 \text{ g mole}}{22.4 \text{ L}} (2{,}000 \text{ ft}^3/\text{min})(28.3 \text{ L/ft}^3)$$

$$= 0.253 \text{ g mole/min}$$

$$= 8.09 \text{ g/min} = 485 \text{ g - S/h}$$

2. Use EC_{max} value from Figure 10.6.

$$EC_{max} = 130 \text{ g - S/m}^3 \cdot \text{h}$$

$$V = \frac{M_S}{EC_{max}} = \frac{485 \text{ g - S/h}}{130 \text{ g - S/m}^3 \cdot \text{h}}$$

$$= 3.73 \text{ m}^3 = \text{volume of packing required}$$

3. In actual practice, one would design a system with a 30 to 70 percent larger volume.

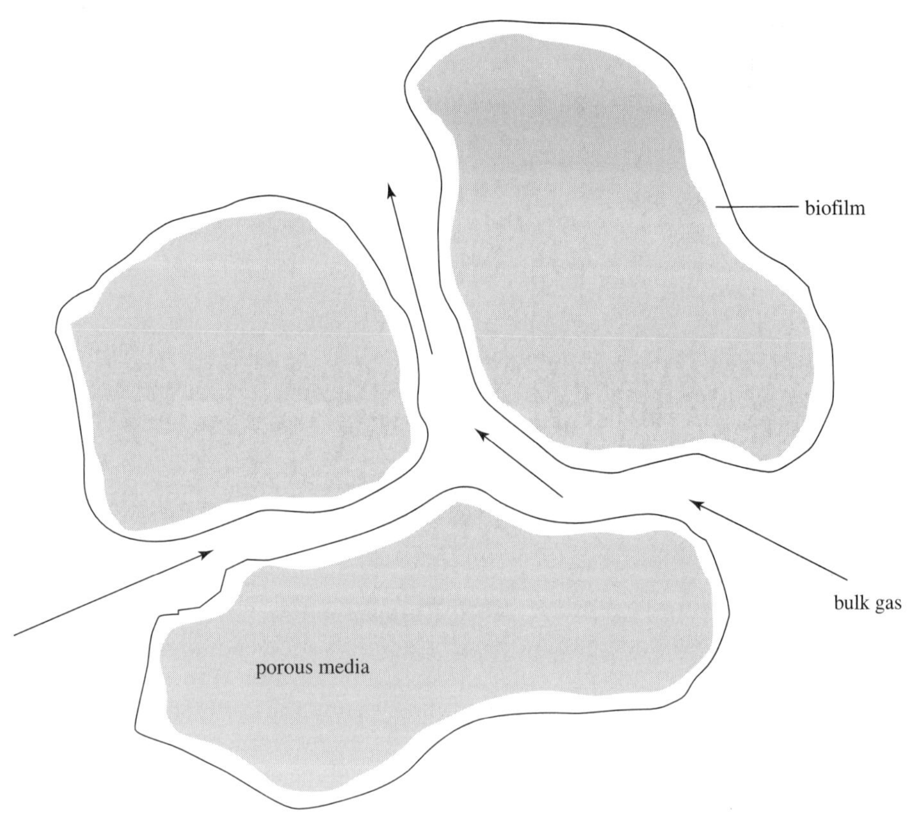

FIGURE 10.7
Microscopic section through the porous bed.

MICROSCALE PROCESSES

A representation of a microscopic section of biofilter media is shown in Figure 10.7. Air containing contaminants follows a tortuous path through the packing particles which are surrounded by a liquid film containing microorganisms. Soluble compounds in the gases partition into this biologically active film, or biofilm, where they are available for biodegradation. Each substrate (electron donors and electron acceptors) and each nutrient required for microbial metabolism must be transported from either the bulk gas phase or the solid phase to the biofilm where the reaction takes place. All products (e.g., CO_2, metabolites), except biomass, must be transported out of the biofilm. The phenomena involved are bulk gas-phase advection, gas-film diffusion, liquid-film diffusion, biofilm diffusion, adsorption, and biodegradation.

Theoretical Modeling of Biofilter Performance

Theoretical models have been developed to aid in understanding mass transport and biodegradation of contaminants in biofilters. Most models in use build on approaches used in earlier models of transport and biodegradation in biofilms

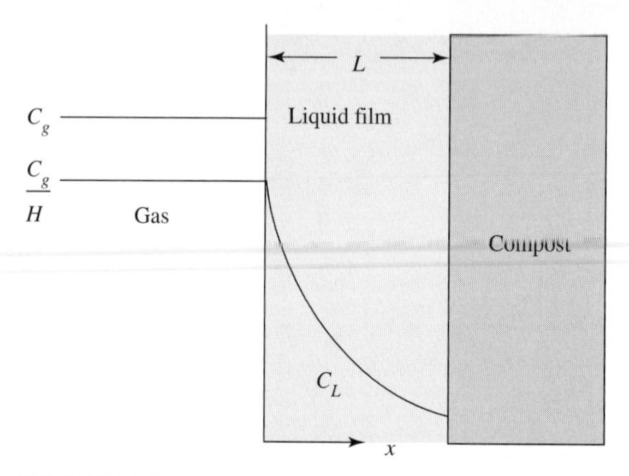

FIGURE 10.8
Schematic representation of the phases in a biofilter.

(Jennings et al., 1976; Härremoes, 1978; Rittman and McCarty, 1980). The approach of both biofilter and biofilm models is to solve the one-dimensional mass balance equations over a microscopic cross section of a biofilm assuming simplified reaction kinetics. By using simplifying assumptions of biodegradation kinetics, three different solutions can be obtained: (1) zero-order reaction/ reaction limitation, (2) zero-order reaction/diffusion limitation, and (3) first-order reaction. Closed-form solutions for the macroscopic gas-phase mass balance for a given VOC are obtained by volume averaging the biofilm equations over biofilter media. Solutions for each of the three cases are developed and compared in the remainder of this chapter.

A conceptual model of the three phases of the biofilter is shown in Figure 10.8. At the interface between the gas and liquid phases, the liquid-layer concentration is assumed to be in equilibrium with the gas-phase concentration. The compound diffuses through the liquid film where it is biodegraded. The differential equation describing the mass balance for a compound C_L in the liquid layer of the biofilter is

$$\frac{\partial C_L}{\partial t} + \frac{\partial N_L}{\partial x} + r_L = 0 \tag{10.11}$$

where C_L = liquid-phase compound concentration, g/m^3
N_L = liquid-phase compound flux, g/m$^2 \cdot$ s
r_L = reaction rate due to biodegradation, g/m$^3 \cdot$ s

The liquid-phase flux N_L is described by Fick's first law of diffusion for the transport of a compound in a quiescent fluid:

$$N_L = -D\frac{\partial C_L}{\partial x} \tag{10.12}$$

where D = liquid-phase diffusion coefficient, m^2/s

The resulting differential equation becomes

$$\frac{\partial C_L}{\partial t} - D\frac{\partial^2 C_L}{\partial x^2} + r_L = 0 \tag{10.13}$$

Biodegradation rate is commonly described by the Monod substrate utilization expression:

$$r_L = -k\rho_b \frac{C_L}{K_S + C_L} \tag{10.14}$$

where k = maximum specific removal rate of the contaminant, s^{-1}
K_S = substrate saturation constant, g/m^3
ρ_b = biomass density, g/m^3

The net growth rate of the biomass can be assumed to follow the following expression:

$$\frac{\partial \rho_b}{\partial t} = Yr_L - k_d\rho_b \tag{10.15}$$

where Y = biomass yield per unit of substrate utilized, g/g
k_d = biomass specific decay coefficient, s^{-1}

An assumption can be made that a quasi-steady-state condition exists whereby the total amount of biomass is just equal to that which can be supported by the substrate flux (Rittman and McCarty, 1980); i.e., the rate of cell growth is equal to the rate of energy expended for cell maintenance:

$$Yr_L = k_d\rho_b \tag{10.16}$$

and the net growth rate $\partial\rho_b/\partial t$ is equal to zero. A second justification for this steady-state assumption even during periods of biofilm growth is that biological growth processes are slow relative to system residence times. By making the assumption that a steady-state condition exists in the biofilm and that mass does not accumulate in the system after some initial period, Equation (10.13) reduces to the steady-state total differential equation

$$D\frac{\partial^2 C_L}{\partial x^2} = k\rho_b \frac{C_L}{K_S + C_L} \tag{10.17}$$

Analytical solutions to Equation (10.17) using zero- and first-order simplifications of the Monod expression are discussed in the following sections.

Zero-order reaction

At liquid-phase substrate concentrations much greater than the half-saturation constant $C_L >> K_S$, Equation (10.17) reduces to a zero-order rate expression

$$D\frac{\partial^2 C_L}{\partial x^2} = K \tag{10.18}$$

where K = zero-order reaction rate, $g/m^3 \cdot s$, is equal to $k\rho_b$. Zero-order reaction kinetics are used when the substrate is not rate-limiting either because of high inlet concentrations or because another compound is serving as a primary substrate.

Two examples of zero-order kinetics are commonly described. In the first case, the compound fully penetrates a biofilm of thickness L. This is referred to as the reaction-limiting case. In the second case, gas-phase compound concentration is lower and the compound penetrates the biofilm to some distance λ which is less than L, where the rest of the biolayer is assumed to be inactive. The latter case is referred to as the diffusion-limiting case. Solutions for these two cases are given below.

For the case where the compound fully penetrates the biofilm, or the reaction-limiting case, Equation (10.18) can be solved with the right-hand boundary condition

At $x = L$
$$\frac{\partial C_L}{\partial x} = 0 \qquad (10.19)$$

which assumes that the flux into the solid phase is zero. This assumption has the interpretation that the flux into the solid phase due to adsorption is equal to the flux out of the solid phase due to desorption. The left-hand boundary condition is given as

At $x = 0$
$$C_L = \frac{C_g}{H} \qquad (10.20)$$

where H = dimensionless Henry's law coefficient
C_g = bulk gas-phase concentration, g/m^3

Use of the left-hand boundary condition assumes that the gas and liquid phases at the interface are at equilibrium, a condition that can be described by Henry's law as discussed above. The solution to the reaction-limiting case is

$$C_L = \frac{K}{D}x - \frac{KL}{2D}x^2 + \frac{C_g}{H} \qquad (10.21a)$$

or in dimensionless form:

$$C^* = 1 + \frac{\phi^2}{2}(\sigma^2 - 2\sigma) \qquad (10.21b)$$

where $C^* = \dfrac{C_L H}{C_g}$

$\sigma = \dfrac{x}{L}$

$\phi = \sqrt{\dfrac{KHL^2}{C_g D}}$

The parameter ϕ, referred to as the Thiele number, is used to describe the ratio between the reaction rate and the diffusion rate.

For the case where the penetration thickness is less than the biofilm thickness, the diffusion-limiting case, the right-hand boundary condition changes to

At $x = \lambda$
$$\frac{dC_L}{dx} \rightarrow 0 \qquad (10.22)$$

Note that as $x \rightarrow \lambda$, $C_L \leq K_S$ and the assumption of zero-order kinetics is violated. The solution to Equation (10.18) in dimensionless form becomes

$$C^* = 1 + \frac{\phi^2}{2}\left(\sigma^2 - 2\sigma\frac{\lambda}{L}\right) \tag{10.23}$$

The penetration thickness λ can be calculated by setting C^* equal to zero at $\sigma = \lambda/L$, yielding

$$\lambda = \sqrt{\frac{2C_g D}{KH}} \tag{10.24}$$

First-order reaction

At liquid-phase substrate concentrations much less than the half-saturation constant, $C_L << K_S$, the Monod expression reduces to a first-order rate expression. Equation (10.17), with the first-order rate expression, becomes

$$D\frac{d^2C_L}{dx^2} - kC_L = 0 \tag{10.25}$$

With the boundary conditions (10.19) and (10.20), the solution in dimensionless form is

$$C^* = \cosh\phi' \operatorname{sech}(\phi'\sigma + \phi') \tag{10.26}$$

where $\phi' = \sqrt{\dfrac{kL^2}{D}}$

The term ϕ' is also referred to as a Thiele number because it is a ratio of biodegradation rate to diffusion rate. Note that the terms in ϕ (10.21) are different from those in ϕ'.

Gas-Phase Mass Balance

A representation of a biofilter bed and coordinate axis is shown in Figure 10.9. The differential equation that describes the steady-state mass balance for the gas-phase compound at a position z in the filter bed assuming negligible along-axis dispersion is

$$v_{\text{pore}}\frac{dC_g}{dz} + N_g A_S = 0 \tag{10.27}$$

where v_{pore} is the gas velocity (m/s) through the porous media, or Darcy velocity, which is obtained by dividing the superficial gas velocity by the media void fraction, A_S is the surface area per unit volume of the porous media (m^{-1}), and N_g is the constituent flux from the gas phase into the biofilm layer (g/m$^2 \cdot$ s). Gas flow through the biofilter bed is assumed to be of the plug-flow type.

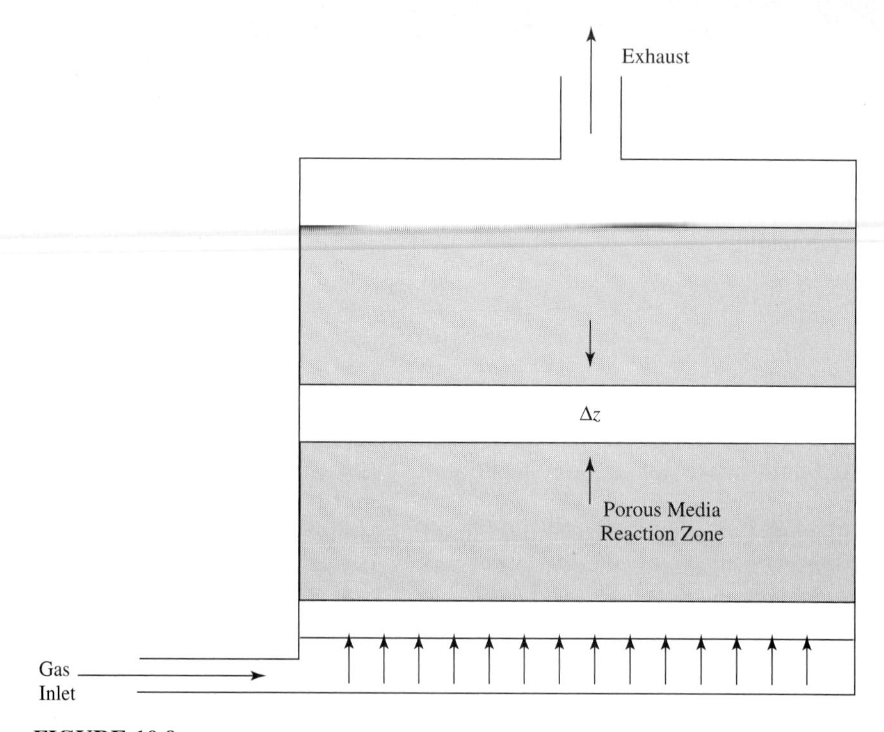

FIGURE 10.9
Schematic representation of a biofilter bed.

The flux into the biolayer N_g can be determined from Equations (10.21*b*), (10.23), and (10.26) for zero-order/reaction limitation, zero-order/diffusion limitation, and first-order biodegradation kinetics, respectively. For the zero-order/reaction limitation case the flux is determined from (10.21) as

$$N_g = -D\left(\frac{dC_L}{dx}\right)_{x=0} = KL \tag{10.28}$$

For the case of zero-order biodegradation kinetics with diffusion limitation the flux is determined from Equation (10.23) as

$$N_g = -D\left(\frac{dC_L}{dx}\right)_{x=0} = K\lambda \tag{10.29}$$

For the case of first-order biodegradation kinetics the flux is determined from Equation (10.26) as

$$N_g = -D\left(\frac{dC_L}{dx}\right)_{x=0} = \frac{kC_g L}{H}\frac{\tanh\phi}{\phi} \tag{10.30}$$

Using the assumption that the substrate half-saturation constant, maximum specific growth rate, and biomass density are not functions of position in the reactor, Equation (10.27) may be solved analytically for the three cases with the

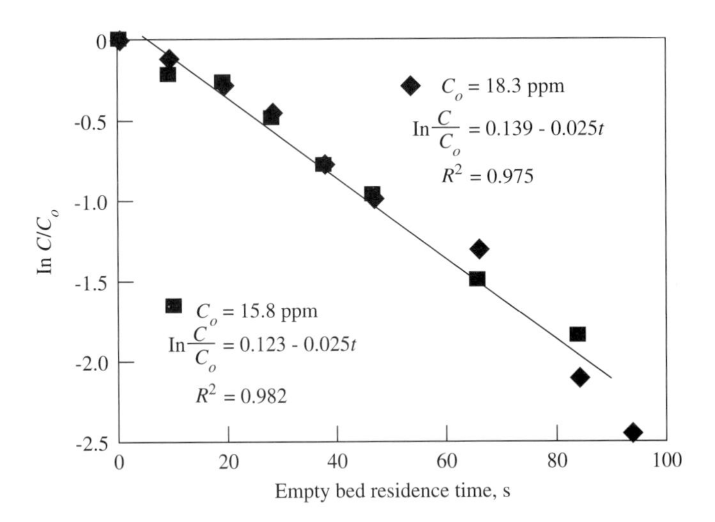

FIGURE 10.10
Log-linear concentration profile for butane in a laboratory-scale biofilter indicating first-order kinetics. (*From Kardono and Allen, 1994.*)

boundary condition, $C_g = C_o$ at $z = 0$, where C_o is the inlet concentration, to yield the following solutions:

Zero-order, reaction limitation:

$$\frac{C_g}{C_o} = 1 - \frac{K_z A_s L}{v_{\text{pore}} C_o} \tag{10.31}$$

Zero-order, diffusion limitation:

$$\frac{C_g}{C_o} = \left(1 - \frac{z A_s}{v_{\text{pore}}} \sqrt{\frac{KD}{2HC_o}}\right) \tag{10.32}$$

First-order:

$$\ln \frac{C_g}{C_o} = -\frac{kzA_s L}{Hv_{\text{pore}}} \frac{\tanh \phi'}{\phi'} \tag{10.33}$$

Equation (10.31), the zero-order/reaction-limitation solution, predicts that the curve of concentration versus height in a biofilter would be a straight line and that higher inlet gas-phase concentrations would require a deeper bed for the same removal efficiency. Figure 10.10 presents the results of an investigation by Kardono and Allen (1994) of the control of butane emissions using a compost biofilter and illustrates first-order reaction behavior.

The results of toluene-removal experiments conducted by Ottengraf and Van Den Oever (1983) in a laboratory-scale biofilter were shown to agree with model predictions (Figure 10.11) when the assumption of zero-order kinetics with diffusion limitation was made, as in Equation (10.32). Note the inlet concentration, given as 0.84 g/m^3, is approximately 200 ppm$_v$ toluene.

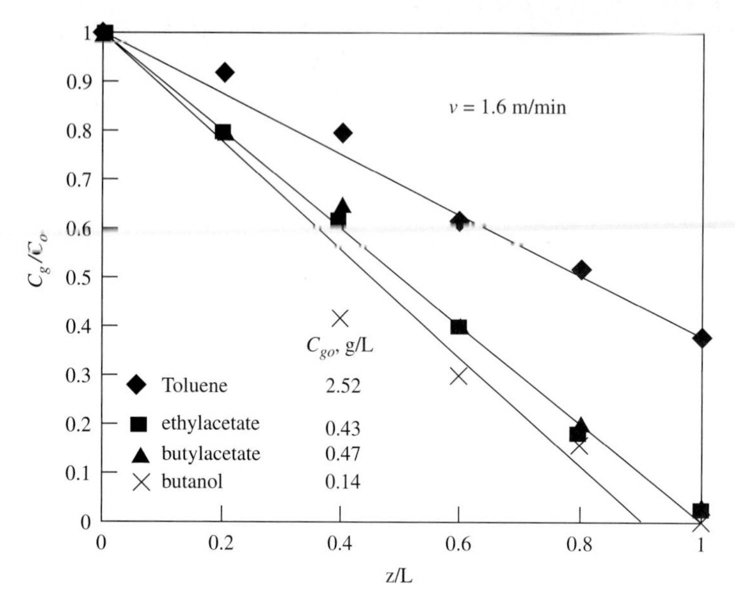

FIGURE 10.11
Concentration profile for toluene in a laboratory-scale compost biofilter showing diffusion limitation in the biofilm. (*From Ottengraf and Van Den Oever, 1983.*)

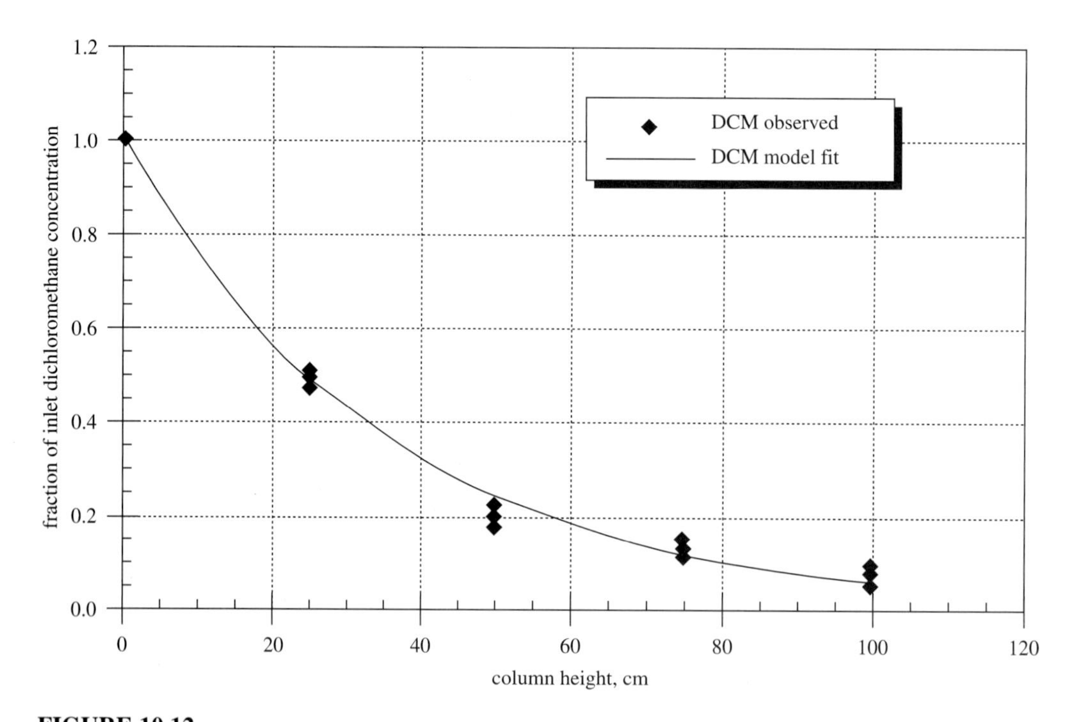

FIGURE 10.12
Concentration profile for dichloromethane in a laboratory-scale compost biofilter showing first-order biodegradation kinetics. (*From Ergas et al., 1994.*)

At a lower inlet concentration, the results of dichloromethane-removal experiments conducted by Ergas et al. (1994) in a laboratory-scale biofilter were shown to agree with model predictions [Figure 10.12 first-order kinetics was made, as in Eq (10.33)]. The data, which are in good agreement with the predicted values, indicate that the model is a useful tool for predicting performance of compost biofilters treating low-concentration VOC emissions. In these experiments, the inlet dichloromethane concentration was 3 ppm_v.

PROBLEMS AND DISCUSSION QUESTIONS

10.1. Estimate the depth of the liquid film in a compost biofilter. Assume the water is distributed uniformly over the solid surface, including the internal pores, and that the water content on a wet-mass basis is 50 percent. The specific surface area of the compost-bulking agent mixture is 8 m^2/g. Density of the dry solids is 500 kg/m^3, and the void ratio of the bulk media is 0.25.

10.2. Estimate the depth of the liquid film for a biofilter in which packing particles have no internal porosity. Assume an average packing-particle diameter d_p of 1 mm and a moisture content on a wet-mass basis of 50 percent, as in Problem 10.1. Density of the dry particles is 500 kg/m^3.

10.3. A biofilter contains a porous medium with pore sizes ranging from 0.01 to 0.1 μm. Will the capillaries be filled with liquid water at a relative humidity of 99 percent if it is assumed that the medium is ideally wetted by the water and that the properties of the liquid are those of water? At what relative humidity will the porous medium become dry assuming no hysteresis effects occur? The Kelvin equation can be used to describe the vapor pressure for curved surfaces:

$$\ln\left(\frac{P_v}{P_o}\right) = \frac{2M\gamma}{RT\rho r}$$

where P_v = vapor pressure, atm
P_o = ambient pressure, atm
M = molecular weight, g/mole
γ = surface tension, 78 dynes/cm
R = gas constant, 0.082 L · atm/mole · K
T = temperature, K
ρ = density of water, g/cm^3
r = capillary radius

10.4. Pilot biofilter experiments were performed with a 1 m^3 (1 m depth × 1 m length × 1 m width) pilot compost biofiltration system. The system was used to control toluene (MW = 92) emissions from a soil vapor-extraction system. Inlet gas-flow rates (Q, m^3/min), inlet toluene concentrations (C_i, ppm_v), and removal efficiencies (e, percent) over a 4-month period of operation are shown in the following table.

Q, m³/min	C_i, ppm$_v$	e, %	Q, m³/min	C_i, ppm$_v$	e, %
1.00	16	99	1.00	366	57
0.50	164	99	0.67	551	55
0.67	151	99	0.70	518	52
0.33	366	98	0.80	523	50
1.10	122	96	1.10	380	48
0.67	228	94	0.90	451	46
0.80	213	92	0.80	579	44
0.70	277	90	0.85	585	43
0.75	290	88	0.67	762	41
0.67	357	83	0.85	643	39
0.75	384	75	0.67	819	36
0.45	681	67	0.67	909	34
0.85	389	60			

a. Calculate VOC loading rate (mg/m³ · min) and elimination capacity (mg/m³ · min) for the system at each data point.
b. Plot elimination capacity vs. VOC loading rate. Discuss your graph with respect to the first- and zero-order kinetic models presented.
c. Design a full-scale biofiltration system to treat emissions from a similar site. Assume a gas flow rate of 3 m³/min and an average inlet toluene concentration of 500 ppm$_v$.
d. What factors should be taken into consideration when recommending this system?

REFERENCES

Apel, W. A., J. M. Barnes, and K. B. Barrett (1995): "Biofiltration of Nitrogen Oxides from Fuel Combustion Gas Streams," *Proceedings of the 88th Annual Meeting and Exhibition, Air and Waste Management Association,* San Antonio, TX, June 18–23.

Barnes, J. M., W. A. Apel, and K. B. Barrett (1995): "Removal of Nitrogen Oxides from Gas Streams Using Biofiltration," *Journal of Hazardous Materials,* vol. 41, pp. 315–326.

Carlson, D. A., and C. P. Leiser (1966): "Soil Beds for the Control of Sewage Odors," *Journal of the Water Pollution Control Federation,* vol. 38, no. 5, pp. 429–440.

Davidova, Y., E. D. Schroeder, and D. P. Y. Chang (1997): "Biofiltration of Nitric Oxide," *Proceedings of the 90th Annual Meeting and Exhibition, Air and Waste Management Association,* Toronto, Canada, June 8–13.

Deshusses, M. A., and G. Hamer (1992): "Methyl Isobutyl and Methyl Ethyl Ketone Biodegradation in Biofilters," in *Biocatalysis in Non-Conventional Media,* J. Tramper, ed., Elsevier Science Pub.

Devinny, J. S., D. S. Hodge, A. N. Chang, and F. E. Reynolds, Jr. (1995): "Biofiltration of Gasoline Vapors from a Soil Vapor Extraction System," *Innovative Technologies for Site Remediation and Hazardous Waste Management,* ed. by R. D. Vidic and F. R. Pohland, American Society of Civil Engineers, New York, pp. 481–488.

Diks, R. M. M., and S. P. P. Ottengraf (1991): "Verification Studies of a Simplified Model for the Removal of Dichloromethane from Waste Gases Using a Biological Trickling Filter," Parts I and II, *Bioprocess Engineering,* vol. 6, pp. 93–99, 131–140.

du Plessis, C. A., K. A. Kinney, E. D. Schroeder, D. P. Y. Chang, and K. M. Scow (1996): "Anaerobic Activity in an Aerobic, Aerosol-Fed Biofilter," *Proceedings,* 1996 Conference on Biofiltration, University of Southern California, Oct. 24–25, pp. 158–163.

Ergas, S. J. (1993): "Control of Low Concentration Volatile Organic Emissions Using Biofiltration," Ph.D. Dissertation, Department of Civil and Environmental Engineering, University of California, Davis.

Ergas, S. J., K. Kinney, M. E. Fuller, and K. M. Scow (1994): "Characterization of a Compost Biofiltration System Degrading Dichloromethane," *Biotechnology and Bioengineering,* vol. 44, pp. 1048–1054.

Ergas, S. J., E. D. Schroeder, D. P. Y. Chang, and R. Morton (1995): "Control of VOC Emissions from a POTW Using a Compost Biofilter," *Water Environment Research,* vol. 67, pp. 816–821.

Eweis, J. B., D. P. Y. Chang, E. D. Schroeder, K. M. Scow, R. L. Morton, and R. Caballero (1997*a*): "Meeting the Challenge of MTBE Biodegradation," *Proceedings of the 90th Annual Meeting and Exhibition, Air and Waste Management Association,* Toronto, Canada, June 8–13.

Eweis, J. B., E. D. Schroeder, D. P. Y. Chang, K. M. Scow, J. R. Hanson, and R. L. Morton (1997*b*): "MTBE Biodegradation in Field and Pilot-Scale Reactors: Culture Characteristics and Removal in the Presence of Other Substrates," *Proceedings of the 1997 Pacific Conference on Chemistry and Spectroscopy,* Irvine, CA, October 21–25.

Hammervold, R. E., T. J. Overcamp, B. F. Smets, and C. P. L. Grady, Jr. (1995): "Experimental Study of the Sorptive Slurry Bioscrubber for Acetone Emissions," *Proceedings* 88th Annual Meeting of the Air and Waste Management Association, San Antonio, TX.

Härremoes, P. (1978): "Biofilm Kinetics," in *Water Pollution Microbiology,* vol. 2, Wiley, New York, pp. 71–109.

Hartenstein, Hans U., and Eric R. Allen (1986): *Biofiltration, an Odor Control Technology for a Wastewater Treatment Plant,* Report to Department of Public Works, City of Jacksonville, FL.

Holubar, P., C. Andorfer, and R. Braun (1995): "Prevention of Clogging in Trickling Filters for Purification of Hydrocarbon Contaminated Wastewater Air," *Proceedings,* USC-TRG Conference on Biofiltration, University of Southern California, October 5–6.

Jennings, P. A. (1976): "Theoretical Model for a Submerged Biological Filter," *Biotechnology and Bioengineering,* vol. 18, p. 1249.

Kampbell, D. H., J. T. Wilson, H. W. Reed, and T. T. Stocksdale (1987): "Removal of Aliphatic Hydrocarbons in a Soil Bioreactor," *J. Air Pollution Control Association,* vol. 37, pp. 1236–1240.

Kardono, and E. R. Allen (1994): "Biofiltration Control of Volatile Hydrocarbon Emissions: *n*-Butane," *Proceedings of the 87th Annual Meeting of the Air and Waste Management Association,* Cincinnati, OH.

Kinney, K. A. (1996): Spatial Distribution of Mass and Activity in a Directionally Switching Biofilter, Doctoral Dissertation, Department of Civil and Environmental Engineering, University of California, Davis.

Kinney, K. A., C. du Plessis, E. D. Schroeder, D. Chang, and K. M. Scow (1996): "Optimizing Microbial Activity in a Directionally-Switching Biofilter," *Proceedings,* 1996 Conference on Biofiltration, University of Southern California, Oct. 24–25, pp. 150–157.

Lehtomaki, J., M. Torronen, and A. Laukkarinen (1992): "A Feasibility Study of Biological Waste-Air Purification in a Cold Climate," in *Biotechniques for Air Pollution Abatement and Odour Control Policies,* A. J. Dragt and J. van Ham, eds., Elsevier Science Pub.

Leson, Gero, D. S. Hodge, F. Tabatabai, and A. Winer (1993): "Biofilter Demonstration Projects for the Control of Ethanol Emissions," *Proceedings of the 86th Annual Meeting of the Air and Waste Management Association,* Denver, CO.

Morgenroth, E., E. D. Schroeder, D. P. Y. Chang, and K. M. Scow (1995): "Nutrient Limitation in a Compost Biofilter Degrading Hexane," *Proceedings of the 88th Annual Meeting and Exhibition, Air and Waste Management Association,* San Antonio, TX, June 18–23.

Ottengraf, S. P. P., and A. H. C. Van Den Oever (1983): "Kinetics of Organic Compound Removal from Waste Gases with a Biological Filter," *Biotechnology and Bioengineering,* vol. 25, pp. 3089–3103.

Peters, D., G. T. Hickman, J. G. Stephanoff, and M. B. Garcia (1993): "Laboratory Assessment of Biofiltration for Fuel-Derived VOC Emissions Control," *Proceedings of the 86th Annual Meeting of the Air and Waste Management Association,* Denver, CO.

Pomeroy, R. D. (1957): "Deodorizing Gas Streams by the Use of Microbiological Growths," U.S. Patent 2,793,096.

Pomeroy, R. D. (1982): "Biological Treatment of Odorous Air," *Journal of the Water Pollution Control Association,* vol. 54, pp. 1541–1545.

Rittman, B. E., and P. L. McCarty (1980): "Model of Steady State Biofilm Kinetics," *Biotechnology and Bioengineering,* vol. 22, pp. 2343–2357.

Shareefdeen, Z., B. Baltzis, O. Young-Sook, and R. Bartha (1993): "Biofiltration of Methanol Vapor," *Biotechnology and Bioengineering,* vol. 41, pp. 512–524.

Smith, F. L., G. A. Sorial, M. T. Suidan, P. Biswas, and R. C. Brenner (1995): "Management of High-Rate Trickle Bed Biofilters, "*Innovative Technologies for Site Remediation and Hazardous Waste Management,* R. D. Vidic and F. G. Pohland, eds., ASCE, pp. 489–496, American Society of Civil Engineers, New York.

Sorial, G. A., F. L. Smith, M. T. Suidan, and P. Biswas (1995*a*): "Evaluation of a Trickle Bed Biofilter Media for Toluene Removal," *Journal of Air and Waste Management Association,* vol. 45, pp. 801–810.

Sorial, G. A., F. L. Smith, A. Pandit, M. T. Suidan, P. Biswas, and R. C. Brenner (1995*b*): "Performance of Trickle Bed Biofilters under High Toluene Loading," *Proceedings of the 88th Annual Meeting and Exhibition, Air and Waste Management Association,* San Antonio, TX, June 18–23.

Togna, A. Paul, and B. R. Folsom (1992): "Removal of Styrene from Air Using Bench-Scale Biofilter and Biotrickling Filter Reactors," *Proceedings of the 85th Annual Meeting of the Air and Waste Management Association,* Kansas City, MO.

U.S. Environmental Protection Agency (1986): *Handbook of Control Technologies for Hazardous Air Pollutants,* EPA Report: 625/6-86/014, Research Triangle Park, NC.

Utgikar, V., R. Govind, Y. Shan, S. Safferman, and R. C. Brenner (1991): "Biodegradation of Volatile Organic Chemicals in a Biofilter," in *Emerging Technologies in Hazardous Waste Management,* American Chemical Society.

van Lith, C. (1989): "Design Criteria for Biofilters," *Proceedings of the 82nd Annual Meeting of the Air and Waste Management Association,* Anaheim, CA.

Wright, W. F., E. D. Schroeder, D. P. Y. Chang, and K. Romstad (1997): "Performance of a Compost Biofilter Treating Gasoline," *Journal of Environmental Engineering,* ASCE, vol. 123, no. 6, pp. 547–555.

Yang, Y., and E. R. Allen (1994): "Biofiltration Control of Hydrogen Sulfide 1. Design and Operational Parameters," *Journal of Air and Waste Management Association,* vol. 44, pp. 863–868.

APPENDIX A

Symbols

Symbol	Definition
a	Langmuir isotherm sorption coefficient, unitless
a	surface area per unit volume, m^{-1}
A	area, m^2
b	Langmuir isotherm saturation coefficient, g/m^3
B	solid-phase cell concentration, g/g
C_A	mass concentration of chemical species A, g/m^3
C_{Ae}	equilibrium concentration of chemical species A, g/m^3
C_D	coefficient of drag, unitless
C_G	gas-phase contaminant concentration, g/m^3
C_L	liquid-phase contaminant concentration, g/m^3
C_S	liquid-phase concentration in equilibrium with gas phase, g/m^3
CASRN	Chemical Abstract Service Registry Number
d	diameter, m
d	characteristic soil grain size, m
D	coefficient of dispersion, m^2/s
D	coefficient of molecular diffusion, m^2/s
D_x	coefficient of dispersion in x direction, m^2/x
D_i	impeller diameter, m
E_o	electrode potential, volts
EC	elimination capacity, $g/m^3 \cdot s$
f_{oc}	fraction of organic carbon in soil, mass carbon/mass dry soil
f_s	degree of saturation, unitless
F	Faraday's constant, 96,630 joules/volt
g	gravitational constant, 9.81 m/s^2
G	free energy, J/mole
$G°$	standard free energy, J/mole
$G°_f$	standard free energy of formation, J/mole

H	dimensionless Henry's coefficient, unitless
H_r	relative humidity, unitless or percent
J_i	flux of chemical species i, $g/m^2 \cdot s$
k	reaction rate coefficient, variable units
k_d	cell maintenance coefficient, d^{-1}
K	equilibrium coefficient
K_C	hydraulic conductivity, m/s
K_F	Freundlich adsorption coefficient, variable units
K_L	liquid-gas mass transfer rate coefficient, m/s
K_{OW}	octanol-water partition coefficient, g/g
K_S	saturation coefficient, mg/L
K_W	ionization constant of water
K_{SD}	soil-water distribution coefficient, m^3/g
m	Henry's coefficient, atm-m^3/mol
m	mass of solid phase, g
M_i	mass of chemical species i, g
M_L	contaminant mass in solution, g
M_S	contaminant mass sorbed, g
n	empirical coefficient or fitting parameter, unitless
n	number of electrons transferred in redox reaction, unitless
N	number, or number concentration, of cells, m^{-3}
N	rotational speed of impeller, s^{-1}
N_g	gas-phase mass flux, $g/m^2 \cdot s$
N_{Re}	Reynolds number, unitless
P	pressure, atm or Pa
P_v	pure component vapor pressure, atm
q	diffusive flux, $g/m^2 \cdot s$
q_L	volumetric flow rate, m^3/s
r_A	rate of transformation of chemical species A, $g/m^3 \cdot s$
r_g	rate of microbial growth, $g/m^3 \cdot day$
r_{max}	maximum biodegradation rate, $g/m^3 \cdot s$
r_N	rate of natural decay, $g/m^3 \cdot s$
r_o	rate of organic removal, $g/m^3 \cdot day$
r_{O_2}	rate of oxygen uptake $g/m^3 \cdot day$
r_T	mass transfer rate, $g/m^2 \cdot s$
r_{TA}	rate of interphase mass transfer of chemical species A, $g/m^3 \cdot s$
R	universal gas constant, 8.2057×10^{-5} atm-m^3/mol-K
R	retardation coefficient, unitless
s	solubility, g/m^3
s	mass solute sorbed per unit mass of soil, g/g
t	time
t_r	empty bed contact time, s
T	temperature, K
v_x, v_y, v_z	velocity in the x, y, or z direction, m/s
V	volume, m^3
V_a	volume of air, m^3
V_w	volume of water, m^3

V_s	volume of dry soil, m^3
V_t	total volume of solids, air, and water, m^3
w	moisture content of soil on a weight basis (g/g soil)
x_A	mass of *A* sorbed onto the solid phase, g
X	mass concentration of cells, g/m^3
Y	microbial yield, mass cells produced/mass contaminant transformed
α_z	empirical coefficient accounting for tortuosity and mixing, m
Θ	volumetric water content of soil, m^3/m^3
θ_H	hydraulic residence time, days
λ	penetration depth, m
μ	liquid viscosity, $N \cdot s/m^2$
μ	microbial specific growth rate, d^{-1}
ρ	density, kg/m^3
ρ_b	bulk density, kg/m^3
ρ_{wb}	wet bulk density
ρ_{gas}	density of air or other gas, kg/m^3
ϕ	porosity, void volume/total volume, m^3/m^3
ϕ	Thiele number, unitless
ω	fractional slurry solids content, kg solids/kg liquid

APPENDIX B

Glossary

acclimation period the period of time between inoculation with microbial culture and observation of activity (such as biodegradation).

advection the process by which a dissolved substance is transported through the bulk motion of the moving fluid.

aerobic an organism which is able to grow in the presence of oxygen. Aerobes may be facultative (able to grow both in the presence and absence of oxygen) or obligate aerobe (can only live in the presence of oxygen).

anaerobic an organism which is able to live and grow in the absence of free oxygen.

anoxic respiration oxidative process similar to aerobic respiration but which utilizes nitrate or another inorganic compound as a terminal electron acceptor.

anthropogenic material or contaminant that results from human activity. Anthropogenic pollutants are the results of discharges or spills rather than natural events such as forest fires.

aquifer an underground layer of porous rock or sand containing water. Generally the term aquifer is applied only to regions from which extraction of the water is economical.

ATP adenosine triphosphate, the main energy carrier in living organisms.

autotroph organism which obtains carbon from the reduction of inorganic compounds, mainly carbon dioxide.

bacteriophage virus which infects bacteria.

bioaugmentation the addition of microbial inoculum to initiate and/or increase the rate of contaminant biodegradation.

biofouling the accumulation of biomass in undesirable places, such as around a screened pumping well, and which may result in coating of surfaces and clogging, thus retarding, or inhibiting proper operation of the well.

biomass the mass of microorganisms in a certain volume of water, sediment, or soil.

biosurfactants surfactants produced naturally by microorganisms.

biotransformation the biologically induced structural transformation of a compound.

BOD biochemical oxygen demand. A measure of the amount of oxygen required to biologically degrade organic material in a given sample. The standard test is carried over an incubation period of 5 days and a temperature of 20°C.

BOD_U ultimate biochemical oxygen demand. The quantity of oxygen required for complete conversion of carbonaceous material to carbon dioxide and other mineral materials.

BTEX benzene, toluene, ethylbenzene, and xylenes, volatile aromatic hydrocarbons of particular concern in gasoline because of their solubility and toxicity.

CASRN Chemical Abstract Service Registry Number

COD chemical oxygen demand. A measure of the total amount of oxygen required to oxidize organic matter to carbon dioxide and water. In a standard test, potassium dichromate in an acidic solution is used to oxidize organic material. Because it is based on chemical oxidation, the COD test does not differentiate between biodegradable, and nonbiodegradable material.

coenzyme a chemical of low molecular weight which binds loosely to an enzyme, and participates in an enzymatic reaction by acting as an intermediate carrier of things such as electrons and functional groups.

colony a group of cells growing on solid nutrient medium, resulting from the reproduction of a single cell. A colony forming unit (cfu) is assumed to be equivalent to a single cell at the beginning of incubation.

cometabolism process by which a substrate is metabolized by a cell while the cell is using another substrate as its carbon or energy source.

competitive inhibitor compound which has a significant affinity for the active site of an enzyme that catalyzes a desired substrate's degradation reaction.

conservative tracer a chemical that does not degrade biologically, but has transport properties similar to those of the biodegradable target compound.

consortia a mixture of two or more groups of organisms or bacteria in which one group may benefit from the other.

denitrifying conditions anaerobic conditions in which nitrate, rather than oxygen, is the main electron acceptor. Denitrification results in the conversion of nitrate to nitrogen gases, leading to loss of nitrogen to the atmosphere in an open system.

DNAPL dense (density greater than water) nonaqueous-phase liquid.

diauxic metabolic control which operates in such a way that enables organisms to select the substrate that allows them to grow at the highest rate.

dioxygenase enzyme that catalyzes the addition of two atoms of molecular oxygen to a molecule.

DO dissolved oxygen.

doubling time the time required for a population of cells to double in number; sometimes referred to as the generation time.

electron acceptor a compound that accepts electrons in an oxidation-reduction reaction. An electron acceptor is also an oxidizing agent.

electron donor a compound that donates electrons in an oxidation-reduction reaction. An electron donor is also a reducing agent.

enzyme protein which both lowers the activation energy and directs the metabolic pathway taken by chemical reactions in an organism.

eukaryote organism characterized by having a nucleus surrounded by a membrane.

eutrophic having an abundance of nutrients, often in reference to nitrogen and phosphorus.

ex situ Latin, meaning removal of material from its natural or original place.

extracellular outside the outermost layer of the cell.

facultative anaerobe organism which grows in the presence or absence of oxygen.

fermentation degradative pathway in which an organic compound serves as both the electron donor and acceptor.

free product pure contaminant; usually used to describe nonaqueous liquids which have accumulated in the soil pores.

GAC granular activated carbon, a material often used as an adsorbent because of its high porosity (high surface to volume ratio) and high surface activity.

half-life the time required for the initial concentration of a compound to be reduced by half, often used in reference to radioactive decay, in situ biodegradation, and other reactions.

heterogeneous mixed; nonuniform in structure and/or composition.

HDPE high density polyethylene.

heterotroph organism which requires an organic form of carbon.

homogeneous having the same structure or composition throughout.

hydraulic conductivity the rate of flow of water through a unit cross-sectional area in the subsurface, under a unit hydraulic gradient ($m^3/m^2 \cdot$ day, or m/day).

hydraulic gradient the change in total head (water pressure) per unit length in a given direction, the principal direction of water flow.

hydrophilic Greek for water-friendly or water-loving. Having a high affinity for water (high solubility).

hydrophobic Greek for water-fearing. Having a low affinity for water. An example is oil which has a low solubility and tends to form a separate liquid phase when mixed in with water.

hydrolysis cleavage of a molecule by reaction with water.

hypha a fungus thread.

inoculum a material which acts as a source of microbial cells, such as activated sludge, soil, and sediment.

in situ Latin meaning in its natural or original place.

lag period the time passing after inoculation with a microbial culture, and before growth is observed.

lignin a naturally occurring organic material with a complex aromatic polymer structure. It is a major component in wood.

lithotrophs bacteria that use inorganic compounds as energy sources.

LNAPL light (density less than water) nonaqueous-phase liquid.

LTU land treatment unit.

lyse breaking apart of the cell membrane.

lysimeter a device used for measuring the percolation of water through soil, and for determining the soluble constituents removed through drainage.

maximum concentration limit (MCL) maximum concentration of a chemical allowed by a regulatory agency in soil, water, or air.

mesophile organism which grows best at temperatures between 15 and 45°C.

metabolism processes through which living organisms grow and obtain energy.

meta position position on an aromatic molecule separated from the point of reference by one carbon position.

micelle with reference to surfactants: an aggregate of molecules in which the hydrophobic ends are lined up on the inside, while the hydrophilic end is lined up to the outside. A micelle can help mobilize hydrophobic compounds by allowing them to partition in the hydrophobic center of the micelle itself.

mil a unit of length equal to 1/1000 inch.

mineralization the complete biodegradation of an organic compound into carbon dioxide, water, inorganic ions and molecules, and possibly cellular material.

monooxygenase enzyme that catalyzes the addition of one atom of molecular oxygen to a molecule.

mutation alteration of a genetic message.

NAPL nonaqueous-phase liquid.

oligotroph deficient in nutrients, often in reference to nitrogen and phosphorus.

organotroph organism that utilizes organic compounds as an energy source for respiratory metabolism.
ortho position position on an aromatic molecule that is adjacent to the point of reference.
PAH polynuclear aromatic hydrocarbon.
para position position on an aromatic molecule that is separated from the point of reference by two carbon positions.
PCB polychlorinated byphenyl.
pH negative log of the hydrogen ion concentration.
phototroph organism that obtains energy from light.
plasmid small circle of DNA that is extrachromosomal and replicates autonomously.
ppm parts per million. For solutes in water, the term is often used as the equivalent of concentrations in mg/L.
ppm_v parts per million on a volume basis. Typically used for concentrations in the gas phase.
ppb parts per billion. For solutes in water, the term would be equivalent to concentrations in mg/m^3.
ppb_v parts per billion on a volume basis. Typically used for concentrations in the gas phase.
prokaryote organism characterized by lacking a nuclear membrane.
psychrophile organism that grows best at temperatures below 20°C.
recalcitrant (refractive) compound that does not undergo biodegradation.
substrate compound that can be used as a carbon, energy, or nutrient source for microbial metabolism.
surfactant a chemical that reduces the surface tension of the liquid in which it is dissolved. Surfactants typically have one hydrophobic and one hydrophilic end. The most common forms of surfactant are detergents and soaps.
thermophile organism that grows best at temperatures above 50°C.
TOC total organic carbon.
TPH total petroleum hydrocarbons.
transposon a segment of DNA that can be moved from one area on a chromosome to another.
UST underground storage tank.
vadose zone also referred to as the unsaturated zone, it is the zone of soil which is below the ground surface, but above the water table. The soil pores within that zone are either partially, or largely filled with air.
VOC volatile organic compound, nominally compounds having a dimensionless Henry's law coefficient greater than 0.01.
water table the upper limit of that part of the subsurface that is completely saturated with water.
xenobiotic strange to life, compound for which metabolic degradation pathways have not evolved in the natural environment.

APPENDIX C

Constants and Conversion Factors

PHYSICAL CONSTANTS

Avogadro's number N_A = 6.02283 molecules/g · mole

Faraday's constant F = 96,630 Joules/volt

Latent heat of fusion of water (0°C, 1 atm) = 333.6 J/g

Latent heat of vaporization of water (100°C, 1 atm) = 2,258 J/g

Molecular mass of dry air = 28.97 g/mol

One angstrom A = 10^{-10}m

One bar = 10^5 N/m^2

One torr (0°C), 1 mmHg = 133.322 kN/m^2

Atmosphere (standard) = 101.325 kPa (kN/m^2)
= 10.333 m of water
= 760 mmHg

One hectare = 10^4 m^2

Specific heat of water C_p(0°C) = 4.2174 J/g · °C
C_p(15°C) = 4.1855 J/g · °C
C_p(20°C) = 4.1816 J/g · °C

Universal gas constant R = 0.082057 L · atm/g-mol · K
R = 8.3144 J(abs)/g-mol · K
R = 1.9872 cal/g-mol · K

CONVERSION FACTORS: SI TO U.S. CUSTOMARY

Multiply SI unit		**by**	**To obtain U.S. customary unit**	
Name	**Symbol**		**Symbol**	**Name**
Acceleration	m/s^2	3.2808	ft/s^2	feet per second squared
Area				
square meter	m^2	10.7639	ft^2	square feet
square kilometer	km^2	0.3861	mi^2	square mile
hectare	ha	2.4711	acre	acre
square centimeter	cm^2	0.1550	in^2	square inch
Energy				
kilojoule	kJ	0.9478	Btu	British thermal unit
kilojoule	kJ	2.7778×10^{-4}	kW-h	kilowatthour
Joule	J	0.7376	ft-lb_f	foot-pound (force)
Joule	J	1.0000	W-s	watt second
Joule	J	0.2388	cal	calorie
Flow rate				
cubic meters per second	m^3/s	35.3147	ft^3/s	cubic feet per second
cubic meters per second	m^3/s	15,850.3	gpm	gallons per minute
cubic meters per second	m^3/s	22.8245	mgd	million gallons per day
Force				
newton	N	0.2248	lb_f	pound force
Length				
centimeter	cm	0.3937	in	inch
kilometer	km	0.6214	mi	mile
meter	m	3.2808	ft	foot
Mass				
gram	g	15.42	gr	grain
gram	g	0.0353	oz	ounce
gram	g	0.0022	lb	pound
kilogram	kg	2.2046	lb	pound
megagram	Mg	1.1023	ton	ton (short: 2,000 lb_f)
Pressure				
Pascal	Pa(N/m^2)	1.4504×10^{-4}	lb/in^2	pounds per square inch
Pascal	Pa(N/m^2)	2.0885×10^{-2}	lb_f/ft^2	pounds per square foot
kilopascal	kPa(kN/m^2)	0.0099	atm	atmosphere (standard)
Temperature				
degree Celsius	°C	1.8 (°C) + 32	°F	degree Fahrenheit
Kelvins	K	1.8 (K) – 459.67	°F	degree Fahrenheit
Velocity				
kilometers per second	km/s	2,236.9	mi/h	miles per hour
meters per second	m/s	3.2808	ft/s	feet per second
Volume				
milliliter	mL	0.0610	in^3	cubic inch
cubic meter	m^3	35.3147	ft^3	cubic foot
cubic meter	m^3	264.172	gal	gallon
liter	L	0.2642	gal	gallon

APPENDIX D

Physical Properties of Water

Temperature, °C	Density ρ, kg/m^3	Dynamic viscosity $\mu \times 10^3$, N · s/m^2	Surface tension σ, N/m	Vapor pressure P_v, kN/m^2
0	999.8	1.781	0.0765	0.61
5	1000.0	1.518	0.0749	0.87
10	999.7	1.307	0.0742	1.23
15	999.1	1.139	0.0735	1.70
20	998.2	1.002	0.0728	2.34
25	997.0	0.890	0.0720	3.17
30	995.7	0.798	0.0712	4.24
40	992.2	0.653	0.0696	7.38
50	988.0	0.547	0.0679	12.33
60	983.2	0.466	0.0662	19.92
70	977.8	0.404	0.0644	31.16
80	971.8	0.354	0.0626	47.34
90	965.3	0.315	0.0608	70.10
100	958.4	0.282	0.0589	101.33

Adapted from J. K. Venard and R. L. Street (1975), *Elementary Fluid Mechanics,* 5th ed., Wiley, New York.

APPENDIX E

Properties of Common Contaminants

TABLE E.1
Properties of selected chemicals commonly found in hazardous wastes

Compound	Molecular formula	Molecular weight	Solubility, mg/L	VP, mmHg	BP, °C	H (25°C)	log K_{ow} (25°C)
Halogenated volatiles:							
Bromoform	$CHBr_3$	253	1,000	5	148	0.0246	
Carbon tetrachloride	CCl_4	154	800	91	77	0.981	2.73
Chlorobenzene	C_6H_5Cl	113	1,000	8.8	132	0.158[g]	2.84[g]
Chloroethene (vinyl chloride)	C_2H_3Cl	62.5	2,792	2,580	–13.4	0.916	0.60
Chlorodibromomethane[1]	$ClCHBr_2$	208	4,750	50	116–122	0.45	2.09
Chloroethane	C_2H_5Cl	65	6,000	1,064	12.2		
Chloroform	$CHCl_3$	119	8,000	160	61	0.163	1.93
Chloromethane	CH_3Cl	51		3,648	–24	0.391	0.91
1,1-Dichloroethane[1]	CH_3CHCl_2	99	7,840	297	57	0.246	1.79
1,2-Dichloroethane[1]	$ClCH_2CH_2Cl$	99	8,690	61	83.5	0.0438	1.47
1,1-Dichloroethene[1]	CH_2CCl_2	97	5,000	500	31.9	1.068	1.84
Dichloromethane	CH_2Cl_2	85	20,000	350	40	0.105	1.15
1,2-Dichloropropane[1]	$CH_3CHClCH_2Cl$	113	2,600	41.2	96.4		2.0
Hexachloroethane	CCl_3CCl_3	237	50	0.22	189	0.33	4.6
1,1,2,2-Tetrachloroethane	$CHCl_2CHCl_2$	168	2,900	8	146	0.438	2.42
Tetrachloroethene[1]	Cl_2CCCl_2	166	160	15.6	121	1.12	2.88
1,1,1-Trichloroethane[1]	CH_3CCl_3	133	4,400	100	74	0.797	2.48
1,1,2-Trichloroethane	$CHCl_2CH_2Cl$	133	5,000	19	113		2.47
Trichloroethene[1]	$ClCHCCl_2$	131.5	1.1[a]	60	86.7	0.395	2.38
Halogenated semivolatiles:							
Bis(2-chloroethyl)ether	$(ClCH_2CH_2)_2O$	143	11,000	0.4	178		1.5
2-Chlorophenol[2]	C_6H_5ClO	129	Miscible	1[b]	174.5		
1,2-Dichlorobenzene	$C_6H_4Cl_2$	147	150	1.2	180	0.126	3.38
1,4-Dichlorobenzene	$C_6H_4Cl_2$	147	80	0.4	174	0.0916	3.38
Hexachlorobenzene[4]	C_6Cl_6	285	6,000[a]	0.00001	323–326	0.0619	5.50
Pentachlorophenol	C_6Cl_5OH	266	20	0.0002	311		5.0
1,2,4-trichlorobenzene[2]	$C_6H_3Cl_3$	181	1[c]	213		0.086	4.3
2,4,5-trichlorophenol[2]	$C_6H_3Cl_3O$	197	1[d]	252		0.113	4.00
2,4,6-trichlorophenol[2]	$C_6H_3Cl_3O$	197	800[a]	400[a]	244.5		3.72
Nonhalogenated volatiles:							
Acetone	CH_3COCH_3	58	Miscible	266[a]	56	0.00085	–0.24
Benzene	C_6H_6	78	1,800	75	80	0.224	2.13
Carbon disulfide	CS_2	76	2,000	300	46		2.0
Ethyl acetate	$CH_3COOC_2H_5$	88	87,000	76	77		0.73
Ethyl benzene	$C_2H_5C_6H_5$	106	150	7.1	136	0.325	3.15

Ethyl ether	$C_2H_5OC_2H_5$	74	75,000	442	35		
2-Hexanone	$CH_3CO(CH_2)_3CH_3$	100	14,000	3	128		
Isobutanol	$(CH_3)_2CHCH_2OH$	74	87,000	9	108		0.76
Methanol	CH_3OH	32	Miscible	97	64		–0.77
Methyl isobutyl ketone	$CH_3COCH_2C_3H_7$	100	19,000	15	–84		1.19
n-Butyl alcohol	$CH_3CH_2CH_2CH_2OH$	74	77,000	4.2	118		0.88
Styrene	$C_6H_5CHCH_2$	104	300	4.5	145		2.95
Toluene	$C_6H_5CH_3$	92	515	22	110.6	0.276	2.69
o- xylene	C_8H_{10}	106.2	575	6.8	144	0.210	3.26
p- xylene	C_8H_{10}	106.2	589	8.9	138	0.283	3.18
Nonhalogenated semivolatiles:							
Anthracene[5]	$C_{14}H_{10}$	178	0.045[a]	2×10^{-4}	340	0.000937	4.54
Benz(*a*)anthracene[5]	$C_{18}H_{12}$	228	0.0094[a]	1×10^{-5}	400	0.000235	5.91
Benzidine	$C_{12}H_{12}N_2$	184	400[b]	11[g]	402		1.30
Benzo(*a*)pyrene[5]	$C_{20}H_{12}$	252	0.001[a]	6×10^{-9}[a]	312		6.06
Benzo(*g,h,i*)perylene[5]	$C_{22}H_{12}$	276	0.0007[a]	1×10^{-10}[a]	—		6.51
Chrysene[5]	$C_{18}H_{12}$	228	0.002[a]	6×10^{-9}[a]	448		5.61
Dimethylphthalate	$C_{10}H_{10}O_4$	194	400	1[f]	285		
Fluoranthrene[5]	$C_{16}H_{10}$	202	0.21[a]	5×10^{-6}[a]	367		4.9
Isophorone	$C_9H_{14}O$	138	12,000	0.2	215		2.22
Methyl *tert*-butyl ether	$C_5H_{12}O$	88	48,000	240		0.0329	
Naphthalene	$C_{10}H_8$	128	30	0.05	218	0.0174	3.36
Nitrobenzene	$C_6H_5NO_2$	123	2,000	<<1	211		1.85
Phenanthrene[5]	$C_{14}H_{10}$	178	1[a]	6.8×10^{-4}	339	0.00105	4.57
Phenol	C_6H_5OH	94	84,000	0.36	182	1.67×10^{-5}	1.45
Pyrene[5]	C_6H_{10}	202	0.13[a]	2.5×10^{-6}[a]	404		4.88
Pyridine	C_5H_5N	79	Miscible	18	115		0.66
Pesticides and herbicides (sol. at 25°C)							
Alachlor[3]	$C_{14}H_{20}ClNO_2$	270	240	2×10^{-5}[a]			
Atrazine[3]	$C_8H_{14}ClN_5$	216	32	6.8×10^{-7}[a]			2.75
Bromacil[3]	$C_9H_{13}BrN_2O_2$	261	820	2×10^{-7}[a]			
Carbofuran[3]	$C_{12}H_{15}NO_3$	221	320	8×10^{-6}[a]			1.60
Chlorpropham[3]	$C_{10}H_{12}ClNO_2$	214	89	9.8×10^{-6}[a]	149[u]		
DDT	$C_{14}H_9Cl_5$	355	0.00001	~0	[w]		
Diazinon[3]	$C_{12}H_{21}N_2O_3PS$	304	40	0.0001[a]	[w]		
Dicamba[3]	$C_8H_6Cl_2O_3$	221	4,500	0.00038			

(Continued)

TABLE E.1 ***(concluded)***

Compound	Molecular formula	Molecular weight	Solubility, mg/L	VP, mmHg	BP, °C	*H* (25°C)	log K_{ow} (25°C)
Pesticides and herbicides (sol. at 25°C): (continued)							
Dieldrin	$C_{12}H_8C_{16}O$	381	0.02	~0	[w]	4.58×10^{-4}	5.48
Diuron[3]	$C_9H_{10}Cl_2N_2O$	233	37	1.6×10^{-7} [a]			
EPTC[3]	$C_9H_{19}NOS$	189	370	0.008	127[v]		
Heptachlor[3]	$C_{10}H_5C_{17}$	373	0.0056	0.00017	[w]		
Lindane[3]	$C_6H_6Cl_6$	291	7.5	6×10^{-5} [a]		1.32×10^{-4}	3.78
Linuron[3]	$C_9H_{10}C_{12}N_2O_2$	249	75	8×10^{-6}			
Malathion	$C_{10}H_{19}O_6PS_2$	330	145	0.00004	[w]		2.89
Metolachlor[3]	$C_{15}H_{22}ClNO_2$	284	530	1.3×10^{-5}			
Monuron[3]	$C_9H_{11}ClN_2O$	199	260	1.7×10^{-7} [a]			
Parathion	$(C_2H_5O)_2PSOC_6H_4NO_2$	291	0.00002	0.0004	707	1.55×10^{-5}	3.81
Picloram[3]	$C_6H_3Cl_3N_2O_2$	241	430	5×10^{-9}			
Prometon[3]	$C_{10}H_{19}N_5O$	225	750	6×10^{-6}			
Simazine[3]	$C_7H_{12}ClN_5$	202	5	1.5×10^{-8} [a]			2.18
Triallate[3]	$C_{10}H_{16}Cl_3NOS$	305	4	0.0002			
Trifluralin[3]	$C_{13}H_{16}F_3N_3O_4$	335	0.3	0.0001	139[s]		5.34

Symbols:
Solubility in water at 20°C unless specified otherwise.
VP: vapor pressure in mmHg at 20°C unless specified otherwise.
BP: boiling point in °C at 760 mmHg unless specified otherwise.
[a]at 25°C.
[b]at 12.1°C.
[c]at 38.4°C.
[d]at 72°C.
[e]at 76.5°C.
[f]at 100°C.
[g]at 176°C.
[s]at 4.2 mmHg.
[t]at 10 mmHg.
[u]at 2 mmHg.
[v]at 20 mmHg.
[w]decomposes at 120°C.

Sources:
[1]Lang, R., T. Herrera, D. Chang, and G. Tchobanoglous, and R. Spicher, *Trace Organic Constituents in Landfill Gas,* University of California, Davis, 1987.
[2]Sax, N. I., and R. J. Lewis, Sr., *Dangerous Properties of Industrial Materials,* 7th ed., van Nostrand Reinhold, New York, 1989.
[3]Taylor, A. W., and W. F. Spencer, *Pesticides in the Soil Environment: Processes, Impacts, and Modeling,* edited by H. H. Cheng, Soil Science Society of America, Wisconsin, 1990.
[4]*The Merck Index: an Encyclopedia of Chemicals, Drugs, and Biologicals,* 11th ed., edited by S. Budavari, Merck & Co., Inc., New Jersey, 1989.
[5]U.S. EPA Health Effects Assessment for Polycyclic Aromatic Hydrocarbons, Ohio, 1984, EPA/540/1-86/013.
All other data compiled from:
U.S. Department of Health and Human Services, *NIOSH Pocket Guide to Chemical Hazards,* Washington, DC, 1985.

APPENDIX F

Composition of Gasoline

TABLE F.1
Approximate composition (mass fractions) of gasoline

Compound name	MW, g	Gasoline*	
		Fresh	Weathered
Propane	44.1	0.0001	0.0000
Isobutane	58.1	0.0122	0.0000
n-Butane	58.1	0.0629	0.0000
trans-2-Butene	56.1	0.0007	0.0000
cis-2-Butene	56.1	0.0000	0.0000
3-Methyl-1-butene	70.1	0.0006	0.0000
Isopentane	72.2	0.1049	0.0069
1-Pentene	70.1	0.0000	0.0005
2-Methyl-1-butene	70.1	0.0000	0.0008
2-Methyl-1,3-butadiene	68.1	0.0000	0.0000
n-Pentane	72.2	0.0586	0.0095
trans-2-Pentene	70.1	0.0000	0.0017
2-Methyl-2-butene	70.1	0.0044	0.0021
2-Methyl-1,2-butadiene	68.1	0.0000	0.0010
3,3-Dimethyl-1-butene	84.2	0.0049	0.0000
Cyclopentane	70.1	0.0000	0.0046
3-Methyl-1-pentene	84.2	0.0000	0.0000
2,3-Dimethylbutane	86.2	0.0730	0.0044
2-Methylpentane	86.2	0.0273	0.0207
3-Methylpentane	86.2	0.0000	0.0186
n-Hexane	86.2	0.0283	0.0207
Methylcyclopentane	84.2	0.0083	0.0234
2,2-Dimethylpentane	100.2	0.0076	0.0064
Benzene	78.1	0.0076	0.0021
Cyclohexane	84.2	0.0000	0.0137
2,3-Dimethylpentane	100.2	0.0390	0.0000

Compound name	MW, g	Gasoline*	
		Fresh	Weathered
3-Methylhexane	100.2	0.0000	0.0355
3-Ethylpentane	100.2	0.0000	0.0000
n-Heptane	100.2	0.0063	0.0447
2,2,4-Trimethylpentane	111.4	0.0121	0.0503
Methylcyclohexane	98.2	0.0000	0.0393
2,2,-Dimethylhexane	114.2	0.0055	0.0207
Toluene	92.1	0.0550	0.0359
2,3,4-Trimethylpentane	114.2	0.0121	0.0000
3-Methylheptane	114.2	0.0000	0.0343
2-Methylheptane	114.2	0.0155	0.0324
n-Octane	114.2	0.0013	0.3000
2,4,4-Trimethylhexane	128.3	0.0087	0.0034
2,2-Dimethylheptane	128.3	0.0000	0.0226
Ethylbenzene	106.2	0.0000	0.0130
p-Xylene	106.2	0.0957	0.0151
m-Xylene	106.2	0.0000	0.0376
3,3,4-Trimethylhexane	128.3	0.0281	0.0056
o-Xylene	106.2	0.0000	0.0274
2,2,4-Trimethylheptane	142.3	0.0105	0.0012
n-Nonane	128.3	0.0000	0.0000
3,3,5-Trimethylheptane	142.3	0.0000	0.0000
n-Propylmenzene	120.2	0.0841	0.0117
2,3,4-Trimethylheptane	142.3	0.0000	0.0000
1,3,5-Trimethylbenzene	120.2	0.0411	0.0493
1,2,4-Trimethylbenzene	120.2	0.0213	0.0140
n-Decane	142.3	0.0000	0.0140
Methylpropylbenzene	134.2	0.0351	0.0170
Dimethylethylbenzene	134.2	0.0307	0.0289
n-Undecane	156.3	0.0000	0.0075
1,2,4,5-Tetramethylbenzene	143.2	0.0133	0.0056
1,2,3,4-Tetramethylbenzene	143.2	0.0129	0.0704
1,2,4-Trimethyl-5-ethylbenzene	148.2	0.0405	0.0651
n-Dodecane	170.3	0.0230	0.0000
Naphthalene	128.2	0.0045	0.0076
Methylnaphthalene	142.3	0.0023	0.0134
Total		1.0000	1.0000

*Source: *Ground Water Management Review,* Spring, 1990, p. 167.

APPENDIX G

Moisture and Energy Balances

The purpose of this appendix is to provide the reader with some background on how to perform moisture and enthalpy balances needed for biofilter design. The appendix begins with some definitions of moisture variables, then an enthalpy balance is performed for an adiabatic evaporation of water into air. Such a balance is the fundamental basis for the psychrometric chart that is used for subsequent calculations, and such a chart is quite handy in performing moisture balances.

There are several measures of humidity, each having slight differences in its definition. The absolute humidity H is the fraction of the mass of water vapor in the air to the mass of dry air:

$$H = \text{absolute humidity} = \frac{18P_w}{29(1-P_w)} \qquad \text{(gr/lb or g/kg)} \tag{G.1}$$

Note that gr is for grain.

The relative humidity H_r is the ratio of actual partial pressure of water P_w to the vapor pressure of water P_v at the same temperature:

$$H_r = \text{relative humidity} = \frac{P_w}{P_v} \qquad (\%) \tag{G.2}$$

The percent humidity H_p is the ratio of actual absolute humidity to the saturation humidity at the same temperature:

$$H_p = \frac{H}{H_s}100 = H_r\frac{1-P_v}{1-P_w} = \text{percent humidity } (\%) \tag{G.3}$$

Now consider an enthalpy balance on an adiabatic saturator through which air enters at some value H_r but that exits saturated (100 percent relative humidity) at temperature T_s. Water is also supplied to the saturator at temperature T_s as shown in Figure G.1. This arrangement approximates that of a "wet-bulb" thermometer.

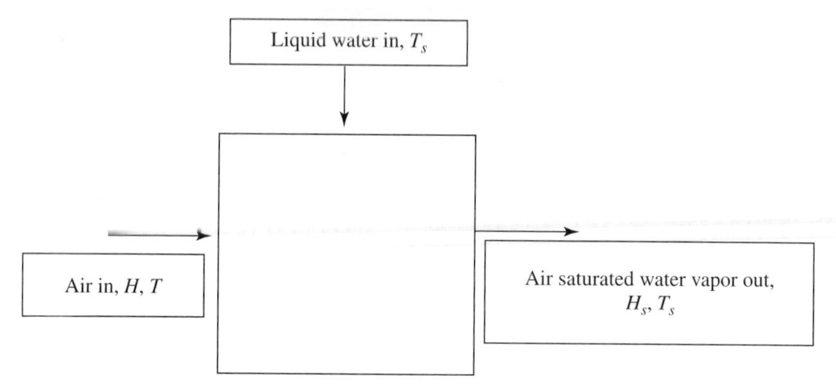

FIGURE G.1
Schematic of adiabatic saturator.

In such a situation all of the energy that evaporates water in the "saturator," i.e., the wetted sock around the wet-bulb thermometer, comes from the entering (ventilating) gas stream. In that case it can be shown that

$$\frac{H - H_s}{T - T_s} = -\frac{C_{pa} - HC_{pv}}{\lambda_s} \tag{G.4}$$

where λ_s is the latent heat of vaporization, C_{pa} is the heat capacity of air, C_{pv} is the heat capacity of water vapor, and T_s is the temperature at which the saturated air leaves the saturator. For a given value of T_s, the quantities C_{pa}, C_{pv}, λ_s, and H_s are known and therefore H is a function of T_s. The relationship between H and T_s is plotted on a psychrometric chart such as that shown. The lesser sloped straight lines running diagonally from lower right to upper left are lines of constant enthalpy (also called "wet-bulb" lines) and satisfy the above equation.

On a psychrometric chart, typically the "dry-bulb" temperature is on the abscissa and the lines of constant enthalpy are sloped as described above. The lines of constant relative humidity form a family of curves running from lower left to upper right. The intersection of the constant-enthalpy lines with the curved line corresponding to 100 percent relative humidity is marked as the "wet-bulb" temperature at the intersections. Lines of constant absolute humidity are horizontal (values read on the right-hand scale). Finally the more steeply sloped lines that run from lower right to upper left are lines of constant volume of air per unit mass of dry air.

A typical application of a psychrometric chart is to determine the relative humidity from a pair of "wet-bulb" and "dry-bulb" temperature measurements. Suppose that the measured dry-bulb temperature is 80°F and the corresponding wet-bulb temperature is 67°F. Following the diagonal wet-bulb line to its intersection with the vertical dry-bulb temperature indicates that the relative humidity of the air is approximately 52 percent. Following a horizontal line to the right-hand side of the chart shows that the absolute humidity is 80 gr/lb$_m$ of dry air. Following the same horizontal line (constant absolute humidity) to the left-hand intersection with the 100 percent relative humidity curve gives the "dew-point" temperature as 61°F, i.e., the temperature at which the air would have to be cooled before condensation of water becomes incipient.

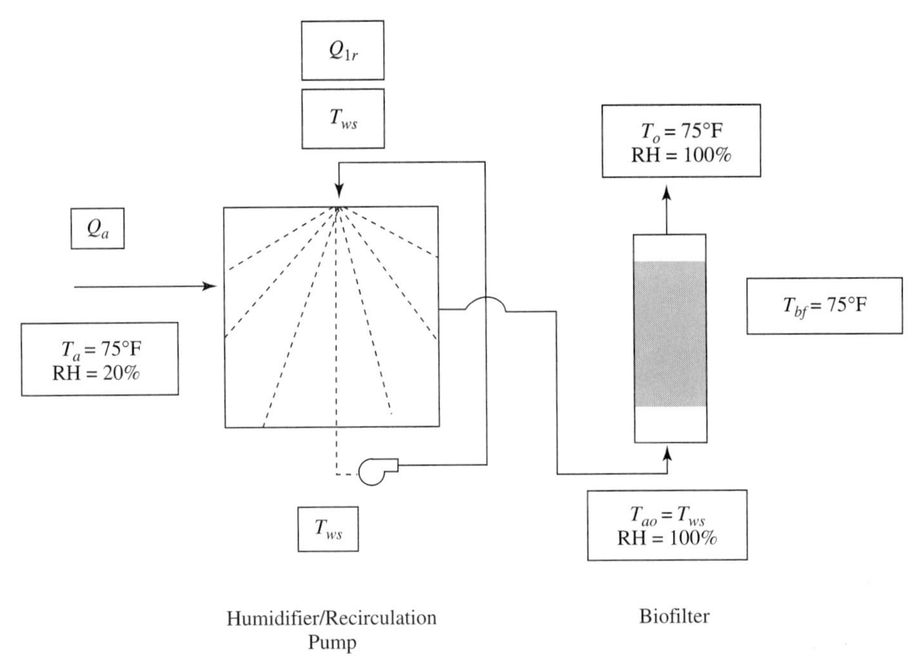

FIGURE G.2
Schematic diagram of a spray chamber humidifier—biofilter system.

HUMIDIFICATION EXAMPLE PROBLEM. Consider the schematic diagram of a spray chamber humidifier (Figure G.2) that is feeding into a biofilter. Water sprayed into the chamber is recirculated (adiabatically—an assumption). The air exiting the humidifier T_{ao} is saturated (also adiabatically—a good assumption) at the same temperature as the water sprayed into the chamber T_{ws}. The inlet air enters the humidifier at a temperature of 75°F, 20 percent relative humidity, and a flow rate of 70 ft^3/min. The biofilter is warmed by the sun and has reached a condition where the air exiting the biofilter is at a temperature of 75°F. Find the following items:

1. The temperature of air exiting the humidifier T_{ao}.
2. The quantity of water evaporated into the inlet airstream (expressed as lb$_m$-H_2O/lb$_m$-dry air)
3. The relative humidity of the humidified air at the temperature of the biofilter assuming that no additional evaporation occurs
4. The rate at which water is evaporated from the biofilter if air leaves saturated
5. The amount of heat that should be supplied to the spray chamber if the air entering the biofilter is to be saturated at the conditions of the biofilter

To solve the problem, use the psychrometric chart on page 287 (Figure G.3). Next locate the intersection of the inlet-air dry-bulb temperature and the 20 percent relative humidity curve (A). Then move along a wet-bulb line (constant enthalpy) to 100 percent relative humidity (B) and read the "wet-bulb" temperature. You should see that T_{ao} = 53°F, relative humidity = 100 percent. Determine the absolute humidity of the inlet air by moving horizontally from point A along a line of constant absolute humidity to the right-hand scale of diagram C to read absolute humidity:

$$H_s(T_a, \text{RH}_{20\%}) = 26 \text{ gr-H}_2\text{O/lb}_m\text{-dry air}$$

Determine the amount of water vapor that is in air saturated at T_{ao} by moving horizontally from point *B* along a line of constant absolute humidity to the right-hand scale of the diagram *D:*

$$H_s(T_{ao}, RH_{100\%}) = 60 \text{ gr-H}_2\text{O/lb}_m\text{-dry air}$$

Determine the amount of water that is added to saturate the air by determining the difference and multiplying by the dry air mass moving through the biofilter.

$$H_s(T_{ao}, RH_{100\%}) - H_s(T, RH_{20\%}) = 34 \text{ gr-H}_2\text{O/lb}_m\text{-dry air}$$

Note that the volume of 1 lb_m of dry air at condition *A* can be read by interpolating from a line of constant specific volume through point *A* (steeper diagonal lines running from lower right to upper left):

$$V_{\text{specific}} = 13.55 \text{ ft}^3\text{/lb}_m\text{-dry air}$$

Therefore, the mass flow rate of dry air corresponding to a 70 actual cubic feet per min (acfm) volumetric flow rate at condition *A* is

$$M_{da} = (70 \text{ ft}^3\text{/min})/(13.55 \text{ ft}^3\text{/lb}_m\text{-dry air}) = 5.17 \text{ lb}_m\text{-dry air/min}$$

and hence the amount of makeup water that must be added to the humidifier is

$$M_{ws} = (M_{da})(34 \text{ gr-H}_2\text{O/lb}_m\text{-dry air})(60 \text{ min/h})(1 \text{ lb}_m\text{/7,000 gr}) = 1.48 \text{ lb}_m\text{-H}_2\text{O/h}$$

Note that the air leaving the humidifier is undersaturated with respect to the outlet of the biofilter (75°F) and hence moisture will be lost from the biofilter. The degree of undersaturation can be determined from the absolute humidity of the air leaving the biofilter (point *B*) and finding the intersection with the dry-bulb temperature of 75°F (point *E*) then moving vertically to saturated conditions at 75°F (point *F*). Thus it can be seen that saturated air at 75°F should have an absolute humidity of 130 gr-H_2O /lb_m-dry air (point *G*). The water removed from the biofilter per hour would be given by

$$M_{w\,bio} = M_{da}(130 - 60) \text{ gr-H}_2\text{O/lb}_m\text{-dry air}(60 \text{ min/h})(1 \text{ lb}_m\text{/7,000 gr}) = 3.19 \text{ lb}_m\text{-H}_2\text{O/h}$$

The lesson to be learned is that a spray humidification unit alone will be effective only if additional heat energy is supplied either from a hotter gas stream entering the unit or heating the water that is sprayed. The amount of heat that would be needed can be determined from the difference in specific enthalpy (on a dry-air basis) from reading values from the chart at conditions *A* and *F* (follow the wet-bulb lines through these points to the specific enthalpy scale on the left side of the chart).

$$H_{\text{needed}} = M_{da}(38.6 - 22) \text{ Btu/lb}_m\text{-dry air}(60 \text{ min/h}) = 5{,}149 \text{ Btu/h}$$

Note also that it would not take much time to dry a biofilter out in hot weather conditions.

NUTRIENT REQUIREMENTS EXAMPLE. A 70 actual ft^3/min airstream at 25°C is contaminated with toluene at a concentration of 50 ppm. It is to be controlled using a biofilter, i.e., assume essentially complete removal of the toluene. Assuming that 40 percent of the input carbon evolves as carbon dioxide for an initial period of 2 months before reaching steady-state operation, calculate the contaminant loading rate and minimum nutrient requirements for cell growth. Use the data in Table G.1 as typical elementary composition of bacterial cells.

Solution

Find the loading rate of the contaminant expressed as on a mass basis of carbon, g-C/day. The volume occupied by 1 mole of gas at 25°C is given by the gas law:

$$V/n = (0.082 \text{ L} \cdot \text{atm/K} \cdot \text{g-mol})(298 \text{ K})/(1 \text{ atm}) = 24.45 \text{ (L/g-mol)}$$

TABLE G.1
Elemental makeup of bacterial cells

Element	Percent of dry weight
Carbon	50
Oxygen	20
Nitrogen	14
Hydrogen	8
Phosphorus	3
Sulfur	1
Potassium	1
Sodium	1
Calcium	0.5
Magnesium	0.5
Chlorine	0.5
Iron	0.2
Sum of trace elements	0.3

Thus the molar flow rate of toluene is given by

$$N_{\text{tol}} = (70\ \text{ft}^3/\text{min})(28 \cdot 3\ \text{L/ft}^3)(1440\ \text{min/day})(50\ \text{ppm})/(24.45\ \text{L/g-mol})$$

$$= 5.8\ (\text{g-mol tol/day})$$

However, there are 7 carbon atoms in each toluene molecule and each carbon atom has a molecular weight of 12 Daltons, so expressed as (g-C/day)

$$N_{\text{carb}} = 490\ (\text{g-C/day})$$

Of the 490 (g-C/day) some 60 percent by mass of biomass is produced. Thus the quantity of biomass formed per day expressed as (g-C/day) is given by

$$N_{\text{biomass}} = (490)(0.6) = 294\ (\text{g-C/day})$$

At the end of 2 months, some 1,760 (g-C) of new biomass would have been formed. From the ratios, one can see that the quantity of the various other nutrients required would be as follows

$$\text{N} = (294\ \text{g-C/day})(14\ \text{g-N/g-mol})/(50\ \text{g-C/g-mol}) = 82.3\ (\text{g-N/day})$$

$$\text{P} = 17.6\ (\text{g-P/day})$$

$$\text{S and K} = 5.9\ (\text{g-X/day})$$

$$\text{Ca, Mg, Cl} = 2.9\ (\text{g-X/day})$$

$$\text{Fe} = 1.7\ (\text{ge-Fe/day})$$

Each of the values above should be multiplied by 60 days to determine the minimum nutrient requirement for the 2-month period.

One can see that if the carbon is not mineralized to carbon dioxide, then plugging of the biofilter with biomass would occur in a relatively short period of time.

BIOFILTER ENTHALPY EXAMPLE—METABOLIC HEAT GENERATION. A 1,000 ft³/min airstream at 120°F and 20 percent relative humidity is treated to remove "aged" gasoline vapors from a soil vapor extraction (SVE) system. Experience has shown that provision of a residence time of about 2 min is satisfactory for removal of highly volatile straight-chain alkanes such as pentane and hexane in a compost-type biofilter. Suppose that an average vapor concentration of 100 ppm total petroleum hydrocarbons (TPH) is to be treated. (Consider the TPH to be equivalent to the same concentration of hexane for stoichiometric purposes.) The soil vapor also contains 10 ppm of dichloromethane (DCM) as a cocontaminant.

Consider an enthalpy balance on the biofilter/humidifier to determine water requirements. Assume that the biofilter will be insulated to reduce external heat loads. Calculate the temperature to which the biofilter will rise at continuous steady-state operation, i.e., treat the biofilter as if it has adiabatic walls, that the exit gas is saturated, and that the energy of metabolic processes is released as heat into the gas. Provide a spray humidification system at the inlet to the biofilter and assume that the liquid temperature entering the humidification chamber is at the same temperature as the air exiting the chamber, i.e., specify water usage and the minimum flow rate to the spray nozzles to effect the desired cooling.

Enthalpy Balance Example Solution/Discussion

A 1,000 ft³/min airstream at 120°F and 20 percent relative humidity has the following properties obtained from a psychrometric chart (Figure G.3):

$$Q_{a,in} = 1{,}000\ ft^3/min \qquad m_{a,H_2O,in} = 0.015\ lb/lb \qquad h_{a,in} = 45.2\ Btu/lb$$

$$MW_{dry\ air} = 29 \qquad MW_{H_2O} = 18$$

$$MW_{avg} = \left(\frac{1 - m_{a,H_2O,in}}{MW_{dry\ air}} + \frac{m_{a,H_2O,in}}{MW_{H_2O}}\right)^{-1} \qquad MW_{avg} = 29$$

$$\dot{m}_{a,in} = \frac{Q_{a,in}}{\left(\frac{1 - m_{a,H_2O,in}}{MW_{dry\ air} \cdot lb} + \frac{m_{a,H_2O,in}}{MW_{H_2O} lb}\right)\frac{0.73\ atm\ ft^3}{R}(460 + 120)R} 1 \cdot atm$$

$$\dot{m}_{a,in} = 1.1 \cdot lb \cdot s^{-1}$$

$$\dot{m}_{a,dry,in} = \dot{m}_{a,in}(1 - m_{a,H_2O,in}) \qquad \dot{m}_{a,dry,in} = 1.1 \cdot lb \cdot s^{-1}$$

Since the inlet liquid water and the outlet liquid water are at the same temperature as the air leaving the humidifier, no heat is gained from the water and the air can be treated as if it becomes saturated adiabatically; i.e., move along a wet-bulb line (constant enthalpy). Reading from Figure G.3,

$$m_{a,H_2O,out} = 0.0238\ lb/lb \qquad h_{a,out} = h_{a,in}$$

Therefore, the humidifier makeup water requirement is

$$\rho_{H_2O} = 62.4\ lb/ft^3$$

$$H_2O_{makeup} = \dot{m}_{a,dry,in}(m_{a,H_2O,out} - m_{a,H_2O,in}) \qquad H_2O_{makeup} = 0.01 \cdot lb \cdot s^{-1}$$

$$vol_{H_2O,makeup} = \frac{H_2O_{makeup}}{\rho_{H_2O}} \qquad vol_{H_2O,makeup} = 1.6 \times 10^{-4} \cdot ft^3 \cdot s^{-1}$$

In gal/h this is

$$Q_{H_2O} = vol_{H_2O\ makeup}\left(\frac{3{,}600\ s}{h}\right)\left(\frac{7.481\ gal}{ft^3}\right)$$

$$Q_{H_2O} = 4.2\ gal/h$$

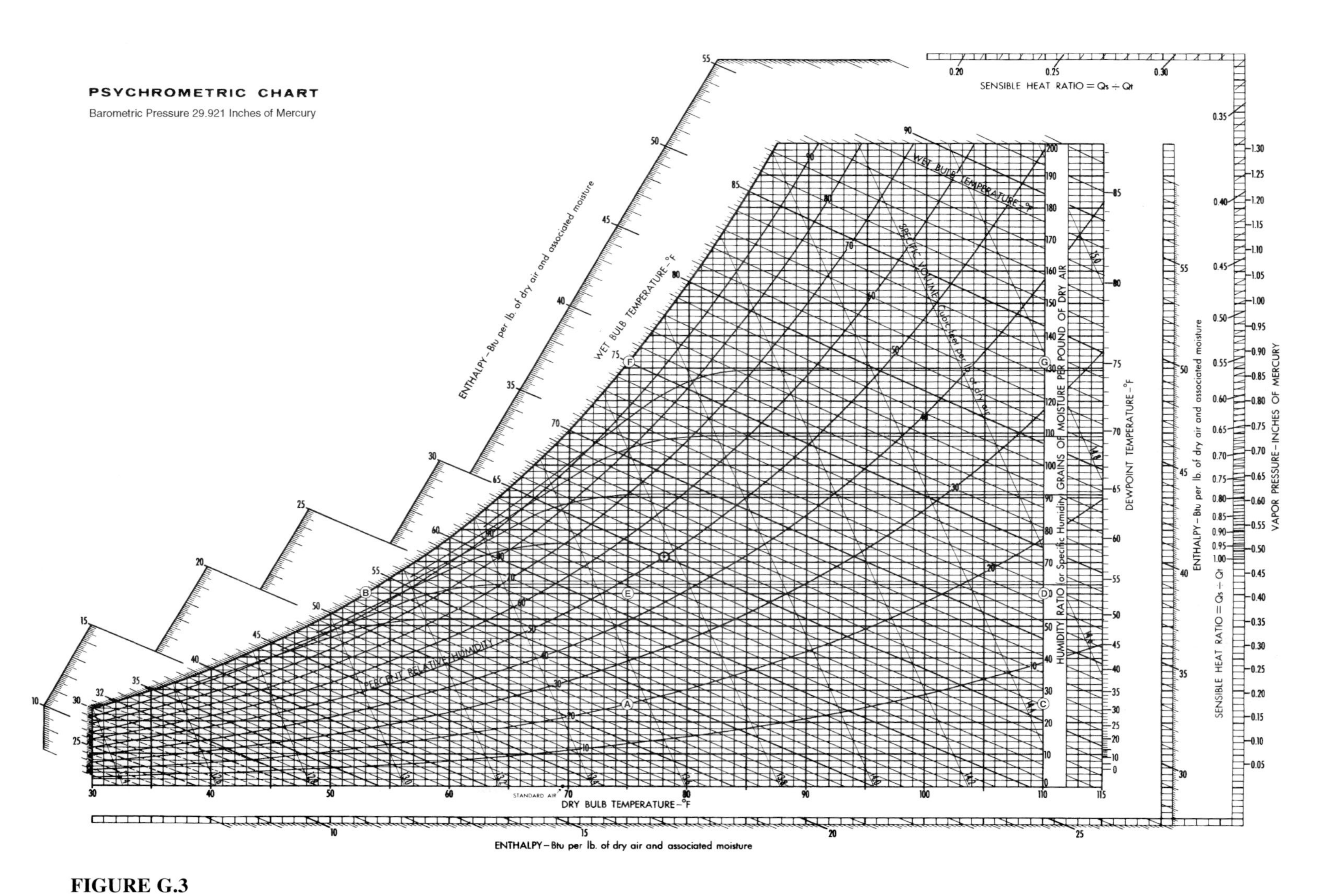

FIGURE G.3
Psychrometric chart. (*Courtesy of Trane Company, LaCrosse, WI*)

The air leaving the humidifier is at a temperature of 82°F or 542°R. Treating the air containing hexane and DCM as inert, then the heat released into the air when the hexane and DCM are degraded to CO_2, H_2O, and HCl, respectively, is given by the following:

$$MW_{hex} = 6 \times 12 + 1 \times 14 \qquad MW_{hex} = 86$$

$$MW_{DCM} = 12 \times 1 + 2 \times 35.5 + 1 \times 2 \qquad MW_{DCM} = 85$$

$$\dot{m}_{hex} = \dot{m}_{a,in} \frac{100}{1 \times 10^6} \frac{MW_{hex}}{MW_{avg}} \qquad \dot{m}_{hex} = 3.4 \times 10^{-4} \cdot lb \cdot s$$

$$\dot{m}_{DCM} = \dot{m}_{a,in} \frac{10}{1 \times 10^6} \frac{MW_{DCM}}{MW_{avg}} \qquad \dot{m}_{DCM} = 3.3 \times 10^{-5} \cdot lb \cdot s$$

The heat of combustion of hexane and DCM are good approximations to the heat released by their oxidation by the microorganisms in the biofilter under steady-state operation. One can find heats of combustion in various references including Perry's *Chemical Engineers' Handbook:*

$$\Delta Hc_{hex} = 45{,}090 \times 10^3 \text{ J/kg}$$

The DCM oxidation will have negligible effect since its heat of combustion, compared to a hydrocarbon like hexane, and its concentration are smaller. The heat of combustion is distributed into the mass of air that accompanies the 1 kg of hexane.

$$\Delta h_{air} = \Delta Hc_{hex} \frac{\dot{m}_{hex}}{\dot{m}_{a,dry,in}} \Delta h_{air}$$

$$\Delta h_{air} = 5.9 \text{ Btu/lb-dry air}$$

From the psychrometric chart the temperature would rise from saturated at 82 to 87°F. The corresponding increase in water would be from 163 to 197 gr/lb-dry air. Note that if not for the latent heat of vaporization, the dry air temperature would rise much more. The amount of H_2O added from the oxidation of hexane would be

$$\dot{m}_{H_2O} \text{ formed} = \dot{m}_{hex} 6 \frac{MW_{H_2O}}{MW_{hex}} \qquad \dot{m}_{H_2O} \text{ formed} = 4.3 \times 10^{-4} \cdot lb \cdot s^{-1}$$

Therefore, the mass of H_2O formed per lb of dry air is

$$\text{Ratio } H_2O \text{ formed per dry air} = \frac{\dot{m}_{H_2O} \text{ formed}}{\dot{m}_{a,dry,in}}$$

$$\text{Ratio } H_2O \text{ formed per dry air} = 3.8 \times 10^{-4}$$

If the biofilter were operated in this manner, then additional water would be needed to prevent the biofilter from drying out. At 87°F, water content per lb-dry air is given by

$$m_a \text{ biofilter}_{out} = \frac{197 \text{ gr-}H_2O\text{/lb-air}}{7{,}000 \text{ gr/lb}}$$

$$m_a \text{ biofilter}_{out} = 0.028 \text{ lb/lb}$$

A condenser at the outlet of the biofilter might be possible, or more conventionally, the water could come from periodic water addition at the top of the biofilter. Therefore, the additional water makeup requirement is

$$\text{Additional } H_2O = \dot{m}_{a,dry,in}(m_a \text{biofilter}_{out} - m_{a,H_2O,out})$$

$$\text{Vol additional } H_2O = \frac{\text{Additional } H_2O}{\rho_{H_2O}}$$

$$\text{Vol additional } H_2O = 7.8 \times 10^{-5} \cdot \text{ft}^3 \cdot \text{s}$$

$$Q \text{ additional } H_2O = (\text{vol additional } H_2O)(7.48 \text{ gal/ft}^3)(3{,}600 \text{ s/h})$$

$$Q \text{ additional } H_2O = 2.1 \text{ gal/h}$$

AUTHOR INDEX

SUBJECT INDEX